T0211567

Lecture Notes in Computer Science   9934

Commenced Publication in 1973
Founding and Former Series Editors:
Gerhard Goos, Juris Hartmanis, and Jan van Leeuwen

Editorial Board

David Hutchison
   Lancaster University, Lancaster, UK
Takeo Kanade
   Carnegie Mellon University, Pittsburgh, PA, USA
Josef Kittler
   University of Surrey, Guildford, UK
Jon M. Kleinberg
   Cornell University, Ithaca, NY, USA
Friedemann Mattern
   ETH Zurich, Zurich, Switzerland
John C. Mitchell
   Stanford University, Stanford, CA, USA
Moni Naor
   Weizmann Institute of Science, Rehovot, Israel
C. Pandu Rangan
   Indian Institute of Technology, Madras, India
Bernhard Steffen
   TU Dortmund University, Dortmund, Germany
Demetri Terzopoulos
   University of California, Los Angeles, CA, USA
Doug Tygar
   University of California, Berkeley, CA, USA
Gerhard Weikum
   Max Planck Institute for Informatics, Saarbrücken, Germany

More information about this series at http://www.springer.com/series/7409

Franco Bagnoli · Anna Satsiou
Ioannis Stavrakakis · Paolo Nesi
Giovanna Pacini · Yanina Welp
Thanassis Tiropanis · Dominic DiFranzo (Eds.)

Internet Science

Third International Conference, INSCI 2016
Florence, Italy, September 12–14, 2016
Proceedings

Springer

Editors

Franco Bagnoli
University of Florence
Florence
Italy

Giovanna Pacini
University of Florence
Florence
Italy

Anna Satsiou
Center for Research and Technology
Thessaloniki
Greece

Yanina Welp
University of Zurich
Zürich
Switzerland

Ioannis Stavrakakis
National and Kapodistrian University
Athens
Greece

Thanassis Tiropanis
University of Southampton
Southampton
UK

Paolo Nesi
University of Florence
Florence
Italy

Dominic DiFranzo
University of Southampton
Southampton
UK

ISSN 0302-9743   ISSN 1611-3349   (electronic)
Lecture Notes in Computer Science
ISBN 978-3-319-45981-3   ISBN 978-3-319-45982-0   (eBook)
DOI 10.1007/978-3-319-45982-0

Library of Congress Control Number: 2016950237

LNCS Sublibrary: SL3 – Information Systems and Applications, incl. Internet/Web, and HCI

© Springer International Publishing AG 2016
This work is subject to copyright. All rights are reserved by the Publisher, whether the whole or part of the material is concerned, specifically the rights of translation, reprinting, reuse of illustrations, recitation, broadcasting, reproduction on microfilms or in any other physical way, and transmission or information storage and retrieval, electronic adaptation, computer software, or by similar or dissimilar methodology now known or hereafter developed.
The use of general descriptive names, registered names, trademarks, service marks, etc. in this publication does not imply, even in the absence of a specific statement, that such names are exempt from the relevant protective laws and regulations and therefore free for general use.
The publisher, the authors and the editors are safe to assume that the advice and information in this book are believed to be true and accurate at the date of publication. Neither the publisher nor the authors or the editors give a warranty, express or implied, with respect to the material contained herein or for any errors or omissions that may have been made.

Printed on acid-free paper

This Springer imprint is published by Springer Nature
The registered company is Springer International Publishing AG
The registered company address is: Gewerbestrasse 11, 6330 Cham, Switzerland

Preface

This volume contains the papers presented at INSCI 2016, the Third International Conference on Internet Science, held on September 12–14, 2016 in Florence.

This conference brought together researchers from across the world to help further develop the emerging discipline of Internet Science. Internet Science is an interdisciplinary field that explores the sociotechnical nature of the Internet through the lenses of Computer Science, Sociology, Art, Mathematics, Physics, Complex Systems Analysis, Psychology, Economics, Law, Political Sciences, and more. Internet Science aims to bridge these different views and theories, in order to create a more holistic understanding of the Internet and its impact on society. In particular, Internet Science asks crucial questions like: How do people behave in the Internet? Are they changing their lifestyle and how? Can the Internet promote sustainability, cooperation, and collective intelligence? Can it support open democracy and policy making? How can the awareness of possibilities and dangers of the Internet be promoted? What about topics like intellectual property, privacy, reputation, and participation? What are the juridical aspects of the Internet? What about arts and humanities in general?

This conference was built on the success of on the First International Conference on Internet Science, which was held on April 9–11, 2013 and the Second International Conference on Internet Science "Societies, Governance, and Innovation" on May 27–29, 2015, which both took place in Brussels, Belgium organized by the FP7 European Network of Excellence in Internet Science - EINS project, with the support of the European Commission.

The organizers of the Third International Conference on Internet Science wish to thank their organizing partners (The University of Bologna, The Centre for Research and Technology Hellas, The University of Florence, and The University of Southampton) for their help and support. We would like to thank the "Collective Awareness Platforms for Sustainability and Social Innovation" (CAPS) initiative for their support of this conference. The CAPS initiative aims at designing and piloting online platforms creating awareness of sustainability problems and offering collaborative solutions based on networks (of people, of ideas, of sensors), enabling new forms of social innovation.

The theme of the Third International Conference on Internet Science was "Openness, Collaboration and Collective Action". This theme aimed to further explore how the Internet can act as a sociotechnical layer to allow people to collaborate and coordinate in open and heterogeneous ways. This theme was further realized by a number of thematic topics:

- Collective Awareness and Crowsourcing platforms
- Collaboration, Privacy, and Conformity in Virtual/Social Environments

- Internet Interoperability, Freedom, and Data Analysis
- Smart Cities and Sociotechnical Systems

July 2016

Franco Bagnoli
Anna Satsiou
Ioannis Stavrakakis
Yanina Welp
Paolo Nesi
Giovanna Pacini
Thanassis Tiropanis
Dominic Difranzo

Organization

Program Committee

Stuart Allen	Cardiff University, UK
Panayotis Antoniadis	NetHood, ETH Zurich, Switzerland
Franco Bagnoli	University of Florence, Italy
Giorgio Battistelli	University of Florence, Italy
Leonardo Bocchi	University of Florence, Italy
Luca Bortolussi	University of Trieste, Italy
Ian Brown	Oxford Internet Institute, UK
Alice Cavaliere	University of Florence, Italy
Jonathan Cave	University of Warwick and UK Regulatory Policy Committee, UK
Tamas David-Barrett	University of Oxford, UK
Claudio De Persis	University of Groningen, Netherlands
Francesca Di Patti	University of Florence, Italy
Dominic Difranzo	University of Southampton, UK
Andreas Fischer	University of Passau, Germany
Patrizia Grifoni	IRPPS-CNR, Italy
Giorgio Gronchi	University of Florence, Italy
Alessio Guarino	Université de la Réunion, France
Andrea Guazzini	University of Florence, Italy
Elisa Guidi	University of Florence, Italy
Giacomo Innocenti	University of Florence, Italy
Georgios Iosifidis	Yale School of Engineering and Applied Science, USA
Konstantinos Kafetsios	University of Crete, Greece
Irene Karapistoli	Democritus University of Thrace, Greece
Yiannis Kompatsiaris	Centre for Research and Technology Hellas, Greece
Bart Lannoo	iMinds - Ghent University, Belgium
Igor Linkov	US Army Engineer RD Center, USA
Meryem Marzouki	CNRS-UPMC Sorbonne University, France
Donald Mcmillan	Mobile Life Centre, Sweden
Sandro Mehic	Centre for Research and Technology Hellas, Greece
Patrizia Meringolo	University of Florence, Italy
Federico Morando	Nexa Center for Internet & Society at Politecnico di Torino, Italy
Paolo Nesi	DISIT Lab, University of Florence, Italy
Heiko Niedermayer	Technische Universität München, Germany
Giovanna Pacini	University of Florence, Italy

Mario Paolucci	Institute of Cognitive Sciences and Technologies, CNR Rome, Italy
Dimitri Papadimitriou	Nokia, Belgium
Symeon Papadopoulos	Centre for Research and Technology Hellas, Greece
Andrea Passarella	Institute for Informatics and Telematics, CNR, Italy
Raul Rechtman	Universidad Nacional Autonoma de Mexico, Mexico
Stefania Righi	NEUROFARBA, University of Florence, Italy
Mark Rouncefield	Lancaster University, UK
Kavé Salamatian	Université de Savoie, France
Panayotis Sarigiannidis	University of Western Macedonia, Greece
Laura Sartori	University of Bologna, Italy
Anna Satsiou	Centre for Research and Technology Hellas, Greece
Ioannis Stavrakakis	National and Kapodistrian University of Athens, Greece
Pietro Tesi	University of Groningen, Netherlands
Thanassis Tiropanis	University of Southampton, UK
Žiga Turk	University of Ljubljana, FGG, Slovenia
Enrico Vicario	University of Florence, Italy
Daniele Vilone	Istituto di Scienza e Tecnologia della Cognizione (ISTC) - CNR, Italy
Stefanos Vrochidis	Information Technologies Institute, Greece
Yanina Welp	Center for Research on Direct Democracy, ZDA, University of Zurich, Switzerland

Contents

Collective Awareness and Crowsourcing Platforms

Incentive Mechanisms for Crowdsourcing Platforms 3
  Aikaterini Katmada, Anna Satsiou, and Ioannis Kompatsiaris

Results of a Collective Awareness Platforms Investigation 19
  Giovanna Pacini and Franco Bagnoli

Debate About the Concept of Value in Commons-Based Peer Production.... 27
  Mayo Fuster Morell, Jorge L. Salcedo, and Marco Berlinguer

Collective Intelligence Heuristic: An Experimental Evidence 42
  Federica Stefanelli, Enrico Imbimbo, Franco Bagnoli,
  and Andrea Guazzini

Collective Awareness Platforms and Digital Social Innovation Mediating
Consensus Seeking in Problem Situations......................... 55
  Atta Badii, Franco Bagnoli, Balint Balazs, Tommaso Castellani,
  Davide D'Orazio, Fernando Ferri, Patrizia Grifoni, Giovanna Pacini,
  Ovidiu Serban, and Adriana Valente

E-Government 2.0: Web 2.0 in Public Administration. Interdisciplinary
Postgraduate Studies Program................................ 66
  Rafał Olszowski

WikiRate.org – Leveraging Collective Awareness to Understand
Companies' Environmental, Social and Governance Performance 74
  Richard Mills, Stefano De Paoli, Sotiris Diplaris, Vasiliki Gkatziaki,
  Symeon Papadopoulos, Srivigneshwar R. Prasad, Ethan McCutchen,
  Vishal Kapadia, and Philipp Hirche

SOCRATIC, the Place Where Social Innovation 'Happens' 89
  Inés Romero, Yolanda Rueda, and Antonio Fumero

Application Design and Engagement Strategy of a Game with a Purpose
for Climate Change Awareness................................ 97
  Arno Scharl, Michael Föls, David Herring, Lara Piccolo,
  Miriam Fernandez, and Harith Alani

Collective Intelligence or Collecting Intelligence? 105
  Richard Absalom, Dap Hartmann, and Aelita Skaržauskiené

Collaboration, Privacy and Conformity in Virtual/Social Environments

Non-trivial Reputation Effects on Social Decision Making
in Virtual Environment   115
  Mirko Duradoni, Franco Bagnoli, and Andrea Guazzini

Small Group Processes on Computer Supported Collaborative Learning.   123
  Andrea Guazzini, Cristina Cecchini, and Elisa Guidi

Perceived Versus Actual Predictability of Personal Information
in Social Networks   133
  *Eleftherios Spyromitros-Xioufis, Georgios Petkos, Symeon Papadopoulos,
  Rob Heyman, and Yiannis Kompatsiaris*

Conformity in Virtual Environments: A Hybrid Neurophysiological
and Psychosocial Approach   148
  *Serena Coppolino Perfumi, Chiara Cardelli, Franco Bagnoli,
  and Andrea Guazzini*

Internet Interoperability, Freedom and Data Analysis

Interoperable and Efficient: Linked Data for the Internet of Things   161
  Eugene Siow, Thanassis Tiropanis, and Wendy Hall

Stable Topic Modeling with Local Density Regularization   176
  *Sergei Koltcov, Sergey I. Nikolenko, Olessia Koltsova,
  Vladimir Filippov, and Svetlana Bodrunova*

An Empirically Informed Taxonomy for the Maker Movement   189
  Christian Voigt, Calkin Suero Montero, and Massimo Menichinelli

Semantic Integration of Web Data for International Investment
Decision Support   205
  *Boyan Simeonov, Vladimir Alexiev, Dimitris Liparas, Marti Puigbo,
  Stefanos Vrochidis, Emmanuel Jamin, and Ioannis Kompatsiaris*

An Analysis of IETF Activities Using Mailing Lists and Social Media   218
  *Heiko Niedermayer, Daniel Raumer, Nikolai Schwellnus,
  Edwin Cordeiro, and Georg Carle*

Assessing Media Pluralism in the Digital Age   231
  Iva Nenadic and Alina Ostling

Responsible Research and Innovation in ICT – A Framework.   239
  Žiga Turk, Carlo Sessa, Stephanie Morales, and Anthony Dupont

End-to-End Encrypted Messaging Protocols: An Overview.   244
  Ksenia Ermoshina, Francesca Musiani, and Harry Halpin

Smart Cities and Sociotechnical Systems

Making Computer and Normative Codes Converge: A Sociotechnical
Approach to Smart Cities   257
  Elena Pavan and Mario Diani

Smart Cities Tales and Trails   278
  Athena Vakali, Angeliki Milonaki, and Ioannis Gkrosdanis

Privacy Through Anonymisation in Large-Scale Socio-Technical Systems:
Multi-lingual Contact Centres Across the EU   291
  Claudia Cevenini, Enrico Denti, Andrea Omicini, and Italo Cerno

The Butlers Framework for Socio-Technical Smart Spaces   306
  Roberta Calegari and Enrico Denti

Public Transportation, IoT, Trust and Urban Habits   318
  Andrea Melis, Marco Prandini, Laura Sartori, and Franco Callegati

Author Index   327

Collective Awareness and Crowsourcing Platforms

Incentive Mechanisms for Crowdsourcing Platforms

Aikaterini Katmada[✉], Anna Satsiou[✉], and Ioannis Kompatsiaris

CERTH-ITI, Thessaloniki, Greece
{akatmada,satsiou,ikom}@iti.gr

Abstract. Crowdsourcing emerged with the development of Web 2.0 technologies as a distributed online practice that harnesses the collective aptitudes and skills of the crowd in order to reach specific goals. The success of crowdsourcing systems is influenced by the users' levels of participation and interactions on the platform. Therefore, there is a need for the incorporation of appropriate incentive mechanisms that would lead to sustained user engagement and quality contributions. Accordingly, the aim of the particular paper is threefold: first, to provide an overview of user motives and incentives, second, to present the corresponding incentive mechanisms used to trigger these motives, alongside with some indicative examples of successful crowdsourcing platforms that incorporate these incentive mechanisms, and third, to provide recommendations on their careful design in order to cater to the context and goal of the platform.

Keywords: Crowdsourcing · Incentive mechanisms · Reputation · Gamification

1 Introduction

Recently, there has been an ongoing interest in crowdsourcing (CS), a practice which emerged with the development of Web 2.0 technologies and capabilities [1]. The term "crowdsourcing" derived from the combination of the words "crowd" and "outsourcing" and is attributed to Jeff Howe [2]. As Howe wrote in 2006, CS is *"the act of a company or institution taking a function once performed by employees and outsourcing it to an undefined network of people in the form of an open call"* [3] Since then, CS has evolved and it can be found in many diverse manifestations leading to various definitions that differ based on the author's perspective. In a recent study examining 40 different definitions on CS, it was concluded that *it constitutes a distributed online process that requires the participation of the crowd for the accomplishment of specific tasks* [4]. CS has already been successfully applied in many areas [5], from business projects [6] to non-profit initiatives [7]. Some examples include "crowdfunding" platforms (e.g. Crowdrise), platforms for civic engagement (e.g. Changemakers), and Open Innovation projects (e.g. Innocentive and Innovation Challenge [8]). CS can also be distinguished in participatory (based on users' directly provided input), as the examples above, and *opportunistic* (based on data indirectly provided by users' mobile devices, e.g. location info), such as Waze and Health Map's Outbreaks Near Me.

Successful CS systems are dependent upon the participation of the users and their continuous involvement. The reasons for participating in CS systems stem from a broad

© Springer International Publishing AG 2016
F. Bagnoli et al. (Eds.): INSCI 2016, LNCS 9934, pp. 3–18, 2016.
DOI: 10.1007/978-3-319-45982-0_1

spectrum of motives, such as altruism, social motivations, and monetary rewards [2]. However, as Zhao & Zhu [1] noted, there is a need for more research to be conducted on the various types of the crowd's motivations, since they vary greatly depending on the CS context. Therefore, a more in-depth understanding of user motives could enable the design of appropriate incentive mechanisms that would eventually promote sustained user engagement in CS platforms. As regards any pertinent research, a number of studies that focus on user motives for participating in various types of CS platforms (e.g. [5, 6, 9–11]), was identified. There are also studies addressing specific incentive mechanisms used in CS, such as reputation systems [12, 13] and gamification [14, 15]. However, there is a lack of studies that offer a holistic overview of all related user incentives and incentives mechanisms commonly applied in CS, as well as impediments and practical implications that should be taken into account when designing these mechanisms.

Towards this direction, this paper aims at: (a) examining user motives and corresponding incentives for participating in CS platforms; (b) investigating the incentive mechanisms that trigger these motives, as well as indicative CS platforms that make use of them; and (c) concluding on some practical design recommendations according to the context and goals of the CS platform. The particular study is of an interdisciplinary nature, since it draws both from the field of Motivational Psychology, examining the concepts of motivation and motivational factors, and Computer Science, researching and presenting various design mechanisms that activate user motives and encourage participation in CS platforms.

The rest of the paper is structured as follows: Sect. 2 presents the methodology of the particular review followed by a thorough analysis of the theoretical framework regarding user motives and incentives. Section 3 delves into the incentive mechanisms that are currently being applied in CS systems, while Sect. 4 highlights the issues that should be taken under consideration in the design of effective incentive mechanisms. Finally, this paper is concluded in Sect. 5.

2 Methodology and Theoretical Framework

2.1 Methodology

For the particular literature review, searches were made in numerous databases, including ACM Digital library, Google Scholar, ResearchGate, JSTOR, IEEE Xplore, etc. The authors first examined relevant research on user motives and incentives for participating in CS systems, and, afterwards, they reviewed and categorized the incentive mechanisms used in CS platforms and mapped their correspondence with relevant motives and incentives. For the needs of this study, a large number of CS platforms was reviewed and categorized according to the incentive mechanisms they incorporate and some prominent examples of successful CS platforms were selected in order to present the incorporated mechanisms. Based on this analysis, the authors concluded on several issues that should be taken into account when designing these incentive mechanisms, followed by useful design recommendations that could improve the overall user experience, as well as increase user loyalty.

2.2   Theoretical Framework

As already mentioned, the success of a CS system depends upon the sustained participation of the users, which, in turn, relies greatly on their motives. For this reason, in this section, we present some fundamental notions regarding user motivation, in order to understand what makes people willing to participate in online CS environments.

In the field of motivational psychology, a person who is activated in order to achieve a goal can be characterized as motivated, whereas a person who is uninspired to act is commonly referred to as unmotivated [16]. Motives can be internal (innate human needs), or external (situations that trigger these needs) [17]. In accordance with the Motive-Incentive-Activation-Behavior Model (MIAB), in a specific situation a suitable incentive will cause an individual's corresponding motive to be activated and lead, as a consequence, to the manifestation of a particular behavior [9].

As regards any specific motives for participating in CS environments, these may vary greatly depending on the participant, the situational context, as well as the system itself. Based on the studies of [2, 9, 18] we identify the following motives relevant to CS environments: (i) *learning/personal achievement*, (ii) *altruism* (iii) *enjoyment/intellectual curiosity*, (iv) *social motives*, (v) *self-marketing*, (vi) *implicit work*, and (vii) *direct compensation*. Learning/personal achievement, altruism, social motives, and enjoyment/intellectual curiosity can be considered intrinsic motives, and are also in line with Maslow's pyramid of needs [4], according to which the two higher needs are self-esteem (including confidence, achievement, and the respect of others), and self-actualization (including creativity, morality, and inner potential) [19].

Following the aforementioned MIAB model, we look into the appropriate incentives that would appeal to the user motives we identified. Actually, each motive can be activated by one or more incentives. A suitable incentive for "learning" would be the *access to the knowledge and feedback of experts or peers*. "Altruism" constitutes the intrinsic motivation to help the community without personal benefit [2], and, thus, can be activated by having the opportunity to *contribute for a good cause*, and by *receiving feedback* concerning the impact of personal contributions. "Enjoyment" refers to the intrinsic motivation to perform an activity simply for the *sheer enjoyment and satisfaction* derived from that action [16] and intellectual curiosity is activated by having the opportunity to meet *new people* and explore *new places and situations*.

"Social motives" can be activated by various incentives, including the will to *attain social status* and *respect by organizers and peers* [9], as well as *present a good social image* according to the values of the online community [20]. Moreover, they may be influenced by the *initial interactions* a newcomer experiences with the online community [21], and can be increased by *presenting personalized social information* to participants [22]. For "self-marketing", *career options* are a decisive incentive especially for volunteers with specialized skills. For example, programmers contributing to open source software may be motivated by career concerns [23], since they have the opportunity to "advertise" themselves by demonstrating their knowledge and skills [9]. "Implicit work" is relevant to the so-called *passive CS*, as it is performed by the user as a *side effect* of accomplishing another task (e.g. the ESP game, reCAPTCHA), or by *contributing information* to third-party websites, even unknowingly (e.g. AdWords, social media [24]). Therefore, it will be excluded from our incentives/incentive mechanisms analysis. Lastly, "Direct compensation" can be differentiated between token and market compensation: the first usually constitutes something desirable, such as a *small monetary prize* or *token*, whereas the second involves *higher payment* [2].

In order to "trigger" the aforementioned incentives, several incentive mechanisms have been designed to elicit and sustain user participation in CS platforms. We have sorted them into four main categories, consisting of: reputation systems, gamification, social incentive mechanisms, and financial rewards and career opportunities. In Fig. 1 we present the correspondence between motives, incentives and incentive mechanisms, as well as several CS platforms that make use of them. More specifically, user motives are placed in the middle inner circle, each one with a different color; suitable incentives for each motive are mentioned in the outer circles, using the same color with the corresponding motive. They are mapped to the incentive mechanisms that sustain them, depicted on the four corners of the image. Examples of CS platforms are strategically placed according to the incentive mechanisms they implement. For example, Waze was placed between Social Incentive Mechanisms and Gamification, since it incorporates both social and gamification elements.

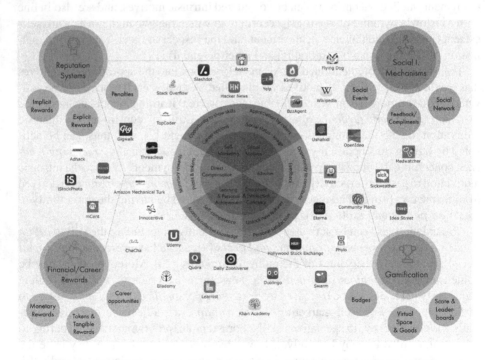

Fig. 1. User motives, incentives & incentive mechanisms (Color figure online)

3 Incentive Mechanisms

3.1 Reputation Systems

Reputation systems are commonly encountered in CS platforms for increasing user participation and quality of contributions. Usually, the platform's users rate other users based on their behavior, and the reputation system combines these ratings to form cumulative assessments of their reputation. Reputation can be measured in discrete or continuous values and the mathematical model (metric) that aggregates ratings can be based upon several different methods, from simple summation and average of ratings to fuzzy logic or probabilistic models [25]. In this section, we describe different reputation metrics that are used according to the CS platforms' goals.

In CS news websites, such as Reddit, Slashdot, and Hacker News, we observe similar reputation metrics with minor but notable differences. On these websites, where the content is mainly user-generated, users accumulate reputation points, called "Karma", based on the ratings of their submissions (posts and comments) and their voting activity on other users' submissions. These reputation systems differ on their levels of "strictness" as regards down-voting and subtracting points from a user's reputation score. Reddit's reputation mechanism sums a user's post and comment karma separately, generating two karma scores based on the number of up-votes minus down-votes. On Hacker News, a user's karma is also calculated similarly. Hacker News differs in that users cannot down-vote posts until they reach a karma score of 500 points, so that they can be considered credible enough to do so. Lastly, on Slashdot, a user's karma is calculated as the sum of up-votes and down-votes on her comments, and is also affected by other things, such as acceptance of her submissions. It also influences the starting score of her comments on the platform: every comment is initially given a score of −1 points for users with low karma, 0 for anonymous users, +1 for registered users, and +2 for users with high karma. An even more strict approach is used by the Q&A platform Stack Overflow; users can build reputation slowly earning reputation points up to a certain daily limit by having their questions and answers voted up, their answers marked as "accepted", etc. [12] and loose reputation points by having their questions and answers voted down, their posts flagged as spam or offensive many times, and even when they vote down on other users' answers. That way, the platform tries to prevent malicious acts and urge users to think twice before down-voting an answer.

Reputation metrics can also take different forms than aggregating ratings. In Amazon Mechanical Turk (AMT), for example, a crowdsourced Internet marketplace, the reputation score of a participant ("worker") is essentially the rate of her approved "Human Intelligence Tasks" (HITS) to those submitted. This rate demonstrates the ability to complete tasks successfully; however, there is no mechanism to detect unfair user scores [13]. Conversely, the reputation of an employer ("requester") is an important motivator for workers to participate and put more effort into their work. For that reason, Turkopticon, a third-party reputation system which gives workers the opportunity to rate requesters based on four aspects of their behavior (Communicativity, Generosity, Fairness and Promptness) [26], was created. On the other hand, in TopCoder, a CS platform which hosts regular contests relevant to design and development, the reputation score of contestants is calculated with a more sophisticated algorithm that takes into account their prior history, their expected performance, as well as their performance as compared to that of other contestants [27].

So far, we have described several approaches to assess a user's reputation. Apart from the reputation metric, reputation systems include some kind of reward for users with high reputation and/or penalties for users with very low reputation. *Explicit rewards* can consist of qualifications upon completion of tasks (e.g. AMT), special badges that prove contribution to the community (e.g. Stack Overflow), privileges in site management (e.g. Stack Overflow), more "moderation points" (e.g. Slashdot), or the right to down-vote (e.g. Hacker News). *Implicit rewards* include respect from the community, career opportunities (e.g. TopCoder), and acknowledgement of their credibility which gives them greater chances of being elected as moderators. Since user rewards often span from gaining badges to social status in the community and career opportunities, reputation systems are often combined with the incentive mechanisms that will be described in the next sections.

On the other hand, *penalties* could consist of blocking users with low reputation from accessing future tasks (e.g. AMT), ban them from posting temporarily (e.g. Slashdot) or posting to specific channels on the platform (e.g. Reddit), and suspend their accounts for a specific period of time that can be increased with subsequent suspensions (e.g. Stack Overflow). Reputation systems can also incorporate mechanisms to prevent malicious behavior, like filtering out posts based on the number of down-votes or user's reputation (e.g., Hacker News, Slashdot), allowing users to post in limited time windows (e.g. Slashdot), and concealing the algorithm used to calculate reputation scores in order to prevent system manipulation [28].

3.2 Gamification

Recently, there has been an increasing interest in the potential of "gamification", which can be defined as "the use of game design elements in non-game contexts" in order to improve user experience and engagement [29]. Such game design elements, also known as "game mechanics", include self-elements, such as points, achievement badges, levels, and time restrictions; social-elements, such as storylines, leaderboards, and interactive cooperation [30]; and can also include the virtual space and goods, as well as virtual gifts [31]. Prandi et al. [14] also mention status (titles that indicate a user's progress) and roles (role playing elements). These game mechanics let users develop their own skills, be creative, and feel competent, while experiencing an often social and enjoyable activity, and motivate them by rewarding their efforts and providing appropriate and timely feedback. Thus, gamification corresponds successfully to intrinsic motives such as enjoyment and social recognition [15, 32].

Gamification is frequently encountered in successful CS platforms and applications. Some notable examples of CS platforms that incorporate the majority of the aforementioned game mechanics include the language learning and crowdsourced translation platform Duolingo, the educational platform Khan Academy, and Foursquare's local discovery and sharing app Swarm. Notable game elements implemented in Duolingo are: (a) levels of progress that sustain user engagement by offering small scale goals;

(b) immediate feedback and helpful tips; (c) clear goals and rules that motivate users to continue; and (d) intuitive and friendly interface that helps creating an immersive experience [33]. Other incorporated game elements are player "lives", scores based on performance, leaderboards and competition between friends, as well as virtual currency which users can use to buy virtual goods or gift it to other users [33]. These virtual gifts promote participation and a sense of privilege and community between users [34]. Khan Academy incorporates achievements, user avatars, badges, levels, content unlocking, and "boss fights" (final tests before leveling up). Notable aspects are that there is no competition between users [32], no social interactions, and user profiles are private by default [15], supporting users' isolation on the platform. Swarm, on the other hand, is a gamified application heavily based on the social interactions between users, indicating that gamification can appeal to social motives and be used in conjunction with social incentive mechanisms as well. Swarm incorporates several game mechanics, such as points ("coins"), badges ("stickers"), and social ranking with leaderboards between friends, motivating users to participate and perform more check-ins [35]. The reason this platform was not claimed under social incentive mechanisms (Sect. 3.3) is that, as argued in [36], Swarm is quite enjoyable even without the social interactions, as a single player "sticker game", offering clear progress and rewards to the users.

Other CS contexts where gamification has been applied include CS platforms in which users participate primarily for altruistic reasons, e.g. civic engagement [37]. Often, these platforms exploit user generated data gathered automatically from sensor-enabled mobile devices (e.g. smartphones). Gamification here can provide extra incentives to participate, apart from the initial intrinsic motivational factors. Waze, for example, is a GPS application for crowdsourced traffic monitoring. Users participate either by sharing traffic and accidents reports or by contributing road data using their smartphones. Waze incorporates gamification elements, such as avatars, points, leaderboards, achievements, levels, badges, and social interactions.

Finally, there are also successful CS *"games with a purpose"*; they are online games which constitute a *"general mechanism for using brainpower to solve open problems"* and can be applied in various, diverse areas, such as computer vision, security and content filtering [38]. Foldit, e.g., is a puzzle game and at the same time a CS platform, in which players try to fold the structures of selected proteins in the best possible way, and researchers then analyze the highest scoring solutions to apply them in real world scenarios. Foldit attracts engaged users through achievement, social interaction, and immersion, supported by several game mechanics [39]. Other CS games with a purpose include the ESP Game [38] and Phylo [40].

3.3 Social Incentive Mechanisms

Social motives often play a major role for participating in CS platforms. For example, it is argued [20] that in online reviewer platforms social image and reviewer productivity are positively correlated, while in online ideas competitions participants want to receive positive reactions regarding their skills [18]. Having a good social image is very important for participants in online communities, who want to be perceived as intelligent, fair, wealthy, and "good", in general [20].

In order to trigger these social motives, many social incentive mechanics that act as enablers of social interactions, giving users the chance to showcase their skills and gain social status in the community can be implemented (e.g. specialized mailing lists, discussion fora, provision of feedback/compliments functionalities, invitations to events, etc.). At the online review community Yelp social interactions and a sense of community contribute greatly in sustaining user interest and participation. Users can connect with friends or meet fellow-minded people, plan events, exchange "compliments" and learn more about a reviewer's personality and taste from her profile page. Yelp members care about presenting a good social image to friends and other Yelp community members by being active and contributing many quality reviews [20].

On the other hand, on the aforementioned Swarm app, users can import their contacts from social media, such as Facebook, meet new friends with similar interests, see their nearby contacts, and exchange messages. The application supports social recommendations through tips, as well as checking-in with friends and adding photos to check-ins. As Cramer et al. [35] mention, these "social-driven" check-ins support friendship, togetherness, and identity. Social incentive mechanisms are also used in Zooniverse, a "cityzen science" platform where users can contribute to novel research in different areas and share the discussion boards with researchers in order to explore and analyze data. Additionally, there are fora, blogs, meme generators and even competitions created by users, which makes participation much more fun [41]. In Wikipedia, a free web-based collaborative encyclopedia, editors claim and receive credibility and recognition in the community, as a reward for their contributions, by displaying lists with articles they have edited on their user pages [42].

3.4 Financial Rewards and Career Opportunities

Another incentive mechanism commonly applied in CS platforms is financial rewards, which trigger extrinsic motives like market and token compensation. Financial rewards are used in order to compensate for the lack of social rewards and intrinsically enjoyable tasks [43], as it not always feasible to replicate situations in which people participate voluntarily, but they can also be used in combination with intrinsic incentives. An example of CS using monetary rewards to incentivize the crowd is InnoCentive, a company that offers cash awards for the best solutions in research and development problems. Here, apart from intrinsic motivations, the desire to win the monetary prize is also a significant motivational factor for the participants [44]. Apart from payment, financial rewards can also comprise small tokens, various prizes, and free access to services and products. For example, the mCent application gives users free Internet access for each sponsored application they download and try out.

Monetary rewards are often encountered in CS platforms combined with reputation systems. One example is the aforementioned Amazon Mechanical Turk, in which workers receive payment upon completion tasks and after approval by the requester or by the platform (automatically). Requesters can also give bonuses in case they are very satisfied with the performance of the workers. Similarly, Gigwalk is a CS mobile application that allows users to find quick jobs in their area posted by retailers and consumer brands; it also matches users with jobs according to their performance score. In other cases, monetary rewards are combined with career and self-marketing opportunities for professionals. In iStockPhotos, an online stock imagery website, users submit their work and receive commission for each sale. Brabham [6] concluded that even though learning and peer recognition are important motivational factors for contributing work to iStock-Photo, the main incentive for participants is the opportunity to sell their work. Similarly, Threadless is an online community of artists, as well as an e-commerce website. Designers can submit their work for public vote by the online community, and receive royalties, cash and gift cards if their designs are selected. Apart from the important financial incentive, users may also participate for self-marketing reasons and higher employability [5]. WiseStep helps employers and recruiters to find high quality talent faster and cheaper, by referrals from the crowd. The participants are, in turn, motivated by having the opportunity to build a strong professional network and win monetary awards by referring their friends. Lastly, an example of CS platform which combines financial rewards and reputation systems is the afore-described TopCoder. Since many technological companies sponsor TopCoder competitions in search for talented developers, and taking into account that reputation in TopCoder is directly linked with performance, it is argued that reputation here is also of important economic value [27].

Finally, it should be mentioned that financial incentives are also important in prediction markets, which give users the opportunity to buy and sell shares based on the outcome of events (e.g. the non-profit Iowa Electronic Markets). The invested amounts are small but they still constitute an important incentive to participate, along with any intrinsic incentives.

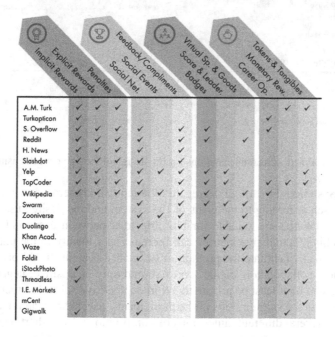

	Implicit Rewards	Explicit Rewards	Penalties	Social Net.	Social Events	Feedback/Compliments	Badges	Score & Leader.	Virtual Sp. & Goods	Career Op.	Monetary Rew.	Tokens & Tangibles
A.M. Turk	✓	✓	✓								✓	✓
Turkopticon	✓										✓	
S. Overflow	✓	✓	✓	✓		✓	✓				✓	
Reddit	✓	✓	✓	✓		✓	✓			✓		
H. News	✓	✓	✓	✓		✓	✓					
Slashdot	✓	✓	✓	✓		✓	✓					
Yelp	✓	✓	✓	✓	✓	✓	✓	✓				✓
TopCoder	✓	✓	✓	✓		✓	✓	✓			✓	✓
Wikipedia	✓	✓	✓	✓	✓	✓	✓			✓	✓	
Swarm				✓		✓	✓	✓	✓			
Zooniverse				✓	✓	✓	✓					
Duolingo				✓		✓		✓	✓			
Khan Acad.						✓	✓	✓				
Waze				✓			✓	✓	✓			
Foldit				✓		✓		✓	✓			
iStockPhoto	✓										✓	✓
Threadless	✓			✓	✓	✓				✓	✓	✓
I.E. Markets											✓	
mCent				✓								✓
Gigwalk	✓			✓							✓	

Fig. 2. Incentive mechanisms elements and CS examples

The aforementioned CS platforms and the most prominent incentive mechanisms characteristics they incorporate are depicted in Fig. 2.

4   Incentive Mechanisms Design Recommendations

Upon examining the design of the aforementioned incentive mechanisms, the authors concluded that there are several issues that should be taken into consideration; the most significant ones are going to be presented here, categorized according to the incentive mechanism they correspond to, and summarized in Fig. 3.

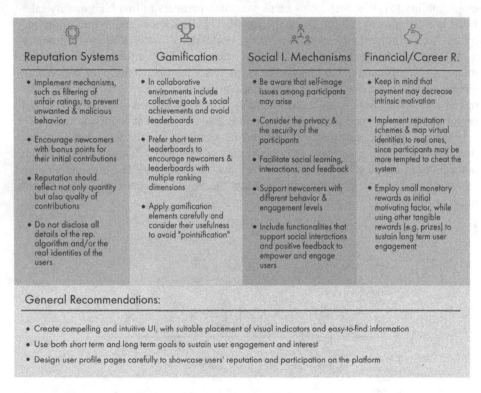

Fig. 3. Practical recommendations for the design of incentive mechanisms

Reputation Systems. First of all, as regards the reputation metric, it is argued that it should be chosen according to the goals of the system and the desired user behaviors [45]. Activity statistics are suitable for building trust between users and supporting member matching, cumulative metrics can increase user loyalty to the platform, and scoring mechanisms facilitate the promotion of quality content. In CS platforms that seek to promote collaboration, public scores should only be positive; negative scores should be avoided or at least be private to avoid competitive spirit. In order to support diversification and the varying skills of the participants, user scores and ranking should be based upon several different dimensions of contributions. Both short term and long term reputation could be included in order to encourage newcomers and provoke their interest with smaller scale goals, while increasing user loyalty [28].

Much attention should be paid to avoid reputation bias due to unfairly positive or negative ratings. In particular, presumed unfair ratings could be excluded based on their statistical properties or the reputation of the rater [45] and domain knowledge filtering methods [25]. Furthermore, they could be cross-checked through meta-moderation schemes, e.g. "rating the raters", or prevented by trying to induce truthful ratings with the use of external rewards, such as financial rewards [25], and the use of anonymous rating schemes [45]. On the other hand, it has also been noticed that malicious users frequently change their identity on the platform and start all over again. One way of preventing that is by mapping virtual identities to the real ones; however, this approach may discourage users from joining the platform in the first place, or, in case their identities are disclosed, giving negative feedback to other users [25], contributing in increasing positive bias. A possible solution could be keeping user identities known only to the reputation system and using community moderation to identify malicious acts and users [25]. In order to support the reputation system's fairness, a common approach is to prevent users from building high reputation very fast (e.g. Stack Overflow), while also allowing them to lose it quickly if at some point they stop contributing or behave maliciously [46]. Lastly, to prevent manipulation of the reputation system, many platforms do not reveal details of their reputation algorithms, even though this practice entails the danger of diminishing the perceived fairness of the system and, consequently, user trust. A middle road could be followed, in which some information regarding reputation score aggregation is disclosed to users and some is unknown (e.g. Reddit, Hacker News).

Gamification. Gamified systems should also be carefully designed, as applying game elements without any consideration for their latent usefulness leads to nothing more than "pointsification" [47]. As a negative consequence, pursuing points may become the primary goal of the participants [41]. Moreover, by merely incorporating game elements into tasks, they would not necessarily become more interesting and engaging. Instead, game mechanics should be implemented very carefully depending on the specific situational context and the targeted users. Leaderboards, e.g., may raise unnecessary competition in CS environments that promote collaboration and target implicit motives. Even in competitive environments, they should be used with caution because they might demotivate newcomers and other low ranked users. Alternatively, short-term leaderboards that allow users to compete on a short time window with the same chances to win by resetting scores every week (e.g. Swarm's weekly leaderboards), or leaderboards including multiple ranking dimensions, could be used. Other alternatives could be customizable leaderboards, allowing participants to choose whom they compete with, or leaderboards that adapt to the user in order to provide optimal motivation levels [37]. As a general recommendation, competition "should be available, but easy to be ignored" [37] (e.g. in Yelp leaderboards are not easily accessible from the main page). Lastly, as regards the inclusion of other social-elements in gamified platforms, the platform's goals should also comprise gamified "collective goals" and not only individual user goals to emphasize on cooperation [41], and highlight social achievements [15].

Social incentive mechanisms. Elements that could increase competitiveness should also be used carefully on platforms that target social motives, since those platforms' primary goal is to support friendly social interactions and collaboration. Design elements that enhance participatory behavior and allow users to connect with others should be included instead. Such elements include online discussion groups, social networks, and functionalities that enable user feedback and the wider distribution of content in the social network. It should also be mentioned that the initial interactions newcomers experience on the platform and the feedback they receive from older members can also affect the levels of their future participation. Indeed, newcomers that have access to their connections' contributions and receive feedback on their activity learn by observation what is feasible and socially acceptable on the platform faster, and they contribute more themselves [48]. Thus, the design of such platforms should facilitate social learning, social interactions such as photo tagging, and support newcomers with different behavior and engagement levels on the platform (e.g. active versus inactive users) [48].

Designers should also consider the fact that security concerns and self-representation issues may arise on social platforms. Swarm (Foursquare) users, for example, are often concerned about the privacy of their check-ins, as well as their social image which, according to their opinion, may be negatively affected by checking in at particular places [36]. A way to deal with these issues is to give users the opportunity to keep their profiles and activity on the CS platform private and/or separate from their social media accounts (e.g. Facebook, Twitter).

Financial rewards/Career opportunities. As regards offering monetary rewards in CS systems the main advantage of this approach is that it is "relatively low cost", since most participants consist of amateurs, scientists, or individuals wishing to apply their skills or pass their free time [49]. However, it should be noted that there is skepticism concerning financial incentives, since extrinsic rewards can decrease people's intrinsic motivation [16]. Moreover, financial rewards may result in short term gain but in the long run they may decrease engagement [39]. Additionally, they do not necessarily lead to better contributions. Indeed, studies regarding participation in Amazon Mechanical Turk have indicated that higher payment had a positive effect on attracting more workers and increasing the quantity of the completed work, but did not lead to increase in its quality and accuracy [43, 50]. Quinn & Bederson [18] also mention that participants may be more tempted to cheat the system in order to increase their reward. A recommended approach for incorporating financial rewards is to employ small monetary rewards as an initial motivating factor, and then utilize other tangible rewards, such as prizes, in conjunction with gamified achievements on the platform to achieve sustained engagement [37].

General design recommendations. Lastly, various other underlying design decisions that appeal to all of the aforementioned incentive mechanisms were also identified. Most of them are relevant to the user interface (UI), including the presentation and placement of incentive mechanisms elements, such as reputation score and visual indicators. For example, profile pages are usually very carefully designed. The profile page of a Stack Overflow user, e.g., contains useful information, such as her reputation score, which is clearly visible on top of the page, recent activity on the platform and even a tag cloud with the subject categories that the particular user is participating in. Suitable placement of information in order to be easily accessible is also an important issue. In Stack Overflow, recent job postings are displayed next to a question, visible to visitors and registered users alike. That way, developers who are interested in displaying their skills to potential employees or they are actively searching for a job can be more incentivized.

Compelling and intuitive UI, as well as appropriate feedback that indicates a user's progress towards mastery, can also encourage and sustain user participation. In both Duolingo and Khan Academy, learners have access to visual indicators of their progress, such as charts and diagrams based on their activity statistics. Their score is always visible on top of the page. In Khan Academy, the contents of a learning topic are displayed in a sequential list with icons that indicate both the type of content (e.g. video, challenge) and the progress of the learner. Positive feedback based on scores that are percentiles of the larger group can also make players feel more empowered and positive about their skills. In the Great Brain Experiment, a CS game in which players participate in experiments that test their cognitive abilities, they might be told that they have better impulse control as compared to 90 % of the population. Lastly, the UI should indicate to the newcomers various ways to contribute, as well as any potential benefits from their participation, in a simple and comprehensive way. In MovieLens, e.g., users are explained that the more ratings they provide the more personalized recommendations they will get, as the system "learns" their preferences.

5 Conclusions

This paper provided a holistic overview of the incentive mechanisms used in CS environments and categorized them under four main directions, i.e., (i) reputation schemes, (ii) gamification practices, (iii) social mechanisms, and (iv) financial rewards and career opportunities. The different incentive mechanisms were analyzed and mapped to different intrinsic and extrinsic user motives and incentives, providing, that way, useful guidelines regarding the selection of incentive mechanisms according to the target users (and their motives) and the context of different CS platforms. Additionally, relevant examples of various CS platforms and applications implementing these mechanisms were discussed and the authors highlighted certain issues that should be taken into consideration for the successful implementation of such mechanisms and concluded on several practical design recommendations.

Acknowledgements. This work has been supported by the EU HORIZON 2020 project PROFIT (contract no: 687895).

References

1. Zhao, Y., Zhu, Q.: Evaluation on crowdsourcing research: current status and future direction. Inf. Syst. Front. **16**(3), 417–434 (2014)
2. Rouse, A.C.: A preliminary taxonomy of crowdsourcing. In: ACIS 2010 Proceedings 76, pp. 1–10 (2010)

3. Howe, J.: Crowdsourcing: A Definition (2006). http://crowdsourcing.typepad.com/cs/2006/06/crowdsourcing_a.html. Accessed 26 Feb 2016
4. Estelles-Arolas, E., Gonzalez-Ladron-De-Guevara, F.: Towards an integrated crowdsourcing definition. J. Inf. Sci. **38**, 189–200 (2012)
5. Brabham, D.C.: Moving the crowd at threadless. Inf. Commun. Soc. **13**, 1122–1145 (2010)
6. Brabham, D.C.: Moving the crowd at iStockphoto: The composition of the crowd and motivations for participation in a crowdsourcing application. First Monday. 13 (2008)
7. Gao, H., Barbier, G., Goolsby, R.: Harnessing the crowdsourcing power of social media for disaster relief. IEEE Intell. Syst. **26**, 10–14 (2011)
8. Crowdsourcing.org. http://www.crowdsourcing.org/
9. Leimeister, J.M., Huber, M., Bretschneider, U., Krcmar, H.: Leveraging crowdsourcing: activation-supporting components for IT-based ideas competition. J. Manage. Inf. Syst. **26**(1), 197–224 (2009)
10. Chandler, D., Kapelner, A.: Breaking monotony with meaning: motivation in crowdsourcing markets. J. Econ. Behav. Organ. **90**, 123–133 (2013)
11. Hossain, M.: Users' motivation to participate in online crowdsourcing platforms. In: 2012 International Conference on Innovation Management and Technology Research (ICIMTR), pp. 310–315. IEEE (2012)
12. Movshovitz-Attias, D., Movshovitz-Attias, Y., Steenkiste, P., Faloutsos, C.: Analysis of the reputation system and user contributions on a question answering website: Stackoverflow. In: 2013 IEEE/ACM International Conference on Advances in Social Networks Analysis and Mining (ASONAM), pp. 886–893. IEEE (2013)
13. Allahbakhsh, M., Ignjatovic, A., Benatallah, B., Beheshti, S., Bertino, E., Foo, N.: Reputation management in crowdsourcing systems. In: 2012 8th International Conference on Collaborative Computing: Networking, Applications and Worksharing (CollaborateCom), pp. 664–671. IEEE (2012)
14. Prandi, C., Salomoni, P. and Mirri, S.: Gamification in Crowdsourcing Applications (2015)
15. Morschheuser, B., Hamari, J., Koivisto, J.: Gamification in crowdsourcing: a review. In: Proceedings of the 49th Annual Hawaii International Conference on System Sciences (HICSS), Hawaii, USA (2016)
16. Ryan, R.M., Deci, E.L.: Intrinsic and extrinsic motivations: Classic definitions and new directions. Contemp. Educ. Psychol. **25**(1), 54–67 (2000)
17. Rani, R., Kumar-Lenka, S.: Motivation and work motivation: concepts, theories & researches. Int. J Res. IT Manage. **8**(2), 12–22 (2012)
18. Quinn, A.J., Bederson, B.B.: Human computation: a survey and taxonomy of a growing field. In: Proceedings of the SIGCHI Conference on Human Factors in Computing Systems. ACM (2011)
19. Maslow, A.H.: A theory of human motivation. Originally Published Psychol. Rev. **50**, 370–396 (1943)
20. Wang, Z.: Anonymity, social image, and the competition for volunteers: a case study of the online market for reviews. BE J. Econ. Anal. Policy **10**(1) (2010). doi:10.2202/1935-1682.2523, ISSN (Online) 1935-1682
21. Joyce, E., Kraut, R.E.: Predicting continued participation in newsgroups. J. Computer-Mediated Commun. **11**, 723–747 (2006)
22. Harper, Y.C.F.M., Konstan, J., Li, S.X.: Social comparisons and contributions to online communities: a field experiment on MovieLens. Am. Econ. Rev. **100**, 1358–1398 (2010)
23. Lerner, J., Tirole, J.: Some simple economics of open source. J. Ind. Econ. **50**(2), 197–234 (2002)

24. Aiello, L.M., Petkos, G., Martin, C., Corney, D., Papadopoulos, S., Skraba, R., Goker, A., Kompatsiaris, I., Jaimes, A.: Sensing trending topics in Twitter. IEEE Trans. Multimedia **15**(6), 1268–1282 (2013)
25. Vavilis, S., Petković, M., Zannone, N.: A reference model for reputation systems. Decis. Support Syst. **61**, 147–154 (2014)
26. Hendrikx, F., Bubendorfer, K., Chard, R.: Reputation systems: a survey and taxonomy. J. Parallel Distrib. Comput. **75**, 184–197 (2015)
27. Archak, N.: Money, glory and cheap talk: analyzing strategic behavior of contestants in simultaneous crowdsourcing contests on TopCoder.com. In: Proceedings of the 19th International Conference on World Wide Web, pp. 21–30. ACM (2010)
28. Dellarocas, C.: Designing reputation systems for the social web. Boston U. School of Management Research Paper (2010)
29. Deterding, S., et al.: From game design elements to gamefulness: defining gamification. In: Proceedings of the 15th International Academic MindTrek Conference: Envisioning Future Media Environments. ACM (2011)
30. Huang, W.H.Y., Soman, D.: Gamification of Education. Research Report Series. Behavioural Economics in Action (2013)
31. Singh, S.P.: Gamification: A strategic tool for organizational effectiveness. Int. J. Manage. **1**(1), 108–113 (2012)
32. Suh, A., Wagner, C., Liu, L.: The effects of game dynamics on user engagement in gamified systems. In: 2015 48th Hawaii International Conference on System Sciences (HICSS), pp. 672–681. IEEE (2015)
33. Rego, I.D.M.S.: Mobile Language Learning: How Gamification Improves the Experience. In: Handbook of Mobile Teaching and Learning, p. 705 (2015)
34. Exton, G., Murray, L.: Motivation: a proposed taxonomy using gamification (2014)
35. Cramer, H., Rost, M., Holmquist, L.E.: Performing a check-in: emerging practices, norms and 'conflicts' in location-sharing using foursquare. In: Proceedings of the 13th International Conference on Human Computer Interaction with Mobile Devices and Services, pp. 57–66. ACM (2011)
36. Lindqvist, J., Cranshaw, J., Wiese, J., Hong, J., Zimmerman, J.: May. I'm the mayor of my house: examining why people use foursquare-a social-driven location sharing application. In: Proceedings of the SIGCHI Conference on Human Factors in Computing Systems, pp. 2409–2418. ACM (2011)
37. Massung, E., Coyle, D., Cater, K.F., Jay, M., Preist, C.: Using crowdsourcing to support pro-environmental community activism. In: Proceedings of the SIGCHI Conference on Human Factors in Computing Systems, pp. 371–380. ACM (2013)
38. Von Ahn, L.: Games with a purpose. Computer **39**(6), 92–94 (2006)
39. Cooper, S., Khatib, F., Treuille, A., Barbero, J., Lee, J., Beenen, M., Leaver-Fay, A., Baker, D., Popović, Z.: Predicting protein structures with a multiplayer online game. Nature **466**(7307), 756–760 (2010)
40. Kawrykow, A., Roumanis, G., Kam, A., Kwak, D., Leung, C., Wu, C., Zarour, E., Sarmenta, L., Blanchette, M., Waldispühl, J.: Phylo: a citizen science approach for improving multiple sequence alignment. PLoS ONE **7**(3), e31362 (2012)
41. Greenhill, A., Holmes, K., Lintott, C., Simmons, B., Masters, K., Cox, J., Graham, G.: Playing with science: gamised aspects of gamification found on the Online Citizen Science Project-Zooniverse (2014)
42. Forte, A., Bruckman, A.: Why do people write for Wikipedia? incentives to contribute to open–content publishing. In: Proceedings of GROUP 5, pp. 6–9 (2005)

43. Mason, W., Watts, D.J.: Financial incentives and the performance of crowds. ACM SigKDD Explor. Newsl. **11**(2), 100–108 (2010)
44. Lakhani, K.R.: InnoCentive.com (A). Harvard Business School Case, 608, p. 170 (2008)
45. Jøsang, A., Ismail, R., Boyd, C.: A survey of trust and reputation systems for online service provision. Decis. Support Syst. **43**(2), 618–644 (2007)
46. Satsiou, A., Tassiulas, L.: Reputation-based resource allocation in P2P systems of rational users. IEEE Trans. Parallel Distrib. Syst. **21**(4), 466–479 (2010)
47. Robertson, M.: Can't play, won't play (2010). http://hideandseek.net/2010/10/06/cant-play-wont-play/. Accessed 01 Mar 2016
48. Burke, M., Marlow, C., Lento, T.: Feed me: motivating newcomer contribution in social network sites. In: Proceedings of the SIGCHI Conference on Human Factors in Computing Systems, pp. 945–954. ACM (2009)
49. Schenk, E., Guittard, C.: Towards a characterization of crowdsourcing practices. J. Innovation Econ. Manage. **1**, 93–107 (2011)
50. Rogstadius, J., Kostakos, V., Kittur, A., Smus, B., Laredo, J., Vukovic, M.: An assessment of intrinsic and extrinsic motivation on task performance in crowdsourcing markets. ICWSM **11**, 17–21 (2011)

Results of a Collective Awareness Platforms Investigation

Giovanna Pacini[1](✉) and Franco Bagnoli[1,2]

[1] Department of Physics and Astronomy and CSDC-Center for the Study of Complex Dynamics, University of Florence, Via G. Sansone 1, Sesto Fiorentino, Florence, Italy
{giovanna.pacini,franco.bagnoli}@unifi.it
[2] INFN, sez, Florence, Italy

Abstract. In this paper we provide two introductory analyses of CAPs, based exclusively on the analysis of documents found on the Internet. The first analysis allowed us to investigate the world of CAPs, in particular for what concerned their status (dead or alive), the scope of those platforms and the typology of users. In order to develop a more accurate model of CAPs, and to understand more deeply the motivation of the users and the type of expected payoff, we analysed those CAPs from the above list that are still alive and we used two models developed for what concerned the virtual community and the collective intelligence.

Keywords: CAPs · Virtual community · Collective intelligence

1 CAPs: Brief Review

Collective Awareness Platforms are important crowdsourcing instruments that may promote cooperation, emergence of collective intelligence, participation and promotion of virtuous behaviours in the fields of social life, energy, sustainable environment, health, transportation, etc. [1, 2].

CAPs do not obey in general to the usual market dynamics: they are developed by volunteers or after a public support (namely, EU projects). Also the participation of the public in CAPs is not due to an immediate return, and there are several motivations, exposed in the following, whose lack of analysis may lead to the failure of the CAP, with an evident waste of effort and public funding.

The core of our investigation is that of examining the motivations for the participation in CAPs based on a model of the individual user based on what is known of human behaviour beyond rationality: human heuristics, emotional components, peers and group influence. In particular, we shall analyse the role of payoff (which in general depends non-linearly on the number of participants), incentives, motivations (reputation, emotional components) and community structure.

2 The Analysis

We analyzed 70 CAPs selected from those financed by EU in the last calls and others involved in European projects. To seek support for modeling the behavior of CAPs the survey followed the principle of group-specific purposive sampling. The CAPs under

© Springer International Publishing AG 2016
F. Bagnoli et al. (Eds.): INSCI 2016, LNCS 9934, pp. 19–26, 2016.
DOI: 10.1007/978-3-319-45982-0_2

examination are extremely varied and therefore we tried to identify a limited set of dimensions to be investigated. The main points that we would like to study are:

- The subject of the action of the CAP: which field/problem/need this platform is addressing.
- The health state of the CAP: is it alive, dead, completed or failed?
- Number of participants, kind of community/group/hierarchical structure that the CAP is promoting.
- Messages and communications among participants, communication network.
- Role and structure of the expected payoff from the point of view of users.

2.1 Applicative Field, Status and Target

We divided the CAPs according to their target fields, as shown in Fig. 1. Sustainability, ITC and sociology cover almost 60 % of the total.

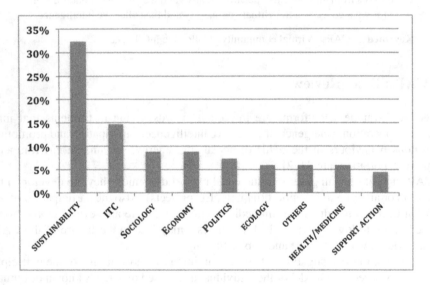

Fig. 1. Application field of CAPs

For what concerns the activity level of CAPS we found that the 75 % of them are alive after at least three years from opening. Some of them may have moved their activity to other media (such as Facebook). What is remarkable (and will be the subject of a further investigation) is that inactive CAPs are, almost all, platforms developed within European projects.

An important aspect of this study is the evaluation of CAPs audience, intended both as a number and as a type of user. For what concerns the geographical target, we found that 50 % have a worldwide audience, 20 % a European one and the others are devoted to local targets. We divided the audience by category of users that may be involved in a CAP; almost half of the CAPS analyzed are addressed to citizens and about a 20 % are dedicated to researchers.

2.2 Social Media Impact

It is very difficult to estimate the number of users from the data obtainable from web site. In many cases we collected data from Facebook and Twitter as reported collectively in Fig. 2.

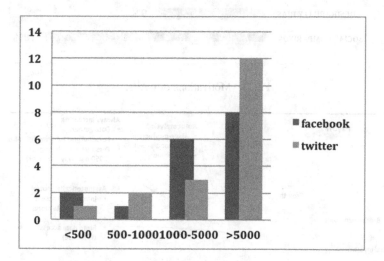

Fig. 2. Social media impact. In the horizontal axis there is the number of likes/followers; in the vertical axis the number of CAPS.

3   Second Analysis

In order to develop a more accurate model of CAPs, and to understand more deeply the motivation of the users and the type of expected payoff, we further analysed those CAPs from the above list that are still alive. Our results are based exclusively on the analysis of documents found on the Internet.

We tried to understand why a user should use a CAP and we highlighted some reasons (there may be several reasons for each CAP). The results are reported in Fig. 3.

Let us now examine the type and scaling of payoff, i.e., the expected return for an user investing time and maybe money into a CAP. We can suppose four different scenarios, see Fig. 4.

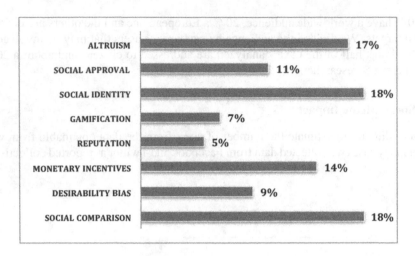

Fig. 3. Motivations of CAPs.

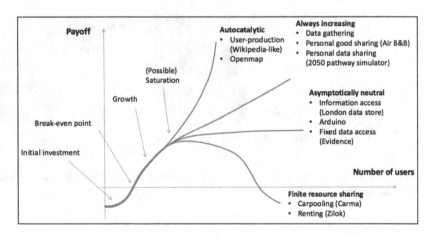

Fig. 4. Payoff versus number of users for different scenarios.

The first one is the autocatalytic one, which can be expected if the value of a CAP is given by the direct user-production (Wikipedia for instance). In this case, the more the users the more the payoff. The second scenario (always increasing) is similar, except that user participate in cataloguing and searching data, not in their production. It is the case for instance of AirB&B. The third scenario, asymptotically neutral, is given by CAPs that provide access to static pieces of information. After that user contributed, for instance by discussing and furnishing support, they may expect to receive a return which does not depend on the number of users. The final scenario, finite resource sharing, is typical of CAPs offering tools for accessing a finite resource, for instance alerting about free park slots. In this case there is an optimum in the number of users, after which the payoff decreases.

This payoff scaling does not mean that users could not be interested in accessing a given resource. As shown in Fig. 5, most of payoff (39 %) is in form of information, money saving for 21 %, social capital for 33 % and skill for 6 %. As typical for this kind of resource, the payoff mainly increases with the number of users (54 %), see Fig. 6.

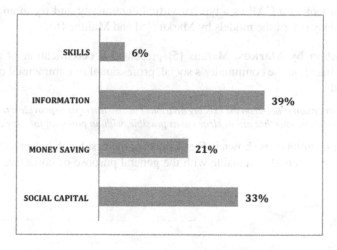

Fig. 5. Type of payoff

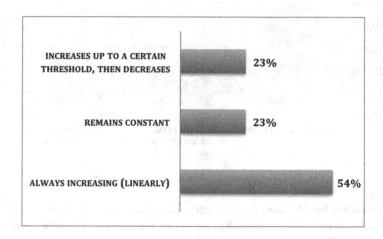

Fig. 6. Payoff scaling.

3.1 CAPs as Virtual Communities

Virtual communities are well know and well studied starting probably from 1985 by Rheingold [3]. We take the definition of *virtual community* from Porter [4]. A virtual community can be defined as

- An aggregation of individuals or business partners
- Who interact around a shared interest,
- Where the interaction is at least partially supported and/or mediated by technology and guided by some protocols or norms.

So we can look at a CAP as a type of virtual community and try to analyse them within the framework of the models by Markus [5] and Malone [6].

Characterization by Markus. Markus [5] presented a classification of the virtual communities based on the community's social, professional or commercial orientation. She explained:

> *This characterization is based on the existing divisions but also attempts to provide a framework for establishing divisions that are as clear cut as possible, without potential for overlaps.*

The CAPs examined, as shown in Fig. 7 only for the first level, have mostly a social orientation, which is understandable with the general purpose of collective awareness platforms.

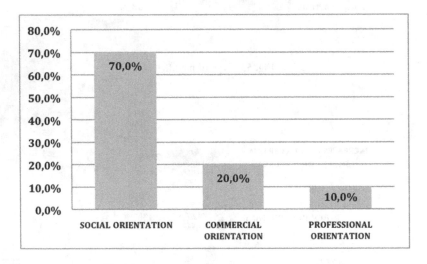

Fig. 7. Orientation of CAPs as virtual communities.

Characterization by Malone. Another very interesting work by Malone *et al.* [6] is based on the concept collective intelligence. Their idea was to describe every system (in particular an IT System) in terms of blocks of collective intelligence, so that one can speak of the "genome" of a collective intelligence system. The identification of the genome is based on four questions:

- What is being done?
- Who is doing it?
- Why are they doing it?
- How is it being done?

The first level can be further specified.

What? This is the first question to be answered for any activity. It is the mission or goal or simply the task. The task can be to *Create* (make something new) or *Decide* (evaluate and select alternative).

Who? The question is about who undertakes an activity. Possible answers are: *Hierarchy*, (someone in authority assigns a particular person or group to perform a task) or *Crowd*, (anyone in a large group who chooses to do so).

Why? This question deals with incentives, the reason for which people take part in the activity. What motivates them? What incentives are at work? The possible answers are: *Money*, where participants earn money from the activity, *Love*, in the sense of intrinsic enjoyment of the activity, the opportunity to socialize, the idea of contributing to something larger than themselves, *Glory* (or reputation), which is the recognition of themselves among their community.

In this our first analyze we did not investigate the question "how" and tried to apply the classification to different roles inside a CAP (user, owners, researcher and so on). The results are shown in Fig. 8.

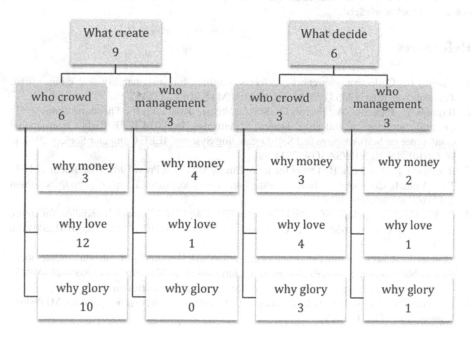

Fig. 8. Genes of collective intelligence of CAPs. The numbers represent how many times we have found the genes.

4   Conclusions

We performed two studies on the aspects that may influence the performances and health status of CAPs. The main lessons are:

- Almost two-third of CAPs are developing applications on sustainability, ITC and sociology.
- Most CAPs do not report on their stakeholders outreach and audiences and therefore it is impossible to evaluate their impact.
- As for potential user involvement, almost half of the CAPs analysed are addressed to citizens and one fifth are dedicated to researchers.
- Most of alive CAPs share information as the payoff for users, and since this resource is in general furnished by users, the expected payoff per user is constant or increases with the number of users themselves, and this is an indicator of a possible further increase of the CAPs audience.

Acknowledgements.   The support of the European Commission under the FP7 Programme Collective-Awareness Platforms under the Grant Agreement no ICT-611299 for the SciCafe2.0 is gratefully acknowledged.

References

1. Sestini, F.: Collective awareness platforms: engines for sustainability and ethics. IEEE Technol. Soc. Mag. **12**, 53 (2012). doi:10.1109/MTS.2012.2225457
2. Bagnoli, F., Guazzini, A., Pacini, G., Stavrakakis, I., Kokolaki, E., Theodorakopoulos, G.: Cognitive structure of collective awareness platforms. In: 2014 IEEE Eighth International Conference on Self-Adaptive and Self-Organizing Systems, IEEE Computer Society 2014, p. 96 (2014). doi:10.1109/SASOW.2014.38
3. Rheingold, H., Landreth, B.: Out of the Inner Circle, Microsoft Press 1985. Also Rheingold, H.: The Virtual Community: Homesteading on the Electronic Frontier. http://www.rhein gold.com/vc/book/intro.html
4. Porter, C.E.: A typology of virtual communities: a multi-disciplinary foundation for future research. J. Computer-Mediated Commun. **10** (2000). doi:10.1111/j.1083-6101.2004.tb0 0228.x
5. Markus, U.: Characterizing the virtual community. SAP Design Guild, 5th edn., (September). Accessible from the WayBack machine. https://web.archive.org/web/20030728104943/, http://www.sapdesignguild.org/editions/edition5/print_communities.html
6. Malone, T.W., Laubacher, R., Dellarocas, C.: The collective intelligence genome. MIT Sloan Manage. Rev. **51**, 21 (2010)

Debate About the Concept of Value in Commons-Based Peer Production

Mayo Fuster Morell[1(✉)], Jorge L. Salcedo[2], and Marco Berlinguer[3]

[1] Berkman Center Harvard, IN3 UOC, Av. Carl Friedrich Gauss, 5,
08860 Castelldefels, Spain
mayo.fuster@eui.eu
[2] Universitat Oberta Catalunya, Av. Carl Friedrich Gauss, 5, 08860 Castelldefels, Spain
jsalcedoma@uoc.edu
[3] IGOP, UAB Edifici MRA, 08193 Bellaterra, Barcelona, Spain
marco.berlinguer@gmail.com

Abstract. We describe a new model of collaborative production called Commons-based peer production (CBPP). This model is frequently supported by digital platforms characterized by peer to peer relationships, resulting in the provision of common resources. Traditionally, it is associated with cases such as Wikipedia or Free Software, but we have recently observed an expansion into other areas. On the basis of an extensive empirical work, we enquired -How does CBPP apply value? and How does value creation function in CBPP? We present an updated version of the meaning of value and sustain the relevance of this debate. After that, we propose how to measure value. We formulate what we call internal and external indicators of value. The first are linked to the internal performance of the CBPP and the second relates to its social value and reputation. Finally we highlight the main features of value that we identified and discuss the limits that we found developing and implementing the proposed diversity indicators.

Keywords: Commons-based peer production · Collaborative economy · Peer to peer production · Value production · Crowd-sourcing

1 Introduction

Several authors have defined CBPP, most importantly Yochai Benkler [8], who partly relying on the work on the traditional commons developed by the 2009 Nobel Laureate Elinor Ostrom [19] systematized a new concept aimed at grasping an emerging and distinctive model of production: Commons-based peer production (CBPP) [7, 8]. Benkler created the term CBPP to describe forms of production in which, with the aid of the Internet, the creative energy of a large number of people is coordinated into large, meaningful projects without relying on traditional hierarchical organizations or monetary exchanges and rewards [8].

But apart from Benkler's initial work, the CBPP concept is still theoretically underdeveloped and is almost nonexistent as an empirically supported theory. After reviewing the previously mentioned characteristics of CBPP, through a questionnaire given to

© Springer International Publishing AG 2016
F. Bagnoli et al. (Eds.): INSCI 2016, LNCS 9934, pp. 27–41, 2016.
DOI: 10.1007/978-3-319-45982-0_3

experts, we have come up with a set of criteria in terms of the delimitation and classification of CBPP (see an extended presentation in "Criteria of Delimitation" [10]). These criteria also define our unit of analysis.

This collaborative production model is frequently enforced or supported through a digital platform, resulting in the provision of common resources. It agglutinates a set of diverse areas of activities and cases that tend to be characterized by peer to peer relationships (in contrast to the traditionally hierarchical command and contractual relationships, with limited mercantile exchange), and/or results in the (generally) open access provision of commons resources that favor open access, reproducibility and derivativeness.

Traditionally, it is associated with cases such as Wikipedia or Free Software, but we have recently observed an expansion into other areas of this production model. For instance, on platforms dealing with car sharing, house sharing, apps exchanging and selling second hand objects or sharing specialized knowledge and notes among university students.

The proliferation and diversity of collaborative platforms is creating significant problems for traditional conceptions of productivity and value. First, because of the growing economic relevance of these types of platforms [4], and secondly due to the problem of how to regulate and reward activities that presently have no market value (e.g. the externalities produced by Free Software for the software industry).

In this vein, the paper addresses these central questions. **How does CBPP apply value?, How does value creation function in CBPP? And what type of value is created?** To answer these research questions, the paper presents the following sections: first, we make a short review of the latest value studies on CBPP and we debate about the relevance of value indicators beyond traditional monetary indicators. We approach the construction of a framework to investigate value within CBPP, providing a set of dimensions of value and applying them empirically. Next, we explain the methods on how we built -to the best of our knowledge- the biggest CBPP database in order to answer, with strong empirical support, our research question. In this section, we also explain the type of statistical analysis that we ran to identify patterns on how CBPP generate value. In the results section we indicate the multiple dimensions of value we developed and test if they are correlated between them. When we present the dimensions of value, we talk about what we called indicators of internal and external value. Finally, we discuss some preliminary conclusions about the generation of value in CBPP and we present further lines of research.

2   The Debate About Value and the Need to Build Value Indicators

The proliferation of collaborative communities is creating significant problems for traditional conceptions of productivity and value. As far back as the 1980s, new forms of collaborative knowledge work were challenging notions of white-collar productivity, rendering the measurement and management of knowledge production problematic [1]. Since the 1990s, questions about the meaning and measurement of value have been raised due to an increasing reliance on socialized forms of collaborative knowledge production in the creative industries [9, 21], in the creation and maintenance of reputation in brand communities [3], in various forms of user-driven innovation [24] and in shared, open, and free forms of productive relations [6, 14, 18]. The ability to measure and define valuable intangible assets—such as brands, intellectual capital and organizational flexibility—remains a pressing problem given the increasing importance of these assets, which are estimated to account for around 70 % of the market value of S&P 500 companies [4]. New definitions of value are necessary to evaluate the contribution of the wide diversity of productive activities.

However, the question of value in collaborative communities is not only an economic one, but also a question of justice. The problem of how to regulate and reward activities that have at present no market value (e.g. the externalities produced by Free Software for the software industry) is contingent on the ability to find a rational and transparent measure of value. The latest developments have emphasized the diversity of notions of value that operate within the information economy. In this paper, we approach the construction of a framework in order to investigate value in CBPP, providing a set of dimensions of value and applying them empirically.

Strategies to quantify the value produced by CBPP by using monetary metrics -for example, quantifying the cost of the work time necessary for the production of its outcomes or by estimating the "consumer surplus" by price experiments- fail to recognize the specificity of these forms of production. Our approach -to a large extent- bypasses the monetary metrics (for a similar strategy, see Wenger et al. [25]). Arguably, without money as a general equivalent, what happens is that the notion of value breaks down into a world of uncertainty, contention and plurality of meanings. However, our choice goes along the growing understanding that "any evaluation exercise should always incorporate a plurality of perspectives on what constitutes value" [17].

2.1 Our Contribution to the Debate About Value

The application of conventional value metrics is increasingly problematic not only in CBPP, but more generally in information and knowledge economics. New definitions of value are necessary in order to evaluate the contribution of the wide diversity of productive activities. We approach the theoretical and empirical foundations for building a framework to investigate value in CBPP by providing a set of dimensions of value, and applying them empirically. There are clearly five different dimensions of value and they have diverse data sources. On the one hand -concerning the dimensions related to community building, objective accomplishment and monetary value-, the data sources were the same CBPP communities we questioned through a survey. From now on, these indicators will be named as "Internal Indicators of value".[1]

---

[1] When building our conceptual framework, we also identified a sixth dimension, which we called Ecological value. However, according to our understanding, this dimension could be quite distinctive and crucial in grasping value within CBPP. We could not find any feasible indicator to operationalize it.

Indicators of Internal Value

Community Building. The ratio, underlying the use of the dimension of the community surrounding the project as a proxy to assess the value generated by it, is that people participation as such is both a sign and a generator of value. On the one hand, the creation of a community is a productive result per se. Additionally, indicators of participation can be considered proxies of productive energies applied to production (and as proxies of the value of the work mobilized). At the same time, participation is an implicit indication of perceived value [13, 25]. Moreover, in many cases participation generates loops of value generation, through network effects [12] and increasing returns [2, 16].

Objective Accomplishment. The dimension of objective accomplishment focuses on a self-defined (indigenous) definition of success, rather than an "objective", universal, external metric. It defines the value achieved, not in terms of monetary value, but in terms of the achievement of substantive missions that motivate the convergence of the stakeholders' efforts. This strategy programmatically desists from identifying a universal, comparable measure among different projects. Rather, it assumes the uniqueness of the features and value programs of each one (along Ostrom's insistence about the singularity of each). Yet, though it recognizes a plurality of definitions/standards/measures of value, at the same time this definition of value allows -to a certain extent-making comparisons, through a level of accomplishment scale of a mission from an applied subjective perspective. Additionally, this approach helps to catch the ad hoc, problem solving, mission-driven logic of many of these collective forms of collaborative action/production. Thus, it potentially accommodates a plurality of organizational configurations and relativizes the importance of the size and duration of the projects.

Monetary Value. In principle, commons and digital commons are not commodified. Thus, the capability of monetary metrics of capturing their core value is very limited. However, in many cases we observe hybrids rather than only "pure commons", including commercial companies developing commons (often playing on the multi-layered outcomes, typical of this form of production). In fact, money can intervene at different moments, through different channels and with different functions within CBPP. For example, it can be a means to cover the costs of the development of the first copy of a resource, afterwards released as a commons; it can contribute in different ways to the sustainability of a project; it can even be the (indirect) core objective of the main developers (e.g. with Google's Android). Moreover, in our society monetary economy dominates at large. Thus, it would be a mistake to completely overlook this dimension. However, at the same time, monetary metrics can distort the effective measure of value in CBPP. Just think about Wikipedia, its overall value could be considered higher than its monetary value.

Apart from what we classified as internal value, for the dimensions related to the social use value and reputation, we relied on proxies and indicators directly accessible by web analytics services (provided by Alexa, Google, Kred, Twitter and Facebook), which we collected automatically through scripts. From now on, these last indicators will be named as external indicators of value.

External Indicators of Value

Social Use Value. Conceptually, the usage or consumption of the resource produced by the community is clearly a measure of the value generated. What is more, we could say that a value to be "realized" requires usage or consumption. Production per se is not a sure indicator of the quantity of value generated. On one side, there is a lot of production that fails to provide utility and is not used. On the other side, there can be small productive communities that produce small resources, which -nevertheless- provide great use value (especially in conditions of non-rivalry, in consumption or usage). To a certain extent, this approach attempts to "objectify" the resulting value and gives a social and objective validation to subjective production (and to producers' potentially biased assessment of its value), and socially validated criteria of success and failure. However, it does not deal with the quantity of the resulting resource per se. Since value can be validated and quantified only through actual use/consumption, these indicators also address a sort of community participation, but mainly through actions of consumption/ use. At the same time, this approach -by recognizing value within consumption independently from price- helps to visualize the social value generated by the practices of open access to resources and the costs implied in the practices of imposing exclusion from the consumption of a non-rival resource.

Reputation. Reputation, on the other hand, is a crucial measure of value and success in contemporary economy (as with brands). From the beginning, the research on CBPP highlighted the importance of reputation as both a motivator for participation and a regulative value within community governance [5, 11, 23]. Reputation embodies the subjective and qualitative evaluation of the relevant stakeholders. It can be considered an indigenous, self-defined criteria of success or value that is not measured by money. However, according to Arvidsson and Peitersen [4], reputation can play a broader function. It can potentially aspire to encompass the fundamental functions of currencies in contemporary production -such as measure, storage and embodiment of value- and, along with the progression of digital connectivity, could potentially provide a synthetic, objective, more democratic (and dynamic) base and measure for a new value regime, different from the exchange value, and more suitable for the challenging characters of CBPP.

In order to operationalize the two dimensions of value -social use and reputation-, we have relied on proxies and used "external indicators of value", that is, web analytics services, collected through scripts. In fact, for social usage, there might be communities that provide data on the social use of the resource produced by the community. However, this is not the case for all communities. Additionally, the data on usage provided by the communities is very diverse and difficult to compare. For these reasons, we preferred to rely on these external indicators.

As with any indicator, we have to recognize the limits in the operationalization of the concept of "value", but beyond this inherent constraint in the creation of any indicator, one of our main caveats is having to use corporate indicators to measure what we called "external indicators of value". The main problem with these indicators is the lack of transparency of the algorithms to calculate them. Nevertheless, they are the most accepted indicators to measure social use and reputation on the Web, and they enabled comparing heterogeneous types of CBPP cases that otherwise would have been really difficult to compare. To summarize, we implemented the following dimensions and indicators of value.

3   Methods

The methodology is based on the statistical analysis of a sample of 302 cases. A "code-book"[2] for data collection -a set of indicators related to the analysis variables- was employed.). To create the sample, the use of a probability or random sample has several advantages. The most important benefit is the possibility to make inferences about the population with a certain degree of confidence. Randomization increases the likelihood that a large sample reflects the characteristics of the underlying population by avoiding assignment or selection based on the value of the variables of interest. However, randomization does not guarantee a representative sample per se. Additionally, random selection involves the risk of "missing relevant cases" [15]. Finally, there are limitations (such as the uncertainty regarding representativeness) to applying randomness to a population that is highly diverse and has an unknown size and boundaries [22]. In other words, using probability samples requires knowledge of the population—for instance, a list or census of the population, or at least a partial list -at some level- of the population. This is not the case in CBPP, which is diverse and whose "universe" is unknown.

Given the lack of adequate conditions and the unsuitability of developing a probability sample of diverse CBPP experiences, as well as the absence of a comparability goal, we used non-proportional quota sampling to build the sample of 302 cases. Our goal does not focus on representation. Rather, the sampling aims to support an analysis that allows us to compare diverse formulas of CBPP (i.e., a comparability goal). Because this sampling aims to guarantee diversity, we expected to be able to talk about even small groups in the universe of CBPP. We ensured the inclusion of a mixed type of CBPP experiences to reflect the heterogeneity of CBPP. From an initial list of cases identified (around 1000), we used different "matching" criteria to ensure the diversity of the sample. Additionally, in order to improve the robustness of our sample, we ensured the systematization of the sampling.

The case selection strategy for the sample was to filter out all the cases that failed to match the definition of CBPP (our unit of analysis). This pertains to the fulfillment of the delimitation criteria of CBPPs we defined, and which dealt with the presence of four features: collaborative production, peer relations, commons and reproducibility.

We included in our sample a diverse range of experiences, some of which are well known and important, in terms of the different dimensions of value that we considered (Table 1), but we also included many experiences that were almost unknown.

---

[2] https://goo.gl/WcGhCi Codebook (23/03/2016).

Table 1. Dimensions and indicators of value

Internal dimensions of value	Community building	Mission accomplished	Monetary value
	How many people– overall - do you estimate participate in the community?	On a scale of 1–10, how far has the project accomplished its mission?	What is the annual turnover (budget) of the project?
	How many registered accounts are there?		
	How many people do you estimate actively contribute to the community?		
External dimensions of value	Social use value of the resulting resource	Reputation	
	Alexa global rank		
	Google pagerank		
	Alexa inlinks		
	Google search results (putting the domain name of the CBPP case between brackets), all times		
	Google search results (putting the domain name of the CBPP case between brackets), last year		
	Facebook likes		
	Twitter followers		
	Kred: influence and outreach		

The data collection was based on four modalities: data from an open directory of CBPP cases (http://directory.p2pvalue.eu/), where we invited members of the CBPP cases -who in a cooperative way helped us to populate the directory-, a survey sent to the cases and web analytics services (data collected through scripts). Finally, during the data collection, "field notes" on general impressions were kept in a field book.

To guarantee the reliability of our sample, another team member (who collected no data on experiences) was assigned exclusively to randomly test almost 30 % of all the cases and verify the data of some outliers. In this way, we controlled the quality of our data. As for the data obtained through scripts, almost 15 % was manually contrasted.

For the statistical analysis of the data, we applied different non-parametric tests. We were aware that non-parametric methods are not as powerful as parametric ones. However, because non-parametric methods make fewer assumptions, they are more flexible, robust, and applicable to non-quantitative (categorical/nominal) variables. Some of the tests that we applied to our dataset were bivariate non-parametric correlations calculated using Spearman's correlation [20].

4   Dimensions of Value and Descriptive Statistics

4.1   Internal Indicators of Value

Community Building. Data suggests that the scale of the communities is extremely variable (Fig. 1). There is not a very frequent range of number of people engaging or/and contributing. 201 to 1000 (or more) is the most frequent range of people that generally participate in the community and the number of registered accounts (although it is "only" around 20 % of cases for both indicators). In contrast, 51 to 200 (or less) is the most frequent range (23 %) of people that actively contribute to the community. It seems rational and in line with a power law dynamics, that the range of very active participants is lower than that of regular participants.

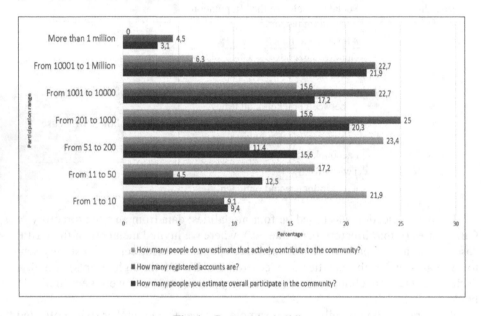

Fig. 1. Community building

Cases do not seem to be composed by very large communities. According to the two first indicators (people that participate and number of registered accounts), 50 % (the median) of cases are below 1000 participants and 60 % of cases (cumulative percent) are below 200 people that participate actively.

Objective Accomplishment. In order to ask the projects to assess their level of mission accomplishment, we asked them to evaluate on a scale of 1–10 how far the project had accomplished its mission. More than 50 % of cases rated their accomplishment from 7 to 10, which could be interpreted as more than medially satisfied in the accomplishment of the mission. The most frequent "score" ranges between 7 and 8 (around 20 % for each score). This suggests that in these cases, participants are quite satisfied (Fig. 2).

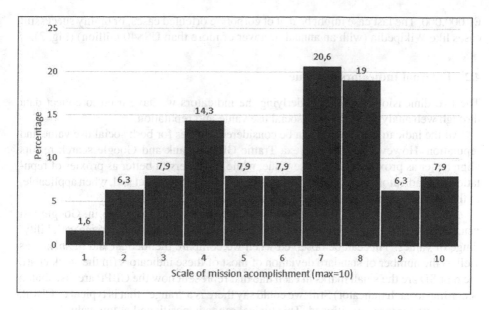

Fig. 2. Mission accomplishment

Monetary Value. In order to have a proxy of the monetary value mobilized around the cases, the survey asked what the annual turnover (budget) of the projects were. What we observed on the dimension of community building is similar to the monetary value of the CBPP communities. The majority of them have an annual budget under 1.000€. The answers obtained showed that 40 % of cases had the lowest turnover level (less than €1.000). This reinforces the idea that CBPP is an activity which has a low level of mercantilization. But around 25 % have more than €100.0000 and 6 % more than

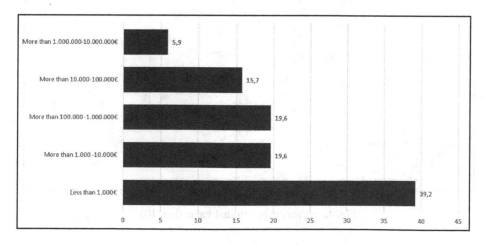

Fig. 3. Monetary value.

€1.000.000. The last case might be that of corporate oriented cases, or highly successful cases like Wikipedia (with an annual turnover of more than US$40 million) (Fig. 3).

4.2   External Indicators of Value

The two dimensions of value underlying the indicators we have used to collect data through web analytics scripts are social use value and reputation.

All the indicators (Table 1) can be considered proxies for both social use value and reputation. However, possibly Alexa Traffic Global Rank and Google search results align better as proxies of social use value, while the others fit better as proxies of reputation. All indicators were applied to the official URL of the project and, when applicable, to the official account of the project on the social networks.

Across most of the indicators (Alexa Global Rank, Alexa Linking in, Google last year, Google all times, Twitter followers, Facebook likes) there is an extreme variability/ range of values. This can be observed when we compare the median and the mean, as well as the number of standard deviation of most of these indicators (in the stock chart, the σ or SD are the small marks in each line that represent how the CBPP are distributed according to each indicator) Still, we could say there is a "range" that is typical of CBPP, where most cases are positioned. This typical range is positioned at low values.

A deviation from the skewed distribution -regarding the concentration of cases in a single range of very low value and very few with very high values- is that of Google Pagerank, the Outreach measure of Kred and the Influence measure of Kred. In these three indicators, 50 % of the observations are near the middle or within the higher range of the scale. The mean and median value of these dimensions suggest (Google Pagerank mean 5.54, median 6 on a scale of 10; Kred Outreach mean 4.37, median 5 on a scale of 10 and Kred Influence mean 694.13, median 727.00 on a scale of 1000) that CBPPs tend to be in the intermediate range of value on the Internet (Fig. 4).

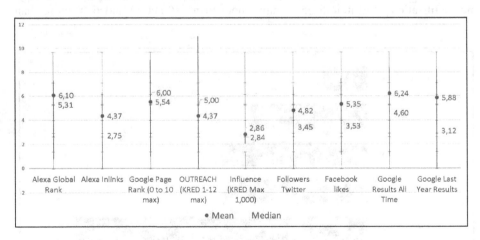

Fig. 4.   Indicators of external value (log_10)

According to the Alexa Traffic Global Rank, 10 % of the sample could be considered as very successful (with a rank lower than 3000). Since the Alexa ranking is applied to the whole universe of Internet websites (the rank goes from 1, the highest value, to more than 6 million), this can be considered as an indicator of the importance of CBPP in the digital economy. Similar conclusions may also be drawn looking at the Google Pagerank.

In regard to Twitter and Facebook, when we analyze the median value of both indicators, it points to a result as high as 50 % of the CBPP, with at least 2,800 followers and more than 3,000 likes. We recognize that we have to contextualize this data, but the majority of CBPPs studied have not been operating for more than 7 years and are relatively young to achieve this high number of followers and likes. The majority of cases tend to be in the middle values of both indicators, so it is frequent for cases to have a considerable number of followers and likes.

5 Correlations Between the Dimensions of Value

In this section, looking at the correlations, we provide data on how the various indicators of value and sub-dimensions of each variable relate to each other. More concretely, the analysis looks at possible explanatory relations between the indicators of internal and external value.

5.1 Between Internal Indicators of Value

Monetary value is moderately correlated (.461** N36) with community building. But it is not correlated with mission accomplishment. This data may suggest that there are communities that just focus on the accomplishment of their mission, neither aiming for nor requiring high monetary turnover or a large engagement of people, but just pursuing the necessary money and people to assure their substantive objective.

5.2 Between External Indicators of Value

We found strong correlations between several indicators of external value (Alexa Traffic Global Rank; Alexa Total Sites Linking In; Google PageRank; Google search all times and last year; Kred1: influence; Kred2: Outreach; Twitter followers; and, Facebook Likes).

We identified that there is a strong correlation between Facebook Likes and Twitter followers (.728** n:175)[3], as well as a very strong correlation between Kred Influence and Twitter followers (.942** n:224), and Kred influence with Facebook likes (.671** n:170). The strong correlation between Kred and Twitter is something expected because of it being the main social medium that this indicator considers when evaluating influence on social media. Nevertheless, we consider it important to highlight how a good reputation on one social medium seems to be related to good reputation on the others.

---

[3] ** => 0.005 and * => 0.05.

We also found a strong correlation (negative, because of the inversion of the scale) between Alexa Traffic Global Rank and Google search results, in its "all times" set (. −790** n:285) and even more in its "last year" set (.−826** n:285). Also, there is a significant correlation between Alexa Total Sites Linking In and Google PageRank (. 725**, n:279). Finally, as it can be expected, the experiences that have a high score in Google search results across all times, also have - in general- a high score in the results limited to last year (corr .806** n:302). This also may mean that a good social value and reputation on one of the external indicators of value reflects a good performance on the others indicators.

5.3 Relationship Between Internal and External Dimensions of Value

We found a low correlation between the index of external dimensions of value (linked to use and reputation) and two of the internal dimensions of value. The index of the external dimensions of value correlate weakly with the index of community building (. 340** n: 64), which would possibly suggest that bigger communities correlate -to a certain extent- with more social use and reputation. The index of the external dimensions of value correlate weakly with monetary value, which would possibly suggest that in order to have visibility and reputation, online monetary power is required (.320* n: 51).

6   Conclusions

As we mentioned at the beginning of this document, value is a complex and practically unexplored concept in the CBPP ecosystem. In this work we propose a framework that considers five dimensions of value, and -with strong empirical support- we identify the main value features of CBPP. We propose to go beyond the monetary formulation of value, considering dimensions of internal and external value. The first dimension is composed of measures such as community building, objective accomplishment and monetary value. The external value dimension is composed by social value and reputation measures, mainly composed by indicators traditionally used on web analytics. These indicators have the advantage that they can be applied on the diverse and heterogeneous cases of CBPP included in our sample.

Regarding what we call internal value, CBPP does not seem to be composed by very large communities. The majority of them are below 200 people that participate actively. In this vein, it seems that the 80−20 work ratio that has been identified in other organizations, applies to CBPPs. That means that a small core of participants -approximately 20 %- assumes the highest level of engagement in comparison with the other members. Nevertheless, it is a hypothesis that we have to explore in depth.

Also, we identified that the majority of CBPPs are satisfied with the accomplishment of their mission, something not necessarily correlated with a high level of community participation or monetary value. According to the indicator of monetary value, the majority of CBPPs have an annual budget under 1.000€. Monetary success does not really seem to be a central motivation.

When we analyzed social use value and reputation we found an extreme variability of values among CBPPs, where there are a few extremely successful cases, but most of them present low values. Nevertheless, on indicators such as Google PageRank, Outreach measure of Kred and the Influence measure of Kred, the CBPPs tend to be in the intermediate range of value on the Internet.

At the moment of testing the different indicators of value, we found a strong correlation between the different indicators of external value, which means that a good performance of CBPP cases on some of the spaces of the Web, for instance social media, also reflects a good reputation and the social value of its web page. However it is something that we have to test on the different types of CBPPs, for instance by means of a cluster analysis.

When we tested if the internal and external indicators of value were correlated, we addressed the fact that the biggest CCBP communities in terms of participation also have a higher social use and reputation, confirming as well something that different community managers know, that is, that to have a higher visibly and better online reputation, it is important to have monetary power.

Regarding some of the limits and necessary improvements of our research, the indicator of objective accomplishment has its own limits. The main problem, in this case, is the subjective assessment of the degree of accomplishment achieved. In respect of the indicators of external value, it is important to say that each indicator, as a proxy, applies differently - with its own problems- to each case configuration. For example, the applicability and reliability of some indicators (like Kred, Twitter, Facebook) depend on the specific use of the social networks given by each project (some do not even use them, while for some of them usage is very marginal). Equally, the values of the Google search results can be more or less distorted, depending on the range of ambiguity that the domain name can generate. More generally, most indicators produce a bias in favor of the projects that are more centralized in their architecture and that are more digitally based. Thus, all of them potentially underestimate the value of projects with a more decentralized architecture and that are less digitally based. Finally, even for the most basic values (like Alexa Traffic Global Rank, Alexa Total Sites Linking, Google PageRank, Number of results by Google search), for a few cases, the values were impossible to collect or plainly wrong: either because the values were too low and the projects came out as not ranked by the web analytics services, or because the websites of the projects were hosted on other platforms (and the measures did not distinguish between the hosted project and the hosting platform).

The origins of the external indicators of value and the control of these by commercial companies clearly expose them to the risk of these services' metrics incorporated biases, and could be influenced by economic interests of providers (e.g. Google metrics could privilege the performance of other Google services, in contrast to the performance of the services of other companies). Additionally, these external value-based indicators on corporate services are not based on FLOSS and the functioning of their algorithms is unknown and non-transparent. That is why they should be used with caution.

An important conclusion from the work undertaken is the need to develop alternative indicators of value (both external and internal to the communities) that are transparent in their functioning. We are also exploring options to adopt Wikipedia visits (if the CBPP has a Wikipedia page) as a potential source of external indicators of value, which is based on FLOSS and is relatively more transparent.

Nevertheless, the limits that the operationalization of value can have on CBPP communities are fundamental to continue improving the indicators and the data available on CBPP, which some authors have denominated a third global model of production.

Acknowledgments. This work is supported by the Framework programme FP7-ICT-2013-10 of the European Commission, through project P2Pvalue (www.p2pvalue.eu) (grant no.: 610961).

References

1. Adler, P.S., Heckscher, C.: Towards collaborative community. In: The Firm as a Collaborative Community: Reconstructing Trust in the Knowledge Economy, pp. 11–105 (2006)
2. Arthur, W.B.: Increasing Returns and Path Dependence in the Economy. University of Michigan Press, Ann Arbor (1994)
3. Arvidsson, A.: Brands: Meaning and Value in Media Culture. Psychology Press, New York (2006)
4. Arvidsson, A., Peitersen, N.: The Ethical Economy: Rebuilding Value After the Crisis. Columbia University Press, New York (2013)
5. Barbrook, R.: The hi-tech gift economy. First Monday 3(12) (1998)
6. Bauwens, M.: Peer to peer and human evolution. Integr. Vis. 15 (2005)
7. Benkler, Y.: Coase's penguin, or, linux and "the nature of the firm". Yale Law J. **112**, 369–446 (2002)
8. Benkler, Y.: The Wealth of Networks: How Social Production Transforms Markets and Freedom. Yale University Press, New Haven (2006)
9. Florida, R.: The Rise of the Creative Class: And How It's Transforming Work, Leisure and Everyday Life. Basic Books, New York (2002)
10. Morell, M.F.: Theoretical synthesis: final theoretical synthesis of WP1, including research reports on data collection. Technical Report #WP1. European Commission (2014)
11. Ghosh, R.A.: Cooking pot markets: an economic model for the trade in free goods and services on the internet. First Monday 3(2) (1998)
12. Katz, M.L., Shapiro, C.: Network externalities, competition, and compatibility. Am. Econ. Rev. **75**(3), 424–440 (1985)
13. Kelty, C.: Two Bits: The Cultural Significance of Free Software. Durham, Duke (2008)
14. Kelty, C.M.: Participation. In: Peters, P. (ed.) Digital Keywords. Princeton University Press, Princeton (2015)
15. King, G., Keohane, R.O., Verba, S.: El diseño de la investigación social: la inferencia científica en los estudios cualitativos. Alianza Editorial, Madrid (2000)
16. Muegge, S.: Platforms, communities, and business ecosystems: lessons learned about technology entrepreneurship in an interconnected world. Technol. Innov. Manage. Rev. 3(2) (2013)
17. Munda, G.: Social multi-criteria evaluation for urban sustainability policies. Land Use Policy **23**(1), 86–94 (2006)
18. O'Neil, M.: Cyberchiefs: Autonomy and Authority in Online Tribes. Pluto Press, London (2009)
19. Ostrom, E.: Governing the Commons: The Evolution of Institutions for Collective Action. Cambridge Univ Pr., Cambridge (1991)

20. Park, H.W., Thelwall, M.: The network approach to web hyperlink research and its utility for science communication. In: Hine, C. (ed.) Virtual Methods Issues in Social Research on the Internet, pp. 171–183. Berg, Oxford (2005)
21. Peck, J.: Struggling with the creative class. Int. J. Urban Reg. Res. **29**(4), 740–770 (2005)
22. Poteete, A.R., Janssen, M., Ostrom, E.: Working Together: Collective Action, the Commons, and Multiple Methods in Practice. Princeton University Press, Princeton (2010)
23. Raymond, E.: The Cathedral & the Bazaar (Revised.). Paperback. O'Reilly Media (2001). Accedido desde. http://biblioweb.sindominio.net/telematica/catedral.html
24. Von Hippel, E.: Democratizing innovation: the evolving phenomenon of user innovation. J. für Betriebswirtschaft **55**(1), 63–78 (2005)
25. Wenger, E., Trayner, B., de Laat, M.: Promoting and assessing value creation in communities and networks: a conceptual framework (2011). http://www.knowledge-architecture.com/downloads/Wenger_Trayner_DeLaat_Value_creation.pdf. Accessed 25 July 2011

Collective Intelligence Heuristic: An Experimental Evidence

Federica Stefanelli[1], Enrico Imbimbo[1(✉)], Franco Bagnoli[2,3], and Andrea Guazzini[1,3]

[1] Department of Science of Education and Psychology, University of Florence, Florence, Italy
{federica.stefanelli,enrico.imbimbo,andrea.guazzini}@unifi.it
[2] Department of Physics and Astronomy, University of Florence and INFN, Florence, Italy
franco.bagnoli@unifi.it
[3] Center for the Study of Complex Dynamics, University of Florence, Florence, Italy

Abstract. The main intrest of this study was to investigate the phenomenon of collective intelligence in an anonymous virtual environment developed for this purpose. In particular, we were interested in studing how dividing a fixed community in different group size, which, in different phases of the experiment, works to solve tasks of different complexity, influences the social problem solving process. The experiments, which have involved 216 university students, showed that the cooperative behaviour is stronger in small groups facing complex tasks: the cooperation probability negatively correlated with both the group size and easiness of task. Individuals seem to activate a collective intelligence heuristics when the problem is too complex. Some psychosocial variables were considered in order to check how they affect the cooperative behaviour of participants, but they do not seem to have a significant impact on individual cooperation probability, supporting the idea that a partial de-individualization operates in virtual environments.

Keywords: Collective intelligence · Crowdsourcing · Cooperation · Social problem-solving · Cognitive heuristics

1 Introduction

During their life, people gradually learn to solve problems of increasing complexity. This ontogenetical evolution was made possible by the corresponding phylogenetic development of species, thanks to which humans have developed the skills to individually learn and to share the knowledge gained [1]. The ability to solve problems in a natural and social environment [3] is know as social problem-solving and it made possible to first humans to confront gradually more complex tasks [4]. All human beings owned the ability to coordinate their activities with that of others in order to solve problems [5]. This can be a results of the role played by natural selection in the evolution of the social nature in

© Springer International Publishing AG 2016
F. Bagnoli et al. (Eds.): INSCI 2016, LNCS 9934, pp. 42–54, 2016.
DOI: 10.1007/978-3-319-45982-0_4

human species [6,7]. Social Darwinism provides a frame for the study of these capacity, not only from a neurophysiology view but also in relation of cognitives activities that underline social-problem solving capabilities [1].

Collective intelligence, is a method of social problem-solving and can be defined as the capacity that a group has to show an intelligence greater than the single members of the group [2]. This capacity is owned by all social species [8] and stabilized evolutionarily as adaptive. Peoples usually turns to groups when they have to solve complex problems because they have better decision-making skills than single individuals since groups can process a larger amount of information, faster and more thoroughly.

Collective intelligence is also a social kind of intelligence because it emerges from cooperative and competitive behaviours [9] acted by members of groups in order to solve a problem [8]. Smith [10] argued that individuals forms collaborative groups because alone they are not able to solve problems perceived as too complex. This could happens either because the single has not all the knowledge and ability needed to solve a task, or because its resolution by a single individual would require him too much time [10]. Given the effectiveness of collective intelligence, one of the most interesting challenges of our time consist in finding a method able to involve all the members of a community in the resolution of common problems and in the generation of innovative ideas.

This process of mass involvement consists in an application of the Collective Intelligence [11,12] and take the name of Crowdsourcing [13,14]. Crowdsourcing demonstrated to be a distributed method of problem-solving [15] extremely effective for a large variety of tasks [16]. It is a web based business model, able to gets more out of creative solution generated by a network of individuals through an open calls for proposals [13]. For Brown and Lauder at the base of Crowdsourcing there is the idea that a group can express an ability in information processing and problem- solving greater than each single member of the group taken alone. This is true especially when the members of the group can strongly interact with each other, and as regards human community, this became possible thanks the diffusion of the latest informatics communication systems like the Internet and, more in general, the World Wide Web [17].

The outdated concept that saw in the single individual the only source of innovation was replaced by a conception that favours a multidisciplinary approach and a collective problem-solving. Three are the dominant ideas: distributed, plural and collaborative [18]. In his book, *The Wisdom of Crowds*, Surowiecki examined several cases in which the solution of a problem depended drastically from the appearing of a vast group of solvers [19]. On the base of this empirical investigations the author concluded that groups can be remarkably intelligent even more their most intelligent member, under appropriate conditions. This wisdom does not depends from the mean of solution proposed by all the members of the group, but by their aggregation and interaction. The web offers the perfect technology able to aggregate millions of people from all over the world, millions of different ideas able to offer intelligent voting system, without disadvantages that came from the excessive communication and the compromise [19]. It is necessary underline that the web is a tool who facilitate the application of this kind of intelligence which already existed before the development of such technologies [19], whatever is the classification that used: success obtained by the application of a distributed intelligence, or the amplification of it [20], or wisdom of crowd [19], or community innovation [21].

Despite of advantages of Crowdsourcing, and Collective Intelligence their true nature still remains elusive. Its knowledge would reveal the basis of their efficacy [22,23]. Huge efforts have been made in order to develop computational and experimental models of social collective behaviour. The standard framework to study the emergency of cooperation in groups interaction environments is the Game Theory, in particular through the use of Public Goods Games and the Prisoner's Dilemma (one-shot or iterated) [24]. Several computational model about the study of decision making in collective behaviour are been created taking into account the size of group. These models showed as cooperative behaviour tends to decrees with the increases of the number of person part of the group [25,26]. An example of recently computational model that, in addition to the dimension of the group, considered also the complexity of the task, has been proposed by Guazzini et al. [1]. It links the Crowdsourcing precess to collective problem-solving. This model was implement both at a computational level, in a evolutionary stochastic system, and at an analytical one, thanks to statistical physics methods. In this study, the authors studied the behaviour of a community composed by 100 virtual agents, that in several different simulations faced the resolution of problems of different complexity divided into groups of different numerosity. The individuals were modelled as perfectly rational agents. The results showed that the maximum fitness was obtained in the condition in which the entire population is divided into two groups of equal size. The cooperation among agents was found to increase with decreasing group size and task difficulty.

The aim of the present study is to test these results with real people, trying to estimate how group size and difficulty of a problem would influence the behaviour of population involved in the resolution of a task. Moreover, the possible implication of psychosocial variables in cooperative behaviours was controlled. To achieve this goal, a multiplayer online game, the Crowdsourcing Game, was developed. In this game, 12 anonymous participants interacted online. Findings, as demonstrated also by mathematical simulations, show that cooperation is more profitable when participants work together to solve simple tasks. The results show also that, unlike rational virtual agents, human beings tend to increase their cooperative behaviour with the increasing difficulty of the task. In psychology the simple, efficient, but not always rational rules that people use to take quick decisions are denoted with the term heuristics. They are mental shortcuts that usually involve focusing on one aspect of a problem ignoring others [27–29].

The present research suggests the existence of a Collective Intelligence heuristics, which promotes cooperation every time people face problems perceived too complex to be solved at the individual level.

Furthermore, according with the SIDE model, this work hypnotize that psychosocial variables of players does not influence in a determinant way their behaviour. The Social Identity Model of Deindividuation Effects (SIDE model) explained how, in condition of computer-mediated communication (CMC), people tend to switch between the salience attributed to their own personal to the social identity [30]. The SIDE model affirms that, through a deindividuation process, people tend to esteem as less salient their own personal identity in favour of an increment of the salience in the importance attributed to the social identity. This could lead to a minimization of differences perceived inside the in-group and the demonstration of behaviours more influenced by social norms who characterize that specific situation [31]. Moreover, the SIDE model said that when the context involves actors as individuals, their personal identity becomes relevant; when the context emphasizes the social identity, actors mostly observe rules associated with the reference group [32,33]. Thus, the CMC modality of interactions could influence the expression of cooperative and altruistic behaviours, if a social norm that encourages such conduct is present in the virtual group.

2 Participants

For the realization of the experiments were recruited 216 participant (150 females, 66 males) aged between 19 and 51 years. The highest level of education achieved by the eighty percent of participants (N = 173) is the high school diploma. The sample was manly composed by singles (48 %) part of at least 3 real social community (M = 3.1; D.S. = 1.34) and members of at least one social network (M = 1.9; D.S. = 1). The recruitment occurred through voluntary census within students and trainees of the School of Psychology and that of Engineering in the University of Florence. Participants were recruited personally by researchers out of the classrooms. To all those who accepted to participate was asked to leave a contact, which was subsequently used in order to inform them of the location and date of the experimental session to which they were assigned. Total anonymity in data processing was guaranteed.

3 Methods Ad Procedures

A multiplayer online game, based on the work of Guazzini et al. [1] exposed above, was developed with the intention to investigate cooperation and competition dynamics in a Crowdsourcing context. To control the possible effect of psychosocial variables in game's behaviour, to participants of the experiments were asked to fill an online preliminary survey composed by a battery of self-reported socio-demographic and psychological questionnaires.

3.1 Materials

The survey was composed by two sections. The first section required to provide information about gender, age, number and members of household, education level, marital status, number and kind of community membership, number and kind of social network used. With the second section are been collected data about personality, honesty, state anxiety, self-efficacy and sense of community of participants, through appropriate psychological validate scales. Specifically, players completed the following scales:

- The Five Factor Adjective Short Test (5-FasT) [34] a personality inventory, used in this work to verify the possible relation between personality factors and the performance obtained in the game. This test consists of five sub-scales each measuring one of the following specific personality trait: Nevroticism, Surgency, Agreebleness, Surgency, Closeness and Conscientiousness.
- The Italian adaptation of Honesty-Humility scale part of the HEXACO personality inventory [35], used to understand if cooperative behaviours could be influenced by the degree of honesty that characterize the person who acts in that way.
- The Italian Adaptation of the State-Trait Anxiety Inventory (STAI) [36], used to verify the existence of a relation between the anxiety of a participant at the moment of playing and his tendency to cooperate.
- The Italian adaptation of General Self-Efficacy Scale [37], used to control the role of self-efficacy in decision-making process
- The Italian adaptation of Social Community scale part of the Classroom and School Community Inventory [38], used to understand if the cooperative behaviour is related to the degree of membership perceived by a player with the anonymous group in which he was insert.

3.2 Software

To perform the experiments and verify the hypotheses of the research, this work foresaw the realization of an online game software that was named *Crowdsourcing Game*. The game was developed through *Google Apps*, a free online service platform, using the *Google Script* programming language based mainly on *JavaSript*. In Fig. 1 is showed the user interface of the game.

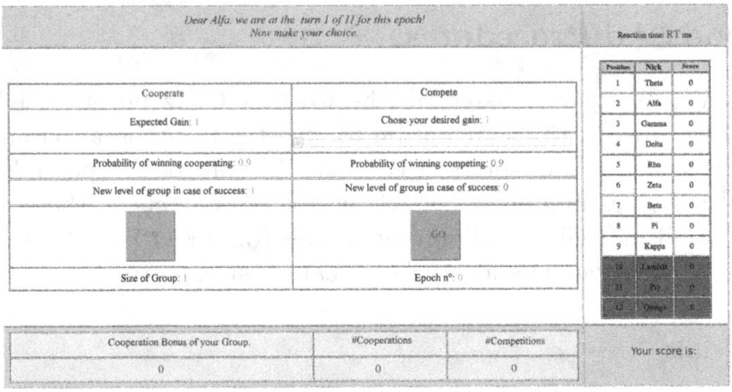

Fig. 1. User interface of the Crowdsourcing Game (original was in Italian language, here is presented the English translation)

3.3   The Crowdsourcing Game's Architecture

Each experimental session involved a N population of 12 players, wich, in four different phases of the match, called epochs, worked divided in n groups of equal numerosity S. The following four different group size have been examined $S = 1$, 3, 6 and 12, dividing respectively the population in 12, 4, 2 and 1 groups. Each experimental session was composed by two match during wich the resolution of a same task, with simplicity quantified by a value R, was assigned to the population. R would assume one of the following values: 0.1, 0.3, 0.5 or 0.9 (R closer to 1 indicates the simplest tasks).

An epoch was composed by 11 rounds where players had to choose either to cooperate or compete with their membership group. Through this choice players could set their expected pay-off. For a cooperator it was set always to minimum value, but the probability of win was fixed at maximum value (R). When a cooperator won, he had to share his gain with all others members of his group, but his expected pay-off was summed to the bonus of his group. The bonus group (group score) increased of one unit, for the next round, every time that in an epoch a cooperator won. Even if more than one cooperator part of same group won, the bonus group would increases of only one unit. Tus, the gain for a winning cooperator was equal to (expected-gain + bonus group)/n. Instead, competitors could choose their expected pay-off from 1 to 10 units, but increasing this latter they decreased the probability of win. Anyway, a competitor did not had to share his gain with any members of the groups and, furthermore, if a member of his group wins cooperating, he receives the shared winnings of all cooperators. Once a player made his choice, the system randomly generated a number between 0 and 1 and compared it with the probability chosen by the player. If the number generated is less than the probability chosen by the player, the exit is a victory for him, otherwise the player looses the round. For every players the goal of the game was to win each epoch individually or at least to not ending in one of the last three positions. No time limit was set for players to make their choice. After 11 rounds the scores were reset to 0 and the size of groups changed.

A total of 18 experiments were performed, in which each participants played two match of the game. All experiments were designed to have a match in which the maximum probability was set to 0.9 and the other match with the maximum probability set to 0.1 or 0.3 or 0.5. The presentation order between the two match was randomized to obtain an equal number of experiments in which the match set to 0.9 preceded or followed the other ones.

The following experimental conditions were obtained:

- 2 experiments in which the session set to 0.9 preceded the session set to 0.1
- 2 experiments in which the session set to 0.9 followed the session set to 0.1
- 3 experiments in which the session set to 0.9 preceded the session set to 0.3
- 3 experiments in which the session set to 0.9 followed the session set to 0.3
- 4 experiments in which the session set to 0.9 preceded the session set to 0.5
- 4 experiments in which the session set to 0.9 followed the session set to 0.5

3.4 Settings

The experiments were conducted within the School of Psychology and the School of Engineering of Florence, in their respective computer labs. Prior to the experimental phase researchers made a series of beta-test for the approximate evaluation of the timing for carrying out an experiment. First, participants completed the preliminary survey. Than, the researchers provided them a detailed description of the rules of the game. Finally, players completed two full sessions of the game. Each experiment lasted approximately an hour and forty-five minutes.

3.5 Data Analysis

To evaluate delineated hypotheses, the experimental log files have been exploited defining three observables as order parameters (i.e. dependent variables). Such variables are the average probability to cooperate of players, the individual score and the group score gained by the participants in each epoch of the game. The control parameters considered in this work (i.e. independent variable) were represented by the size of the groups in which the population was divided, the difficulties of problems asked to being solved and the psychosocial variables collected. Such empirical dimensions are been preprocessed assessing the consistency of the statistical properties required to run the inferential analysis (i.e. distributions' skewness and kurtosis, homogeneity of the examined variances). The analysis of variance method (i.e. ANOVA) was used to analyse the order parameters to reveal statistically significant differences between and among the experimental conditions.

4 Results

The experiments show that the average probability of cooperation decreased with the increase of the size of the groups in which is divided the population of participants, while it increases with the increase of difficulty of the problem. The highest probability to cooperate was seen in small groups who face complex tasks, while the lowest in large groups that address simple tasks (Fig. 2).

The individual score increased with increasing the simplicity of the problem. Moreover, is important to note that a maximum of individual gain, in each condition of difficulty, was observed when the population was divided into two groups of equal numerosity (Fig. 3).

The group score (bonus group) increases as a function of group size and tend to increment with the increase of the simplicity of the problem. There is a maximum-gain point in the situation in which the population is divided into two groups of equal size and the difficulty of the task is 0.5 (Fig. 4).

4.1 Regression Models

In order to predict the individual ad group score of the players in the Crowdsourcing Game and their tendency to cooperate related to psychosocial variables, three linear regression models have been produced (Tables 1 and 2).

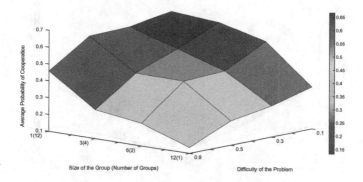

Fig. 2. Average probability to cooperate as function of size of the group and difficulty of the problem

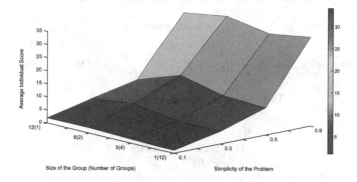

Fig. 3. Individual score as function of size of the group and simplicity of the problem

Table 1. Best model for individual score

R	Adj-R^2	F
.616	.615	1381.822**

$$**p < .0001$$

The linear regression model presented for the individual score explains the 61 % of the variance and can be summarized as follows,

$$0.038(Surgency) + 0.784(Simplicity\ of\ the\ problem) + \varepsilon(error).$$

At the end of each epoch of the game, both energy of players and simplicity of the task resulted significant predictors of the final score of a single participant (Tables 3 and 4).

The linear regression model presented for the group score explains the 32 % of the variance and can be summarized as follows,

$$0.045(Hexaco\ Honesty) + 0.493(Group\ Size) +$$
$$0.267(Simplicity\ of\ the\ problem) + \varepsilon(error).$$

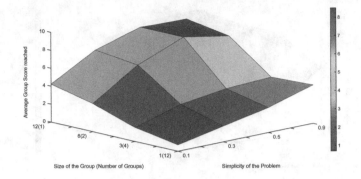

Fig. 4. Group score as function of size of the group and simplicity of the problem

Table 2. Predictors coefficients of the best model for the individual score

Predictor	Stand. coefficient	t	Sig.
Surgency	$\beta_1 = .038$	2.574	$p < .01$
Simplicity of the problem	$\beta_2 = .784$	52.498	$p < .0001$

Table 3. Best model for group score

R	Adj-R^2	F
.317	.315	266.109**

$$**p < .01$$

Honesty of a players, the size of the groups in which the population is divided and the simplicity of the problem faced are significant predictor of the score reached by a group at the and of each epoch (Tables 5 and 6).

The linear regression model presented for the cooperation probability explains the 23 % of the variance and can be summarized as follows,

$$0.056(Neuroticism) - 0.056(Surgency) + 0.048(Conscentiousness)$$
$$+ 0.062(Hexaco\ Honesty) - 0.051(STAI) + 0.049(Sense\ of\ Community)$$
$$- 0.346(Group\ Size) - 0.316(Simplicity\ of\ the\ task) + \varepsilon(error).$$

Table 4. Predictors coefficients of the best model for the group score

Predictor	Stand. coefficient	t	Sig.
Hexaco Honesty	$\beta_1 = .045$	2.265	$p < .05$
Group size	$\beta_2 = .493$	24.782	$p < .0001$
Simplicity of the problem	$\beta_3 = .267$	13.401	$p < .0001$

Table 5. Best model for cooperation probability

R	Adj-R^2	F
.233	.233	65.167**

$**p < .0001$

Table 6. Predictors coefficients of the best model for cooperation probability

Predictor	Stand. coefficient	t	Sig.
Neuroticism	$\beta_1 = .056$	2.200	$p < .05$
Surgency	$\beta_2 = -.056$	-2.507	$p < .01$
Conscientiousness	$\beta_3 = .048$	2.226	$p < .05$
Hexaco Honesty	$\beta_4 = .062$	2.870	$p < .001$
STAI	$\beta_5 = -.051$	-1.992	$p < .05$
Sense of community	$\beta_6 = .049$	2.277	$p < .05$
Group size	$\beta_7 = -.346$	-16.363	$p < .0001$
Simplicity of the task	$\beta_8 = -.316$	-14.931	$p < .0001$

The level of neuroticism of each player, his energy, his conscientiousness, his honesty, the anxiety perceived by him at the moment of the game, his sense of community, the size of the group in which he is insert and the simplicity of the problem to solve are all significant predictors of the probability to cooperate of a player.

5 Discussion

The research presented in this paper attempted to explore collective dynamics involved in Crowdsourcing contexts using an anonymous online community. To do this the computational model of Crowdsourcing, developed by Guazzini et al. [1], was tested with people. The role of groups size, of problem difficulty and of psychosocial variables in cooperative an competitive behaviours was evaluated. First of all, findings showed how people cooperated greatly when faced with problems characterized for their high difficulty of resolution, acting in an irrational way respect virtual agents of the model. The higher trend for cooperation was observed for small groups involved in the resolution of complex tasks. This exhibition of a collective behaviour in solving difficult problems also in an environment who encouraged competitive actions (how showed in the theoretical simulation of the model) could be seen as the manifestation of a Collective Intelligence Heuristic owned by human beings.

Such heuristic could unconsciously bring people to look for a collective action when abilities to solve a task or resources needed to deal with a problem are perceived to be higher than those of the individual. Secondarily, a maximum gain for what concern both individual and group score achieved by participants was observed in the condition which the population was divided into two different competing groups of equal numerosity. In particular, the maximum peak of gain for the group score appeared in the situation in which players were divided into two groups of same size and tried to solve an uncertain problem (i.e. probability being solved $= 0.5$). Finally, the psychosocial variable of participants, controlled to verify their role in shaping behaviour, during the experiments are found have a minor and not so relevant impact than other control parameters. This suggested that in anonymous Crowdsourcing dynamics subjective difference can be omitted to predict the cooperative behaviour of participants of the community.

The main limitation of the study can be found in the sample. Indeed, it was composed mostly by young university students and, for this reason, it was not rappresenatative of the population. Moreover, it was not possible to equaliz it for gender. Also, because of the narrowness of the sample, it was not possible to execute the same number of trials for each difficulty level. Another limitation is constituted by the self report nature of the survey used to collect information about psychosocial variables.

The present research gives a baseline framework for the study of Crowdsourcing dynamics providing some evidence about possible ergonomic variables to keep in consideration in the view of enhancing cooperative, collaborative and collective behaviour. This can improves the efficiency of platform and service that already use this method of problem-solving [39–44]. These findings shown that presenting a problem perceived as challenging to a working community dived into two groups of equal numerosity, it's possible promote the cooperative trend of the members.

Acknowledgement. This work was supported by EU Commission (FP7-ICT-2013-10) Proposal No. 611299 SciCafe 2.0.

References

1. Guazzini, A., Vilone, D., Donati, C., Nardi, A., Levnajić, Z.: Modeling Crowdsourcing as collective problem solving. Sci. Rep. **5**, 16557 (2015)
2. Singh, V.K.: Collective intelligence: concepts, analytics and implications. In: 5th Conferencia; INDIACom-2011. Computing For Nation Development, Bharati Vidyapeeth. Institute of Computer Applications and Management, New Delhi, March 2011. ISBN: 978-93-80544-00-7
3. D'Zurilla, T.J.: Problem-SolvingTherapy: A Social Competence Approach to Clinical Intervention. Srpinger, New York (1986)
4. Molen, M., Snchez-Zapata, J.A., Margalida, A., Carrete, M., Owen-Smith, N., Donzar, J.A.: Humans and scavengers: the evolution of interactions and ecosystem services. BioScience (2014). doi:10.1093/biosci/biu034
5. Dumas, G.: Towards a two-body neuroscience. Commun. Integr. Biol. **4**, 349–352 (2011)
6. Darwin, C.: The Descent of Man and Selection in Relation to Sex. Princeton University Press, Princeton (1871)
7. Van Lawick-Goodall, J.: The behaviour of free-living chimpanzees in the Gombe stream reserve. In: Van Lawick-Goodall, J. (ed.) Animal Behaviour Monographs. Rutgers University Press, Columbia (1968)

8. Luo, S., Xia, H., Yoshida, T., Wang, Z.: Toward collective intelligence of online communities: a primitive conceptual model. J. Syst. Sci. Syst. Eng. **18**(2), 203–221 (2009)

9. Lvy, P.: LIntelligence collective. Pour une anthropologie du cyberespace. La Dcouverte, Paris (1994)

10. Smith, J.B.: Collective Intelligence in Computer-based Collaboration. Lawrence Eribaum, Hillsdale (1994)

11. Arolas, E., Guevara, F.: Towards an integrated crowdsourcing definition. J. Inf. Sci. **38**, 189–200 (2012)

12. Zhao, Y., Zhu, Q.: Evaluation on crowdsourcing research: current status and future direction. Inf. Syst. Front. **16**, 417–434 (2014)

13. Howe, J.: The rise of crowdsourcing. Wired Mag. **14**(6), 14 (2006)

14. Surowiecki, J., Silverman, M.P.: The wisdom of crowds. Am. J. Phys. **75**, 190–192 (2007)

15. Brabham, D.C.: Crowdsourcing as a model for problem solving an introduction and cases. Convergence: Int. J. Res. New Media Technol. **14**(1), 75–90 (2008)

16. Minder, P., Bernstein, A.: CrowdLangfirst steps towards programmable human computers for general computation. In: Workshops at the Twenty-Fifth AAAI Conference on Artificial Intelligence (2011)

17. Brown, P., Lauder, H.: Collective intelligence. In: Schuller, T., Baron, S., Field, J. (eds.) Social Capital: Critical Perspectives, pp. 1–38. Oxford University Press, Oxford (2000)

18. Mau, B., Leonard, J.: Massive Change. Phaidon, New York (2004)

19. Surowiecki, J.: The Wisdom of Crowds: Why the Many are Smarter than the Few and How Collective Wisdom Shapes Business, Economies, Societies, and Nations. Doubleday, New York (2004)

20. Bush, V.: As we may think. Atlantic Mon. **176**(1), 10–18 (1945)

21. Von Hippel, E.: The Sources of Innovation. Oxford University Press, New York (1988)

22. Hackman, J.R.: Collaborative Intelligence: Using Teams to Solve Hard Problems. Barrett-Koehler Publishers Inc., Oakland (2011)

23. Sornette, D., Maillart, T., Ghezzi, G.: How Much Is the Whole Really More than the Sum of Its Parts? 1+1=2.5: superlinear productivity in collective group actions. Plosone **9**(8), e103023 (2014)

24. Sigmund, K.: The Calculus of Selfishness. Princeton University Press, Princeton (2010)

25. Szolnoki, A., Perc, M.: Group-size effects on the evolution of cooperation in the spatial public goods game. Phys. Rev. E **84**(4), 047102 (2011)

26. Hamburger, H., Guyer, M., Fox, J.: Group size and cooperation. J. Conflict Resolut. **19**(3), 503–531 (1975)

27. Lewis, A.: The Cambridge Handbook of Psychology and Economic Behaviour, p. 43. Cambridge University Press, Cambridge (2008). ISBN: 978-0-521-85665-2

28. Harris, L.A.: CliffsAP Psychology. Wiley, New York (2007). ISBN: 978-0-470-19718-9

29. Nevid, J.S.: Psychology: Concepts and Applications, p. 251. Cengage Learning, Belmont (2008). ISBN: 978-0-547-14814-4

30. Lea, M., Spears, R.: Computer-mediated communication, de-individuation and group decision-making. Int. J. Man Mach. Stud. **34**(2), 283–301 (1991)

31. Spears, R., Lea, M.: Panacea or panopticon? The hidden power in computer- mediated communication. Commun. Res. **21**(4), 427–459 (1994)

32. Reicher, S.D., Spears, R., Postmes, T.: A social identity model of deindividuation phenomena. Eur. Rev. Soc. Psychol. **6**(1), 161–198 (1995)
33. Postmes, T., Spears, R., Lea, M.: Breaching or building social boundaries? SIDE- effects of computer-mediated communication. Commun. Res. **25**(6), 689–715 (1998)
34. Giannini, M., Pannocchia, L., Grotto, R.L., Gori, A.: A measure for counseling: the five-factor adjective short test (5-fast). Counseling. Giornale Italiano di Ricerca e Applicazioni **3**, 384 (2012)
35. Lee, K., Ashton, M.C.: Psychometric properties of the HEXACO personality inventory. Multivar. Behav. Res. **39**(2), 329–358 (2004)
36. Spielberg, C.D.: Manual for the State-Trait Anxiety Inventory STAI (Form Y). Consulting Psychologists Press, Palo Alto (1983)
37. Sibilia, L., Schwarzer, R., Jerusalem, M.: Italian adaptation of the general self-efficacy scale. Resource document. Ralf Schwarzer web site (1995)
38. Prezza, M., Pacilli, M.G., Barbaranelli, C., Zampatti, E.: The MTSOCS: a multidimensional sense of community scale for local communities. J. Commun. Psychol. **37**(3), 305–326 (2009)
39. Schwartz, C., Borchert, K., Hirth, M., Tran-Gia, P.: Modeling crowdsourcing platforms to enable workforce dimensioning. In: International Telecommunication Networks and Applications Conference (ITNAC), pp. 30–37. IEEE (2015)
40. Peng, X., Ali Babar, M., Ebert, C.: Collaborative software development platforms for crowdsourcing. IEEE Softw. **2**, 30–36 (2014)
41. Rand, D.G.: The promise of Mechanical Turk: how online labor markets can help theorists run behavioral experiments. J. Theoret. Biol. **299**, 172–179 (2012)
42. Horton, J.J., Rand, D.G., Zeckhauser, R.J.: The online laboratory: conducting experiments in a real labor market. Exp. Econom. **14**(3), 399–425 (2011)
43. Suri, S., Watts, D.J.: Cooperation and contagion in web-based, networked public goods experiments. PLoS One **6**(3), e16836 (2011)
44. Chanal, V., Caron-Fasan, M.-L.: The difficulties involved in developing business models open to innovation communities: the case of a crowdsourcing platform. M@n@gement **13**(4), 318–340 (2010)

Collective Awareness Platforms and Digital Social Innovation Mediating Consensus Seeking in Problem Situations

Atta Badii[1], Franco Bagnoli[3,5(✉)], Balint Balazs[4], Tommaso Castellani[2], Davide D'Orazio[2], Fernando Ferri[2], Patrizia Grifoni[2], Giovanna Pacini[3], Ovidiu Serban[1], and Adriana Valente[2]

[1] Intelligent Systems Research Laboratory, University of Reading, Reading, UK
[2] IRPPS-CNR, Rome, Italy
[3] Department of Physics and Astronomy and CSDC, University of Florence, Florence, Italy
{giovanna.pacini,franco.bagnoli}@unifi.it
[4] Environmental Social Science Research Group (ESSRG), Gödöllő, Hungary
[5] INFN, Sez. Firenze, Florence, Italy

Abstract. In this paper we show the results of our studies carried out in the framework of the European Project SciCafe2.0 in the area of Participatory Engagement models. We present a methodological approach built on participative engagements models and holistic framework for problem situation clarification and solution impacts assessment. Several online platforms for social engagement have been analysed to extract the main patterns of participative engagement. We present our own experiments through the SciCafe2.0 Platform and our insights from requirements elicitation.

Keywords: Collective Awareness Platforms (CAPS) · Digital social innovation · Participatory Engagement models · Problems situation disambiguation · CAPS platforms · UI-REF · SciCafe2.0

1 Introduction

Collective Awareness Platforms (CAPs) are applications based on Internet or mobile communication, scaffolding on social networking for supporting communities by delivering new services, building innovative knowledge, promoting collective intelligence. The final goal of CAPS is the promotion of more sustainable lifestyles and inducing transformative social innovation [1]. Often such 'voluntary model' is conceptualized as a collaborative commons paradigm as it bypasses the capitalist markets and relies on zero marginal cost [2]. Many applications are devoted to real actions, beyond simple knowledge-sharing, for instance by promoting energy saving (e.g., car-pooling, food sharing, buying groups) and essentially harness communication among people [3]. With regard to the optimization and potentials of ICT-enabled spontaneous, massive and collective citizen involvement the concept of crowdsourcing has been recently defined as a process of accumulating the ideas, thoughts or information from many independent participants, with aim to find the best solution for a given challenge [4].

© Springer International Publishing AG 2016
F. Bagnoli et al. (Eds.): INSCI 2016, LNCS 9934, pp. 55–65, 2016.
DOI: 10.1007/978-3-319-45982-0_5

Within the SciCafe2.0 project we are setting up an observatory of crowdsourcing (European Observatory for Crowdsourcing) devoted to participative engagement, in the spirit of the Science Café movement [5, 6]. We are studying the information flow in a participative event, and what the psychological and social components beyond the individual participation are. We are also actively experimenting on such phenomena, through specific set-up and developing a platform for supporting participative actions. Finally, we are supporting the Science Café movement, through specific actions and by means of experiments on mixed real-life and Internet-based participative models. Based on this, we present our approach to requirements elicitation that proved to be helpful co-creating robust, scalable and sustainable solutions; then we turn to social-cognitive patterns of social collaboration; and finally analyze patterns of online participative engagement and tools that help such collaboration.

2   Holistic Framework for Problem Situation Clarification and Solution Impacts Assessment

One of the most challenging obstacles to resolving problem situations is ambiguity in the problem situation that can cloud the issues and make it difficult to identify the causal roots of the problem. Thus the first priority should be to find a way to establish shared sense making about the problem by overcoming the likely barriers such as the conflicting motives of each (sub) group, their sub-languages, metaphors, subjectivity and any myths and causal fantasies. This process of consensus solution seeking requires dialogue and methodologically-guided elicitation and analysis of the values and priorities of implicated stakeholders.

To empower the stakeholders to achieve a more objective insight about the interplay of influences in the problem space we are reminded that things are most likely to be remembered and defended as personal interests worth protecting only in the contexts that they are deemed significant by human beings according to their personal and/or social constructs. Accordingly, to work towards a solution, the contexts of the most valued interests of each implicated sub-group have to be made explicit so as to identify both distinct and shared values and possible trade-offs in specific (sub)contexts. This will pave the way for areas of (inter)subjectivity and (dis)agreement to be de-limited within specific (sub)-contexts so as to facilitate consensus solution building.

UI-REF which stands for User-Intimate Integrative Requirements Elicitation and Usability Evaluation Framework [4] is a normative ethno-methodological framework to support problem situation disambiguation, requirements prioritisation and user–solution usability relationship evaluation. As such UI-REF incorporates other methods and instruments, such as empirical ethnographic approaches, cultural probes, laddering, online self-report, action research, nested-video-assisted situation walkthrough, virtual user and gaming-assisted role-play approaches to help reduce the ambiguities. This is achieved by identifying, and de-limiting the areas of disagreement and conflicts of interest and concluding the contextualised priorities of the stakeholders in the (sub) problem space(s) where consensus solutions can emerge, endure and thrive to pave the way for increasingly more robust scalable and sustainable solutions to be co-creativity established through deeper engagement as mediated by CAPS tools and digital social innovation.

Various methodologies have been proposed for usability requirements and evaluation and impact assessment. The UI-REF methodological framework is outlined here as one of the possible strategies to elicit and prioritise requirements and ensure maximum possible replicability potential for the resulting solution as well as optimal trade-offs to re local/global and immediate/downstream impacts.

As the relationship between the stakeholders, the problem situation and an emerging solution will evolve over time and the solution needs to be re-visited so as to remain dynamically responsive to evolving realities and relationships of the situation, it follows that there is a need for a Dynamic Usability Evaluation and Holistic Impact Assessment Framework, e.g. the Dynamic Usability Relationship-Based Evaluation (DURE) method [7, 8] which takes account of the dynamic relationship that can develop between the stakeholders and the solution.

3 Participative Engagement Models

People collaborate for several evolutionary human biology reasons beyond the acting for themselves. Firstly, the genetic component of collaboration implies that collaborating with others is beneficial: even if it is costly or detrimental for the collaborator. This is the main reason for the collaboration in social insects (and in some other animal) and for kin caring - an effective strategy in a small Neolithic village that may lead to quite surprising effects on a highly connected society like ours. Secondly, sexual selection drives the appearance of ornaments (like the peacock tail), which seems useless or even deleterious for survival, but are fundamental for finding a mate. Thirdly, the origin of our intellectual capacities is based on alliances and power. The way this goal is implemented is through rather sophisticated mechanisms of understanding other's wishes (the theory of mind), which is lacking in social-impaired individuals (notably, autistic or suffering from Asperger syndrome). Fourthly, group selection and natural forms of loyalty to our in-group (accompanied with fierce hatred against out-groups) is limited by our cognitive capacities. We apply different heuristics when facing a chat group (4–5 people) or a small group (up to 10–12 people) or a crowd. All in all, we do not generally act following a deep reasoning, but rather applying "rules of thumb" (heuristics) that were successful in our recent (evolutionary speaking) past.

How these heuristics determine our behavior in the Internet world, the propensity towards collaboration, the importance we assign to privacy and reputation, are among the main subjects of our investigations. In particular, we are interested in how they modify when passing from the "real life" world made of physical contacts to the cyber-world, which is missing many of the non-verbal messages we most often rely on. We rely a tentative classification based on four types and on three functions [7, 8]. As for types, we defined four categories as follows:

1. Tools are components used in online participatory activities;
2. Toolkit is a collection of tools that are used in online participatory methodologies;

3. Technique/application is a tool/toolkits put into action (implemented tool/toolkit);
4. Method is a combination of tools, toolkits, techniques put together to address defined goals.

Functions are related to

1. Telling (receive and provide information);
2. Enacting (Discuss, Deliberate, Propose, Vote);
3. Making (Share projects, Co-design projects, Collective problem solving, Share goods).

Such a classification allows to group on-line participatory platforms basing on their primary functionality, identifying 10 paradigms or building blocks of on-line participation [9]: INIP – Interactive Information Provider; AST – Ask-Tell; CODI – Collective Discussion; DIREP – Discussing for Reaching Power Nodes; REP – Reaching Power Nodes; COST – Consulting Stakeholders; SHAGO – Sharing Goods; MAP – Mapping; CODE – Co-Design; COPS – Collective Problem-Solving. These paradigms are considered as 'bricks' with which real participatory platforms are composed.

4   The SciCafe2.0 Platform

In the SciCafe2.0 project [10] the main goal is to set-up an observatory of crowdsourcing devoted to participative engagement in the spirit of the Science Café movement. In particular, we aim at the promotion of Science café networks through a supporting agency, the extraction of scenarios and best techniques, the use of this or similar methodologies (like the world cafés) beyond science, and the development of a web interface for supporting this type of communication. We are also interested in the cognitive basis of cooperation, participation and the emergence of collective intelligence.

The technological part of the project was devoted to the development of tools for promoting the combination of different kinds of services for online communities. We profited of the PLAKSS (PLAtform for Knowledge and Services Sharing) framework, developed by CNR [11, 12]. Such a platform was devoted to:

- Set and model the community specifying the characteristics of its participants. PLAKSS can model both people and virtual agents (organizations, devices, etc.) as members of the community.
- Support the different types of collaboration that can occur in Web 2.0 integrating external resources like Google and other social networks:
  - content-based: people collaborate sharing content.
  - group-based: people collaborate gathering around an idea or interest.
  - project-based: people work together on a common task or project such as a development project or a book.
- Set inferring rules for acquiring knowledge and for studying interactions between members of the community. The knowledge can support the modelling and management of complex processes.

The framework includes functionalities:

- Create a new community specifying the profiles and the information to manage for the members of the community.
- Instantiate the community managing the information, documents and data of the members.
- Support different activities of the community members, like hangouts meetings.
- Manage digital libraries.
- Manage exchange of information and interaction between members
- Share and propagate knowledge between members.

The PLAKSS framework has been used for instantiating the SciCafe2.0 platform. The SciCafe2.0 platform is conceived as a participatory crowdsourcing platform that allows people and organizations to be active actors, playing both the roles of problem and solution providers. It acts as a multiplier of knowledge and innovation:

- Aggregating and making it possible to easily access and share services, information and knowledge already available via pre-existing tools in an organized and unified manner;
- Enabling users to create their personal repository of services, information and knowledge, that can be shared with other users.

Since the main purpose of the SciCafe2.0 Project is to foster communities dialog and inquiry on specific topics, its members usually need to create a collaborative dialogue and to share knowledge and ideas. For this purpose, the Scicafe2.0 platform implements the dialog:

- Managing Science Café events using the Hangout on-line conference.
- Including in the World Café tool the Hangout on-line conference, and its functionalities such as chats.
- Providing and integrating functionalities allowing users to manage discussions organized in different tables, using and sharing posts, documents, images, videos.
- Providing and integrating functionalities for managing a blackboard for collecting and organizing opinions, or forms (integrated by Google) for managing questionnaires and data collections.

For showing the current development of the platform we introduce how the platform implement the virtual World Café meeting. This kind of virtual meeting is structured in discussion tables specified and configured by the organizer of the World Cafe. More than one table can be defined. A control panel allows to the "table chair" to start (or re-start) the table discussions. When a table starts, a hangout event is open. Users can play different roles in a virtual World Café meeting:

- Organizer of the World Café
- Table chair
- Participant to a table
- Public of the event

Depending on their role, users can have different views on the defined tables and can play different actions. In particular, an organizer manages the different tables; s/he can assign or change assignment of chairs to the tables. A table chair, when a hangout event is open, can start the table inviting participants; s/he can decide the date and time of the table meeting, or can immediately start and follow the table. Each participant to a table can contribute at the discussion directly by voice, by chat, with opinions on post-it, but can also follow discussions of the other tables (changing her/his role, from "participant" to "public" of the event), accessing the different hangouts live transmission by YouTube users.

Each meeting is also recorded in a video registration of the table on YouTube. The table chair can specify the authorized audience for the registration. Different levels of privacy are managed. In fact, a virtual World Café meeting can be public (and followed by all people, connected to YouTube), or the registration can be restricted to a small group as for example the SciCafe2.0 group that represents the authorized audience.

The World Café tool allows users to organize the space of the blackboard in different areas according to the different aspects or objectives that are discussed in the World Café meeting and need to be modelled. All users can write a post-it putting it on the blackboard and when necessary moving it (according to established semantics) in the different areas of the blackboard shared among participants at the meeting. The blackboard and its content can be saved for the inclusion in the activity stream containing the World Café meeting as one of the activities. After 20 min, participants move to another table and add to the content on that table's paper. At the end of the discussion in the table, the documentation (blackboard, forms…videos) of the World Café meeting is automatically collected and recorded into the task having the same title of the World café. Each virtual World café meeting is part of an activity in the SciCafe2.0 platform and more than one virtual World café meeting is usually contained in the same activity. The virtual World café meeting allows users to participate to the stream related to world cafés directly from the activities.

The Citizens' Say Knowledge Exchange is used by the SciCafe2.0 Platform to provide the required access to external knowledge and additional functionalities such as recommendations, Keyword Extraction, Named Entity Recognition, Text Enhancement (Annotation) as well as a parametric description of the way citizens have responded to a participative engagement session – as required and envisaged within the scope of the SciCafe2.0 project. Thus the Citizens' Say Knowledge Exchange provides access to external repositories of information (e.g. DBpedia) and also makes recommendations to the SciCafe2.0 users; suggesting activities/events depending on each user's specific interest (relevant profile) and activities description. This allows the SciCafe2.0 tool to search for individuals, organizations or events that are present in the external Knowledge Repository. One other feature of the Citizens' Say Knowledge Exchange is the Annotation tool, which provides enhanced text information or links, by linking important entities to Wikipedia or DBpedia articles.

5   Community Engagement and Requirements Elicitation

EU-level policy supports engagement. The need for stakeholder engagement and transparent dialogue with citizens is clearly articulated in Article 11 of the TEU and the White Paper on European governance (2001). Being the primary customer for SciCafe2.0 project DG Connect has developed its own inclusive approach to the involvement of stakeholders into policies, programmers and services. SciCafe2.0 project regards stakeholder engagement as a process that encompasses relationships built around one-way communication, basic consultation, in-depth dialogue and working partnerships. SciCafe2.0 also developed a Stakeholder Outreach Reference Document to guide co-development of a content marketing plan, which helps to improve the engagement processes. In the document we distinguish four main stakeholder groups as amplifiers, brokers and the medium of the message to the majority:

- Real Communities with Real problems: the project develops and deploys the SciCafe2.0 platform which can be tested in solving community difficulties.
- Partner Communities such as for example CAPS projects, Network of Science Cafes, Responsible Research & Innovation communities, European Innovation Partnerships, The Living Knowledge Network of Science Shops.
- Gateway Networks such as for example The European Network of Regions (ERRIN).
- Public Administration and Policy Making: e.g. DG Connect, Various other Public Institutions such as e.g. Local Authorities. Municipalities.

Our Citizens' Say Participatory Engagement Tool Stakeholders' Requirements Workshop took place in Brussels on the 27th of March, 2014 organized by the SciCafe2.0 Consortium and DG CONNECT. The workshop offered open parallel sessions with stakeholder sub-groups around small tables discussing their requirements facilitated by members of the SciCafe Consortium. The event was a success with an attendance of approximately 20 persons. The participants included representatives from DG CONNECT, European Regions Research and Innovation Network (ERRIN), Vrije Universiteit Brussels, Responsible Research and Innovation Projects, amongst other persons of interest. The workshop managed to specify the features best valued by the various stakeholders for the Participative Engagement Tool.

From the Citizens' Say Participatory Engagement Tool Stakeholders' Requirements Workshop the SciCafe2.0 Consortium managed to specify stakeholders' requirements for the Participative Engagement Tool. The most important requirements specified from this workshop are that the Tool needs to be simple to use, comforting, the user must be able to set different privacy levels and the tool must also be multilingual.

The SciCafe2.0 Consortium ran a session for practitioners and academics involved and interested in participatory engagement activities. It was held as part of a conference on Innovative Civil Society organized in Copenhagen by the Living Knowledge network of science shops over 9th to 11th April 2014. The session was entitled Scientific Citizenship: Deepening and widening participation and raising the quality of debating and decision making. The objective of this session was to specify more requirements and features best valued by the potential adopters for our Participative Engagement Tool. The workshop was based on Metaplan methodology and aimed at eliciting enablers and barriers from the participants to take part in on-line discussions. The workshop generated a wide variety of insights regarding user requirements; the observation we would like to draw attention to was the repeated emphasis on the social dimensions and constraints: synchronicity, emotions, resonance, collaboration, attendance, reputation and reaching consensus.

From the 1st to the 2nd of July 2014 the SciCafe2.0 Consortium attended the CAPS 2014 Conference held in Brussels. During the first day, 1st of July, a session was held by SciCafe2.0 entitled Citizens' Say: Have Your say! This session was split into two spaces. One was used to present the SciCafe2.0 Project and Citizens' Say Platform and the second one was a "hands-on-session". Both parts of the session as held at the CAPS2014 Conference were attended by a smaller audience than previous workshops but there was an open plan setting and people just dropped in and out so the total audience was larger than that at any time. Participants could explore and experiment with the platform on their computers in the room, assisted by the SciCafe2.0 Consortium members who continued to engage with the audience.

6   A Real Case Study "Science with and for Society Observatory"

Within SciCafe 2.0 project we have made a comparison of existing on-line participatory methodologies [13], we have implemented and edited a Handbook of Online Participatory Methodologies [14] in which some paradigms of on-line crowd-sourcing participatory methodologies are proposed, based on the analysis of online platforms. The emerging results contributed to the development of the SciCafe2.0 platform.

The SciCafe2.0 platform integrating the Citizens' Say Knowledge Exchange Tool provides for stimulating participation/cooperation. Basing on the results of this preliminary work a Delphi-based model of collective participation and knowledge building in the decision making processes was designed and implemented on the SciCafe2.0 Platform. The aim of the model was to explore the potential of CAPS in participative policy-making, directly connecting with real social contexts and including relevant social actors. We implemented the participatory model to a real case study, in our case the "Science with and for Society Observatory" of the Second Municipality of Rome. The participatory process was designed from the beginning as a combination of online and offline activities and It is based on the collective participation of a multitude of different actors, including policy makers, experts, citizens. As a typical Delphi model, our process is divided in different steps, and combines off-line and on-line activities.

This experiment was a success, more of three hundred people have participated at the first plenary meeting of the Observatory, and many of them decided to continue the participatory process participating at the works of the different groups and at the online activities hosted by SciCafe2.0 platform. After the implementation of the participatory process, we performed the validation by means of an online workshop in which we applied the Delphi methodology within the RE-AIM Framework.

The Delphi-based validation workshop involved 10 panelists from different categories (policy makers, researchers, science museums, schools and citizens) who discussed the effectiveness of online participatory decision making as well as the advantages and specifics of the different participative instruments and we obtained a high level validation of the participatory model; in fact we received 8 recommendations regarding the implementation of a participatory model in order for the participatory model to be successful.

In our experience we can say that Scicafe2.0 and its participatory model is a suitable for change in the dynamics of social innovation especially those relating to participatory methodologies aimed to citizens involvement and bottom up actions, it facilitates and entices users to the implementation of participatory methodologies applicable to different fields and walks of life, becoming a powerful tool for public engagement within the broader dynamics of social innovation within the macro changes that new media favor, but also stand out in the social dynamics, economic policies, effectively making them a major catalyst for change.

7   Lessons Learnt and Insights Arising from the Participatory Engagement Sessions

- Human resources are the real limitations in organizing moderator support for the participatory engagement sessions. In order to furnish a satisfactory interaction, people connected from remote need to receive timely responses from the online moderator. In practice, this means that one needs two moderators, one for the online engagement and one for the face-to-face interactions. It was concluded that the availability of some kind of automatic moderation facility should be explored as an added value, for instance using a portable device (a tablet) for online moderation by a single moderator.
- The inclusion of various instruments within the SciCafe2.0 platform, such as Google Hangouts, requires a particular attention to the third parties' policies (e.g. access, data storage, privacy, etc.). Indeed, one can need to visualize copyrighted material (like for instance pieces of movies), the recording of which will be blocked by Google.
- Non-verbal communication (i.e. emotion) was considered as an essential aspect both in written and oral discussions. Therefore, the implementation of some kind of emotional feedback within the SciCafe2.0 platform would be an added value. One possibility is that of adding emoticons, like in WatsApp and Google chat.
- For the Hangout discussion sessions, some capability for regulating the turn taking in speaking was considered as desirable.
- Enhanced support for the users maintaining an overview of the discussion themes and threads, e.g. by way of some graphic tool was also considered as a helpful feature. We already implemented the threaded discussion and the use of colors for distinguishing recent from old messages. A further possibility would be to add separate "rooms" for the discussion and a kind of "wall" where the main ideas arising from the rooms' discussion can be publicly posted. This was part of the original design of the interface, whose implementation was however delayed for technical problems.

8   Conclusions

In this paper the authors have presented an account of their studies in the area of Participatory Engagement models specifically addressing the aspects of Collective Awareness Platforms and Digital Social Innovation Mediating Consensus Seeking in Problem Situations. The paper has explored the various influences at play in societal problem situations including socio-psycho-cognitive, social engagement models, constructs and the situated cultural, and ambiguity challenges of the problem environment as well as the methodologically-guided means of reducing ambiguity, thus reducing and delimiting the contexts where there is disagreement and in doing so increasing agreement including about disagreements - towards consensus solution co-creation. The paper also briefly describes the UI-REF Framework for problem situation disambiguation and Requirements prioritization. The SciCafe platform, including the Citizens' Say tool, is featured as an example of Engagement Platforms and the World Café as an example of a Participatory Engagement Model. The paper concludes with an account of the SciCafe2.0 user requirements elicitation and community engagements and the resulting insights shared.

Acknowledgements. The SciCafe2.0 Consortium wishes to acknowledge the support from the European Community DG Connect under ICT-611299 the Collective Awareness Platforms.

References

1. Sestini, F.: Collective awareness platforms: engines for sustainability and ethics. IEEE Technol. Soc. Mag. **31**(4), 54–62 (2012)
2. Rifkin, J.: The Zero Marginal Cost Society: The Internet of Things, the Collaborative Commons, and the Eclipse of Capitalism. Macmillan, New York (2014)
3. Bagnoli, F., Guazzini, A., Pacini, G., Stavrakakis, I., Kokolaki, E., Theodorakopoulos, G.: Cognitive structure of collective awareness platforms. In: 2014 IEEE Eighth International Conference on Self-Adaptive and Self-Organizing Systems Workshops (SASOW), pp. 96–101. IEEE (2014)
4. Guazzini, A., Vilone, D., Donati, C., Nardi A., Levnajić Z.: Modelling crowdsourcing as collective problem solving. Scientific Reports 5, Article number: 16557 (2015). doi:10.1038/srep16557. http://www.nature.com/articles/srep16557
5. Dallas, D.: The cafè scientifique. Nature **399**, 120 (1999). doi:10.1038/20118
6. Pacini, G., Bagnoli, F., Belmonte, C., Castellani, T.: Science is ready, serve it! Dissemination of Science through Science Cafè, in Quality, Honesty and Beauty in Science and Technology Communication, PCST 2012 book of papers, edited by M. Bucchi and B. Trench, Observa Science in Society (2012)
7. Badii, A.: User-intimate requirements hierarchy resolution framework (UI-REF): methodology for capturing ambient assisted living needs. In: Proceedings of the Research Workshop, International Ambient Intelligence Systems Conference (AmI 2008), Nuremberg, Germany, November 2008
8. Badii, A.: Online point-of-click web usability mining with PopEval-MB, WebEvalAB and the C-assure methodology. In: Proceedings of Americas Conference on Information Systems (AMCIS 2000), University of California, Long Beach (2000)

9. Sanders, E.B.-N., Brandt E., Binder T.: A framework for organizing the tools and techniques of participatory design. In: Proceedings of the 11th Biennial Participatory Design Conference, pp. 195–198. ACM (2010)
10. http://scicafe2-0.european-observatory-for-crowdsourcing.eu/
11. Ferri, F., Ferri, P., Caschera, M.C., D'Ulizia, A.: KRC: knowInG crowdsourcing platform supporting creativity and innovation. AISS Adv. Inf. Sci. Serv. Sci. **5**(16), 1–15 (2013)
12. Grifoni, P., Ferri, F., D'Andrea, A., Guzzo, T., Pratico, C.: SoN-KInG: a digital eco-system for innovation in professional and business domains. J. Syst. Inf. Technol. **16**(1), 77–92 (2014)
13. Castellani T., D'Orazio D., Valente A.: Case studies of on-line participatory platforms, IRPPS Working Papers 67 (2014)
14. Castellani T., Valente A., D'Orazio D., Ferri, F., Grifoni P., D'Andrea A., Guzzo T., Balazs B., Bagnoli F., Pacini G., Badii A.: Handbook of on-line participatory methodologies, SciCafe2.0 Deliverable D4.1 (2014)

E-Government 2.0: Web 2.0 in Public Administration. Interdisciplinary Postgraduate Studies Program

Rafał Olszowski[✉]

Aurea Libertas Institute, ul. Batorego 2, 31-135 Kraków, Poland
`rafal.olszowski@omni.pl`

Abstract. Author is a researcher working in the educational project "E-Government 2.0 in Practice" realized since 2014 under the European Commission's program ERASMUS+. The project developed innovative and interdisciplinary teaching curricula in the area of e-Government 2.0, offering a new quality of university education as regards the content provided and educational tools applied, in the subjects covering: Web 2.0 in public administration, particularly practical use of the ICT tools in the field of e-participation, e-consultations, e-petitions, e-democracy, crowdsourcing, collaborative decision-making, open data re-use etc. The knowledge database, the case studies, the new curricula, teaching methods and didactic tools have been created and are available for a free use for all interested universities. The paper includes the following subjects, reflecting the scope of work carried out in the project: identification of the contemporary political, social, technical and research trends in the e-Government 2.0 area, the key questions outlining an e-Government 2.0 framework, proposed typology of the e-Government 2.0 websites and internet applications, as well as information about access to the results of the project. The project results are available for all the interested educational organizations under creative commons license on the http://www.egov2.eu website. Every university can therefore implement the project results in the educational program.

Keywords: e-Government · Web 2.0 · Postgraduate studies · e-Participation · e-Consultations · e-Democracy · Crowdsourcing · Collaborative decision-making

1 "E-Government 2.0 in Practice" Project Founding

Educational project "E-Government 2.0 in Practice" has generated results that should be considered as an important factor for competence development of the European university alumnae. The project gathered experts specializing in e-Government, political science and civic engagement, as well as specialists in IT implementation and social media. The project was realized by the international consortium of the universities (Tischner European University, Poland, and Tallinn Technical University, Estonia), NGOs (Aurea Libertas Institute, Poland) and ICT enterprises (Friendly Social Ltd, UK).

The implementation of e-Government solutions constitutes a priority in the EU members states. When compared to other EU countries, Poland places itself in an extremely unfavorable position. Low social participation is accompanied by deficits in

© Springer International Publishing AG 2016
F. Bagnoli et al. (Eds.): INSCI 2016, LNCS 9934, pp. 66–73, 2016.
DOI: 10.1007/978-3-319-45982-0_6

the sphere of e-Government. Although great endeavors have been made over the last 10 years to implement change, the lack of system approach and existing barriers in the area seemed to prevail, one of them being a deficit of specialists with practical experience and adequate knowledge of potential solutions. The situation became even worse as there seemed to be no relevant curricula provided by universities that would be practically-oriented and that would use effective teaching methods like workshops, simulations, etc. The project proposed by the consortium has been conceived to fill in this gap and contribute to the increase of knowledge related to e-Government 2.0 and presence of this knowledge in the university studies programs. The most interesting result of the project is an educational method based on the use of online game (three scenarios for the game were developed) and the ICT application that simulates several e-Government 2.0 processes. This method and the technical capabilities of the application will be described in the final section of this text.

2   Identification of Contemporary Political, Social, Technical and Research Trends in the e-Government 2.0 Area

The project focused on identification of contemporary political, social, technical and research trends that really and potentially affect the e-Government 2.0 mechanisms with the use of ICT. Therefore, we identified several issues as key e-Government 2.0 trends. First, we worked on determination of key terms and definitions. We noticed, that due to the lack of interdisciplinary approach to the subject, each of disciplines uses its separate terms, i.e. e-Government 1.0 & 2.0, electronic participation, e-democracy, political crowdsourcing [1], Digital Political Participation [6], Social Media Based Government [11], ICT enabled collaborative governance [5] and others. Than we focused on digital strategies implemented by governments of selected countries and international organizations (European eGovernment Action Plan, US Digital Government Strategy) and connected with them strategies on transparency and access to open public data, for example the declaration of the UN General Assembly (Open Government Declaration) and Open Government Partnership established as a result of the above-mentioned declaration. We analyzed whether these policies meet the guidelines of the OECD reports "Rethinking e-Government Services: User-Centered Approaches" and "Focus on citizens" [19, 20] when it comes to the transformation of public administration; do these policies lead to the structural transformation of public sector; whether the use of ICT might be not only the method to improve "the quality of democracy" but also "the quality of creating public policies through the participation of citizens in them", or perhaps the activities of governments are only cosmetic and don't lead to the transformation of the system.

Another observed trend was crowdsourcing as a systematized method of civic contribution to the creation of legal solutions, based on the examples of Finland and Iceland [1]. As an important phenomenon we also see the new grassroots movements that aim to the radical reconstruction of current political order by proposing direct decision making by citizens through the Internet, including in the decision making process deliberation and multistage delegated voting (e.g. Partido de la Red, The Pirate Party,

DemocracyOS, Liquid Feedback) or deliberations leading to a consensus (e.g. Occupy Movement, Podemos, M15, Loomio and others). We also took a review of the most important implemented research projects regarding E-Government 2.0; including EU priorities concerning the use of collective intelligence in the public sphere "Collective Awareness Platforms", and research on e-participation (CROSSROAD, MOMENTUM etc.) within the EU 7th Framework Programme and Horizon 2020.

Result of this analysis was creation of the typology of projects and initiatives that implement e-Government 2.0 in various public and civic areas, including e-petition, e-consultations, collaborative governance with the participation of social partners, voting with the use of the Internet (Estonia), crowdsourcing, direct democracy, "consumerisation" of public services connected with the "open data" policy.

3   The Key Questions Outlining an e-Government 2.0 Framework

During our work on outlining an e-Government 2.0 framework numerous questions appeared. These questions still require in-depth study and look for answers, however, since the e-Government 2.0 is a new area of research, we left these questions open as a contribution to further work and topics for the future discussion. First of all, we asked, what is the scope of achieving the aims of "Open Government", namely inclusivity, responsibility (for "participation contract") and transparency (the open nature of the process of arriving at decisions) and how does it affect the activity of citizens; are there (according to the systematic of Aitamurto and Landemore [1]) mechanisms that increase the participation of citizens in formulating valuable ideas and increase the quality of arguments in online civil debate (systematizing opinions with the use problem mapping, partial, hierarchical and delegated voting) and involvement of citizens in helping carrying out public tasks, for example through microtasking.

Then, we asked, is it possible to use in e-Government 2.0 projects the method of analysis of the collective intelligence, such as CIDA - Collective Intelligence Deliberation Analytics [23] and how large should citizen groups be in order to carry out interactions that lead to effective decision making; what should be the size of a group, type and quality of the people interested, how they should be recruited to the group, what entry barriers and the methods of evaluation of their conduct should be applied [16]. We also considered the awareness of users when it comes relations between their activity and made public decision: how does it affect the nature of involvement in e-Government 2.0, the scope of involvement, the capacity of citizens to contextualize their involvement, cohesiveness with political cycles and understanding of the role of evaluation?

Therefore, the question arises, does Nielsen 90/9/1 pattern of participation unevenness applies to the analysed cases?, if so, how does it influence the quality of e-Government 2.0, what part of an involved group creates the most relevant content and how to respond to the risk of being dominated by the minority? We also considered other problems and dangers to overcome, for instance the barriers of e-participation, identified by P. Panagiotopulos & T. Elliman: lack of trust as regards the honesty of a public part, populist demands that attract the attention of mass media, the danger of establishing too much lower criteria of access – it could lead to an irrelevant discussion, delay in reaction on the part of administration; partial coverage of the problem and big information pollution [12], problems with systematic evaluation that leads to constructive conclusions [14], the danger of camouflaged lobbying of solutions, superficial participation, without deepen discussion - slacktivism/clicktivism.

4 Typology of the e-Government 2.0 Websites and Internet Applications

Taking into account the described above political, social, technical and research trends in the e-Government 2.0 area, as well as key questions outlining an e-Government 2.0 framework, we proposed typology of the websites and internet applications in the study area. The groups that have been identified are not categories, but merely provide help in indicating the most important common characteristics among e-Government 2.0 applications. The proposed typology includes:

1. INSTITUTIONS 2.0: communication using Web 2.0 tools. This category includes public institutions websites that are not devoted to any specific e-Government 2.0 projects, but are primarily informative and use Web 2.0 technologies. It may also include services that belong to the traditional e-Government area, but feature additional social functions (e.g. allow users to publish comments) or are integrated with external social media (e.g. Facebook or Twitter) and communication between public institution and citizens that utilizes social media. A good example of public institution websites that are characterized by a high level of 2.0 communications are web pages of British local offices. The Enfield website, which received numerous awards, is an excellent example: http://www.enfield.gov.uk/. The Leicestershire (www.leics.gov.uk) and Hounslow (www.hounslow.gov.uk) websites are also worth mentioning.
2. E-CONSULTATIONS/E-OPPINIONS. This category includes websites and applications that allow citizens to take part in online public consultations announced by public institutions. The topics for consultations are submitted by the public sector entity; most of the consultations have a specific schedule and conditions, and the issues include plans that are being prepared or implemented, reforms, new regulations, budget statements, etc. Applications that make it possible to review measures taken by various types of public institutions are also part of this group. Examples of implementations are: Budgetsimulator.com (UK), Regulations.Gov (USA), Pathways 2050 (UK) – consultations concerning CO_2 emission reduction plan; and Opinion Space (USA) – an application developed at the UC Berkeley.
3. E-ACTIVITY/CROWD-LABOR CROWDSOURCING. Applications that belong to this group may be referred to as E-ACTIVITY or, according to the list mentioned earlier, crowdsourcing-oriented initiatives of the crowd-labor type. Citizens' activity may involve performance of micro-tasks such as cataloging, tagging, describing archival pictures, digitizing collections, conducting observations related to scientific research (e.g. by measuring the Internet bandwidth). Examples are: Citizen Archivist, Nature's Notebook, OldWeather and DigiTalkoot - The National Library of Finland Microtask. This group includes also popular applications that use

digitized maps in order to collect data related to a specific geographic location. This sub-category we labelled "GEOLOCATION: applications that use maps". The important implementations are related in civic engagement in managing security or crisis management, which facilitate collection of data related to defects of urban infrastructure, accidents, risk of crime, as well as natural disasters monitoring, elections monitoring, information on war damage, etc. Examples: FixMyStreet, Ushahidi, First to See.

4. COLLABORATIVE GOVERNANCE/CROWD COLLABORATION – participation in making political decisions. This groups includes solutions based on collaborative decision making, aggregation of ideas and solutions to problems. The idea that inspired the developers of such applications is the possible decision making process that would include the broadest group of participants possible, which uses a principle that is analogous to the creation of open source code by programmers. Thus group include crowdsourcing that uses open framework, which results in aggregation of new ideas; "direct democracy" systems based on various forms of voting, systems that implement participative budgets, electronic petition systems, etc. Examples: Liquid Feedback, Democracy OS, Loomio, E-Petitions (UK), We the People (USA).

5. OPEN DATA: re-use of open public data. Implementations related to OPEN DATA includes applications and websites run by non-governmental entities that re-use open data obtained from public administration as the basis for their activity. Private entities develop their own applications by processing published data or using the opportunity to integrate their solutions with the government software using the open API. Examples: Banklocal, Chicago Councilmatic, Doorda, Look at Cook, My Building Doesn't Recycle.

5   E-Government 2.0 Teaching Method

In this final part of the paper I would like to present the most important result of our project that constitute the core of the e-Government 2.0 Interdisciplinary Postgraduate Studies Program: an original method of teaching, assuming the use of online educational games as a key element of the learning process on e-Government 2.0. For this purpose we have prepared IT application which simulates several processes in this area: deliberation combined with voting, budget consultations, ranked voting system, crowdsourcing using geo-location and mobile devices, vote delegation. This application can be used during student classes in accordance with the developed scenarios.

As a part of the created e-learning system three scenarios for an educational game were developed. They concern the implementation of participatory budgeting in a fictional community and public consultations on construction of a waste incineration plant. Some of the participants will play the roles of the local government staff, and others will play the roles of the residents of the city proposing civic projects or participating in consultations. "Employees of the administration" will be responsible for the selection and configuration of ICT tools, which "residents" will use, and for supervising the process of projects development and selection. The "residents' will use ICT tools to refine their ideas, build knowledge bases, debate, vote and realize mutual consultations. All participants of the game will have access to the following features:

1. Performing structured debate, carried out in groups, using the voting mechanism and vote delegation;
2. Visualization tools to consult the budgets structure, allowing the users to propose alternative ways of distributing the available resources;
3. Presentation of the ideas and projects, using: geo-location of their elements on the maps, publishing pictures and other materials (also with the use of smartphones);
4. Sharing and collaboratively editing the documents;
5. Using the possibility of preferential voting;
6. Evaluation of the results of their work.

Within the project it was also created a knowledge base on existing e-Government solutions and initiatives. The base has a form of an online catalogue and includes resources developed through the analysis of the governmental strategic documents, outcomes of research projects and evaluation of selected e-Government 1.0 and e-Government 2.0 implementations. Subsequently curricula along with related courses (and syllabi) were prepared. To complement new curricula, a set of didactic tools was developed including a casebook — a publication collecting model applications in e-Government, e-learning courses, video tutorials, multimedia lectures, presentations and podcasts. The project results are available for all the interested educational organizations under creative commons license on the http://www.egov2.eu website. Every university can therefore implement the project results in the educational program.

Conducted analytical work and educational experiences associated with the preparation of the above-described materials lead us to the conclusion that the way in which the public administration may benefit from Web 2.0 is primarily increase in public participation, which leads to increased citizens' awareness and involvement in public affairs. Social media enable new forms of involvement in common issues and new ways to consult issues and interact with the citizens, including engaging the citizens in making public decisions. "2.0 Projects" that are skillfully implemented facilitate the dialogue, stimulate creativity and involve the people who are interested in cooperation with the administration.

Due to the increased collaboration with citizens, improvement in public services quality is expected: the users may contribute to improving quality of services by proposing innovative solutions, including proposals for new legislative acts. Just as in business, producers and service providers collect and process customers' opinions to enhance their offer, the citizens' opinion are used to make public decisions. What is more, the scope of cooperation between individual central and local government agencies is expected to extend (e.g. by creation of common knowledge bases and inter-institutional expert teams that would communicate via the web), which is supposed to result in better public institutions management.

OECD report "Rethinking e-Government Services: User-Centered Approaches" [20] defines expectations related to the transition of public administration towards model

2.0: if e-Government 1.0 was based on the assumption that information technologies should facilitate the current administrative work, the e-Government 2.0 approach includes implementation of ICT solutions as an element of a much broader phenomenon: structural transformation of the public sector, which means it is a way to develop better governance methods. E-Government becomes the driving force behind the transformation of the entire public sector. In the expected model, the government does not introduce changes, but creates conditions for them, which results in the increase in citizens' involvement in public affairs. The state becomes not the only one, but one of many actors in performing public tasks, and its basic duty is to provide infrastructure and allow non-governmental entities, such as NGOs, businesses and individual citizens to act. All the entities have to actively propose new solutions and take some of the government's duties in such fields as spatial planning, security, health service, education. The citizens will be able to participate in the provision of public services (e.g. by using their smartphones, taking pictures and collecting data). The crucial part of the e-Government 2.0 phenomenon is yet not the development of technology or online communities, but making a fundamental change in the government's manner of operation towards an open collaborative project based on cooperation with citizens that may include open consultations, open data, and shared knowledge.

References

1. Aitamurto, T., Landemore, H.: Five design principles for crowdsourced policymaking: assessing the case of crowdsourced off-road traffic law in Finland. J. Soc. Media Organ. **2**(1), 1–19 (2015)
2. Baumgarten, J., Chui, B.: E-Government 2.0. McKinsey Quarterly, July 2009. http://www.mckinseyquarterly.com/Business_Technology/E-government_20_2408
3. Bertot, J.C., Jaeger, P.T., Grimes, J.M.: Using ICTs to create a culture of transparency e-government and social media as openness and anti-corruption tools for societies. Gov. Inf. Q. **27**, 264–271 (2012)
4. Boughzala, I., Janssen, M., Assar, S. (eds.): Case Studies in E-Government 2.0: Changing Citizen Relationships, p. 215. Springer, Switzerland (2015)
5. Charalabidis, Y., Koussouris, S., Lampathaki, F., Misuraca, G.: ICT for governance and policy modelling visionary directions and research paths. In: Charalabidis, Y., Koussouris, Y. (eds.) Empowering Open and Collaborative Governance: Technologies and Methods for Online Citizen Engagement in Public Policy Making, pp. 263–282. Springer, Heidelberg (2012)
6. De Marco, S., Antino, M., Morales, J.M.R.: Assessing a measurement model for digital political participation. In: Charalabidis, Y., Koussouris, S. (eds.) Empowering Open and Collaborative Governance Technologies and Methods for Online Citizen Engagement in Public Policy Making, pp. 61–78. Springer, Heidelberg (2012)
7. Dixon, B.: Towards e-Government 2.0: an assessment of where e-Government 20 is and where it is headed. Publ. Admin. Manage. **15**(2), 418–454 (2010)
8. Ergazakis, K., Askounis, D., Kokkinakos, P., Tsitsanis, A.: An integrated methodology for the evaluation of ePetitions. In: Charalabidis, Y., Koussouris, S. (eds.) Empowering Open and Collaborative Governance: Technologies and Methods for Online Citizen Engagement in Public Policy Making, pp. 39–59. Springer, Heidelberg (2012)
9. GOV 2020. Deloitte University Press. http://government-2020.dupress.com

10. Grönlund, Å., Horan, T.A.: Introducing e-Gov: history, definitions, and issues. Commun. AIS **15**(1), 713–729 (2005)
11. Khan, G.F.: Social media-based government explained utilization model, implementation scenarios, and relationships. In: Boughzala, I., Janssen, M., Assar, S. (eds.) Case Studies in e-Government 2.0: Changing Citizen Relationships. Springer, Heidelberg (2015)
12. Klein, M., Convertino, G.: A roadmap for open innovation systems. J. Soc. Media Organ. **2**(1), 1 (2015)
13. Levy, P.: L'Intelligence collective: Pour une anthropologie du cyberespace. La Découverte, Paris (1994)
14. Loukis, E.: Evaluating eParticipation projects and lessons learnt. In: Charalabidis, Y., Koussouris, S. (eds.) Empowering Open and Collaborative Governance: Technologies and Methods for Online Citizen Engagement in Public Policy Making. Springer, Heidelberg (2012)
15. Meijer, A.J., Koops, B.J., Pieterson, W., Overman, S., Tije, S.: Government 2.0: key challenges to its realization. Electron. J. eGov. **10**(1), 59–69 (2012)
16. Misuraca, G., Broster, D., Centeno, C., Punie, Y., Lampathaki, F., Charalabidis, Y., Askounis, D., Osimo, D., Skuta, K., Bicking, M.: Envisioning digital Europe 2030: scenarios for ICT in future governance and policy modelling. European Commission, JRC, Spain (2011)
17. Nielsen, J.: 90-9-1 rule for participation inequality: lurkers vs. contributors in Internet communities. Jakob Nielsen's Alertbox (2006). http://www.useit.com/alertbox/participation_inequality.html
18. O'Reilly, T.: What is Web 2.0: Design patterns and business models for the next generation of software. http://oreilly.com/web2/archive/what-is-web-20.html
19. OECD: Organization for Economic Co-operation & Development "Focus on citizens: public engagement for better policy and services – policy brief". OECD, Paris (2009)
20. OECD: Organization for Economic Co-operation & Development. Rethinking e-Government Services: User-Centered Approaches. Paris (2009)
21. Panagiotopoulos, P., Elliman, T.: Online engagement from the grassroots: reflecting on over a decade of ePetitioning Experience in Europe and the UK. In: Charalabidis, Y., Koussouris, S. (eds.) Empowering Open and Collaborative Governance: Technologies and Methods for Online Citizen Engagement in Public Policy Making, pp. 79–94. Springer, Heidelberg (2012)
22. Schudson, M.: Click here for democracy a history and critique of an information based model of citizenship. In: Jenkins, H., Thornburn, D. (eds.) Democracy and New Media. MIT Press, Cambridge (2004)
23. Shum, S.B., De Liddo, A., Klein, M.: DCLA meet CIDA: collective intelligence deliberation analytics. http://catalyst-fp7.eu/wp-content/uploads/2014/05/dcla14_buckinghamshumdeliddo klein.pdf

WikiRate.org – Leveraging Collective Awareness to Understand Companies' Environmental, Social and Governance Performance

Richard Mills[1]([✉]), Stefano De Paoli[2], Sotiris Diplaris[3], Vasiliki Gkatziaki[3],
Symeon Papadopoulos[3], Srivigneshwar R. Prasad[1], Ethan McCutchen[4],
Vishal Kapadia[5], and Philipp Hirche[5]

[1] Department of Psychology, Cambridge University, Cambridge, England
rm747@cam.ac.uk
[2] Abertay University, Dundee, Scotland
[3] CERTH-ITI, Thessaloniki, Greece
[4] Decko Commons e.V., Berlin, Germany
[5] WikiRate e.V., Berlin, Germany

Abstract. WikiRate is a Collective Awareness Platform for Sustainability and Social Innovation (CAPS) project with the aim of "crowdsourcing better companies" through analysis of their Environmental Social and Governance (ESG) performance. Research to inform the design of the platform involved surveying the current corporate ESG information landscape, and identifying ways in which an open approach and peer production ethos could be effectively mobilised to improve this landscape's fertility. The key requirement identified is for an open public repository of data tracking companies' ESG performance. Corporate Social Responsibility reporting is conducted in public, but there are barriers to accessing the information in a standardised analysable format. Analyses of and ratings built upon this data can exert power over companies' behaviour in certain circumstances, but the public at large have no access to the data or the most influential ratings that utilise it. WikiRate aims to build an open repository for this data along with tools for analysis, to increase public demand for the data, allow a broader range of stakeholders to participate in its interpretation, and in turn drive companies to behave in a more ethical manner. This paper describes the quantitative Metrics system that has been designed to meet those objectives and some early examples of its use.

Keywords: Sustainability · Open data · Peer production · Collective Awareness · Crowdsourcing · Corporate Social Responsibility

1 Introduction

Companies are increasingly expected to, and in some cases legally required to, report on their Environmental, Social and Governance (ESG) performance. The voluntary production of Corporate Social Responsibility (CSR) reports is now

© Springer International Publishing AG 2016
F. Bagnoli et al. (Eds.): INSCI 2016, LNCS 9934, pp. 74–88, 2016.
DOI: 10.1007/978-3-319-45982-0_7

commonplace among large companies. Driving this trend is increased stakeholder demand [1], including demand from consumers [7]. Additionally, recently or forthcoming legislation requires:

- companies that trade in the United States and file with the Security and Exchange Commission (SEC) to produce "Conflict Minerals Reports" [22].
- companies that trade in the UK to publish statements about the steps they take to avoid slavery in their supply chains [15].
- Indian companies over a certain size to spend 2 % of their profits on CSR activities [19].
- companies based in the EU with more than 500 employees tp report on ESG performance [10].

The majority of CSR reports are delivered as **PDF** documents (and/or online "Integrated Reports") following a bespoke structure as determined by the reporting company. The company has full control over this document, and freedom to present itself in the best possible light. One of the benefits companies seek when they engage in voluntary CSR reporting is an improvement in their reputation [4]. CSR reports tend to be written in a way that maximises this gain, and in some cases present disinformation or "greenwashing" [14].

To analyse a company's ESG performance based on their reporting, one must first interrogate that reporting and extract concrete information, then contextualise it by, for example, comparing to other companies of a similar size and/or operating in the same industry. This is however a difficult task, as it involves picking the same pieces of information out of the reporting output of every company being assessed.

Wikirate is a Collective Awareness Platform for Sustainability and Social Innovation (CAPS) project [23] funded by the Framework Programme 7, and launched in October 2013 with a mission to design and build a platform for "crowdsourcing better companies". This papers presents research conducted to determine how the WikiRate project could best pursue that goal, and the resulting design which deploys wiki principles to create an environment where the peer production [2] of data on ESG performance can take place. WikiRate offers a public repository where this data can be stored and tools for analysis and critique. Stakeholder demand is often cited as a driver of improved reporting [1], and WikiRate aims to demonstrate and increase the demand for this data by making it available in a usable format. As WikiRate is a peer production effort, all contributors are of equal status and can engage fully in the discourse about what we really want from CSR and the reporting thereof.

2   Corporate Social Responsibility Reporting

One of the major developments in CSR reporting in recent decades has been the establishment of reporting standards, and the adoption of these standards by many large corporations. The Global Reporting Initiative's (GRI) G3, and more recently G4, standards have the greatest levels of adoption by companies.

The G4 defines 58 General Standard Disclosures, and 91 indicators for measuring sustainability impacts. The G4 guidelines state that companies should report on all of the 91 sustainability indicators that they deem "material" (relevant) to their business.

The degree to which companies disclose the information these indicators ask for varies between indicators, industries and companies. Sutantoputra [28] proposed a "social disclosure rating system" for assessing companies' CSR reports. This is based on the GRI's G3 guidelines and awards points based on whether a company reported specific pieces of information, giving more weight to "hard" indicators (where firms could face litigation if they are found to be lying) than "soft" indicators (which tend to be promises about the future). The rationale for this approach is Voluntary Disclosure Theory [27], which posits that while disclosures are voluntary, companies who perceive themselves to be performing well have an incentive to disclose more about their ESG performance.

Corporate Knights Capital [8] analysed the CSR reporting of the world's largest 4,609 listed companies in 2012 to see if seven sustainability indicators were disclosed. These indicators are as follows: Employee turnover: 12 %; Energy use: 40 %; Greenhouse Gas emissions: 39 %; Injury Rate: 11 %; Payroll: 59 %; Waste: 23 %; Water: 25 %. For a member of the public that is interested in the relative injury rates at competing companies in an industry, there is an 11 % chance that they will be able to find information somewhere inside the CSR report of each company they research. These documents are often large, and finding the answer to a specific question involves looking up the indicator in an index (where this is provided) and then scanning a page/section to find the information.

In Corporate Knights' report [8] the seven indicators were each listed with GRI specification points, but also "Bloomberg ESG Fields". Bloomberg offer access to data about companies' ESG performance as part of their "Bloomberg terminal" service [3] which seems to cost around $24,000 per year for a single terminal access point [21]. There are several high-profile social and environmental ratings that are similarly opaque and inaccessible to the public[1]. There is evidence that these kinds of ratings can affect companies' behaviour. Chatterji [6] analysed companies that were covered by the KLD Social Ratings and found that companies with the worst performance in a year showed greatest improvement in subsequent years, more improvement than initial good performers and companies that weren't rated. Sharkey and Bromley [24] explored this further and reported an additional indirect effect whereby an increased number of rated peers led to reductions in toxic emissions, even among companies that were not themselves rated.

Some platforms, like CSRhub.com, mix data from paid-for and public sources together to produce a web-based offering that shows paying subscribers ratings of ESG performance broken down into themes - with an option to see values collected from public but not private sources. There are also ratings based on this

---

[1] e.g. Kinder, Lydenberg, Domini Research & Analytics (KLD) Social ratings, Dow Jones Sustainability indices.

data that are published openly[2], with a description of their scoring methodology that explains it quite clearly. However, ratings based on private-access sources have an in-built limitation on how transparently they can present their scoring methodology – it is difficult to expose the gears of the mathematical apparatus at work without also exposing the data being processed to produce each score.

A useful quantitative record of companies' ESG performance is available for analysis, but only to people who are part of an organisation that can afford access to this resource. This data is a commercial product, the raw material of a CSR analysis industry that uses it to produce reviews and ratings, and sells these products on to investors, analysts, and in some cases ethically minded consumers. Many stakeholders have no access to analysable data on companies ESG performance or proprietary ratings built upon that data. This limits the public's capacity to critique the actions of corporations and the manner in which these are reported. The data from CSR reporting is openly published, but it is also locked away either inside a PDF file or behind a paywall with an expensive key.

The GRI has already moved to encourage machine-readable CSR reports by developing and releasing an XBRL taxonomy – XBRL is a XML-like format used for much mandatory financial reporting. XBRL adoption was not immediately achieved in financial reporting due to organisational constraints such as legacy systems for reporting being perceived as delivering the same functionality [9]. This may also be a factor hindering a more widespread adoption of the standard in CSR reporting, and only a few examples of reports in this format can be found[3].

Reporting standards like G4 are valuable because they ask concrete standardised questions of companies and can actually elicit responses. There are many questions about companies' ESG performance that can only be answered from within, and thus a company's reporting output is the original source for much of what is known about their performance. Reporting standards offer a framework for interpreting this output, posing a set of questions of companies' performance that either have answers or do not.

CSR reporting standards and procedures are still maturing, and issues like a lack of external assurance [11] remain to be resolved. Reading companies' reporting output is by no means the only way that the public can understand their impacts. There are many organisations taking a more active approach to investigating companies' behaviour. For example organisations like the *Business and Human Rights Resource Centre*[4] collect and interpret qualitative information about companies' behaviour from pre-existing external sources.

Organisations like *Amnesty International* and *Global Witness* conduct investigative research that tends to focus on particular themes, using methods

---

[2] Corporate Knights Capital also produce Newsweek's Green Rankings – http://www.newsweek.com/green/worlds-greenest-companies-2014.

[3] https://www.globalreporting.org/services/Analysis/XBRL_Reports/Pages/default.aspx.

[4] http://business-humanrights.org/.

like interviews, secret filming/photography, and close scrutiny of docu-ments/accounts[5]. The purpose of this research is to establish an evidence-base which can be used for advocacy - the focal point being a published report, with a campaign organised around that report. This kind of research sometimes involves the collection of useful company-level data as a by-product. For example, as part of the "Digging for Transparency" report into conflict minerals, *Global Witness* and *Amnesty International* [13] analysed Conflict Minerals Reports[6] of 100 com-panies and found that 79 failed to meet the minimum requirements of the law. The reporting status of these 100 companies was not published as that was not the main thrust of the report, which sought to draw attention to a systemic problem. WikiRate offers a place to display this kind of data transparently and make it available for analysis by others, who can also critique or refine the methodology, and if they find it useful apply it to produce data for additional companies.

A number of organisations (e.g. *Oxfam, Greenpeace*) also produce public rat-ings of corporate performance along certain themes. In conducting this research, these organisations first define a set of indicators through which they measure companies' performance, then collect data to establish how companies perform in relation to those indicators, and produce a formula that turns the raw indicator data for a company into a score. The level of detail provided about the data and scoring methodology varies between projects. Oxfam's Behind the Brands[7], and Ranking Digital Rights' Corporate Accountability Index[8], are both noteworthy as being transparent with regard to the raw indicator data for companies and how this is turned into a score. Even in these cases however, because each organisation is conducting their research independently, and there are no common standards for how to define indicators or represent the data and scoring methodology – it is difficult for other researchers to replicate and build upon this work. The lack of commonalities in how this research is conducted and reported, and the fact that it is distributed between many different sources, makes it difficult for an individual to form a coherent understanding of how companies perform. The fragmentation of this research is also likely to diminish its potential to influence the behaviour of companies.

It is worth noting that there is a strong disconnect between the kinds of ratings that are offered to the public by organisations like Oxfam and Ranking Digital rights, and proprietary ratings such as KLD. NGOs may not have access to analysable data representing companies' reporting output, and proprietary indices may under-utilise publicly available data [5]. Knowing how a company is rated by KLD is restricted to those who have access, and to our knowledge it is not possible to see what the individual indicator values are for a company, or how those have been transformed into a rating. The public has limited insight into

---

[5] https://www.globalwitness.org/en/about-us/.

[6] Mandatory for companies that file with the United States Securities and Exchange Commission (SEC) under the 2010 Dodd Frank Act.

[7] http://www.behindthebrands.org/.

[8] https://rankingdigitalrights.org/index2015/.

these ratings that can influence companies' behaviour [6], and the organisations producing these ratings not accountable to the public.

3   WikiRate Core Concepts

WikiRate is designed as a commons for the peer production [2] of an information resource that can be used to collectively: (1) establish what the impacts of a corporation's activities are and understand which practices or policies are causal; (2) identify the types of data that can be used to track companies' performance; (3) figure out which questions are most important to ask; (4) find the answers to those questions where they are available and (5) push companies to disclose them when they are not available.

Neutrality is one of the key principles around which the platform has been designed. WikiRate is *not pro or anti-companies*, and does not take a position on the relative importance of issues associated with corporate impacts. The only issue WikiRate takes a position on is corporate transparency. The manner in which companies behave and associated "externalities" should be out in the open for all to see. Without this information, it is impossible for stakeholders to form an accurate impression of which companies have net positive or negative impacts. WikiRate wants to enable stakeholders to formulate their judgments in an informed manner, not to dictate the relative importance of issues or metrics of performance.

Neutrality is important for WikiRate because ultimately the best source of information about companies' ESG performance should be the companies themselves. By being fair to companies and distinguishing between hard facts and value judgments, WikiRate offers companies a single place where they can organise and conduct their reporting in future, in a more direct and real-time dialogue with their stakeholders. The lag on CSR reporting has been identified as an issue with using this as an effective monitor of companies' behaviour [26], and WikiRate is well placed to mitigate this issue by allowing companies to enter data on a piece-by-piece basis as it becomes available.

WikiRate's design calls for two broad types of Metric. **Researched Metrics** are containers for storing "raw data" that comes from an external source, every value for these metrics must cite at least one source. **Calculated Metrics** serve the analysis of that data, they perform some mathematical or logical operation on input metrics to produce their output automatically. This fundamental distinction allows for the disassociation of data from analysis, and the easy re-use of data in multiple analyses that are free to interpret it in different ways. Metrics are complemented by structures for textual wiki content (the Wiki part of WikiRate, comprised of Notes and Reviews) that serve interpretation, critique, and information that does not fit neatly into Metric containers. Figure 1 shows a schematic overview of the relationships between WikiRate content types.

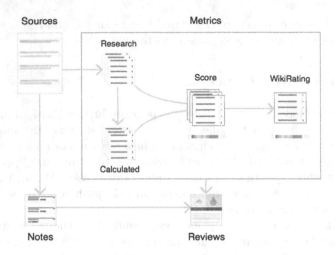

Fig. 1. Overview of the relationships between WikiRate content types

3.1   Researched Metrics

Researched Metrics have been designed to accommodate many different types of information in a standardised format, including

- low-level numerical indicators like quantity of water used
- binary or categorical answers to questions such as whether a company engages in a particular practice or not
- pre-existing ratings of company performance as produced by external research, advocacy and media organisations

A Researched Metric is a container for asking the same **question** of many different companies. Metric Values (data-points) represent the **answer** to that question for a particular company in a particular year. Metrics must also have a short **title** that can be displayed in lists. Each metric must nominate a **Metric Designer** – this is the individual or organisation who formulated the question and defined a methodology for answering it[9]. On the page for a company (Fig. 2) a reader can see all of the relevant metrics and the most recent answers for the company.

Metric pages (Fig. 3) display all of the meta-data associated with a metric, and show how a filterable selection of companies perform. There are two spaces for expanding upon the definition of a metric. The **About** section describes the information being sought, its utility, and/or how it should be interpreted. The **Methodology** section instructs researchers on how/where they can find the answer for a company.

Everything on WikiRate can be discussed and edited, to facilitate a discourse about how information should be interpreted and which questions are

---

[9] A Metric's full name is of the format Metric_Designer+Metric_Title, this allows metrics that share the same name to exist independantly in the system.

Fig. 2. The company page (Metrics tab) for Google Inc. on WikiRate

Fig. 3. The Metric page for PayScale+CEO_to_Worker_pay

most important. Researched Metrics are containers for collecting raw data, but they also incorporate social spaces for critique and interpretation of that data. All researched metric values must have a source, and a reader can easily follow links to see these sources. An individual value for a researched metric has the following properties:

- the **answer** to the question the metric asks
- the **company** and **year** it relates to
- the **source** of the information

Finally, each researched metric has a **Research Policy** that determines who is eligible to add new values. Metrics that invite community members to research and add new values have a **"community assessed"** policy – these metrics tend to involve interrogating published documents like CSR reports or Conflict Minerals Reports to extract the answers to their question. Metrics that represent information like scores or ranks awarded by external entities have a **"designer assessed"** policy because only the designer is in a position to apply that scoring methodology faithfully to additional companies (or to cover additional years).

Researched Metric Data. There are three ways in which WikiRate is gathering data for researched metrics.

1. Organisations that already produce company-level data about ESG performance design metrics and import their data from CSV files.
2. There are many existing publicly accessible and structured sources of company-level data scattered across the websites of different organisations. We are scraping this data and importing it as metric values that each link back to the original source.
3. WikiRate participants are encouraged to research and add new metric values for companies by following the methodology of the Metric they want to research[10].

The first two methods are incorporated in recognition of the fact that a considerable volume of information relevant to assessing companies' ESG performance already exists in structured sources – it would be inefficient to task community members with adding this information on a value-by-value basis when it can be added in bulk at relatively little effort. Our approach is to reserve the time and effort of our community members for tasks which require a human touch.

Data Scraping. An easy-to-use Information Extraction framework, named easIE, was developed to gather ESG data from publicly accessible Web sources and integrate it into WikiRate's database. easIE enables the extraction of ESG data about companies from heterogeneous Web sources in a semi-automatic manner, and organizes the extracted data around three key notions: (i) company: that represents a corporation (or a company) and is related with an id, name, country, a set of aliases and a website, (ii) metric: that represents a piece of information related to a company and (iii) article: the source of the information. A full description of easIE can be found in [12].

An extensive list of Web sources with CSR data was produced as part of they survey the ESG information landscape - by browsing the source lists of established aggregators of CSR data, and through communication with community members and other researchers/advocates. Next, each source was studied and

---

[10] Companies can answer questions about their own performance directly through the same mechanism, but must declare that the account used to add this data is operated by an official company representative.

the sources were ordered by the ease of extraction, the number of companies they covered and the relevance of the data they contained. Then, appropriate extraction rules for easlE were determined and a set of configuration files was built. So far, we have gathered data from 32 different sources. The created database comprises of 466,147 metric values related to 50,074 companies.

Peer Produced Data. To facilitate the extraction of metric value data from unstructured source documents, an interface has been designed which emulates aspects of existing crowd-sourcing approaches (e.g. [20]). Source documents can be displayed in one part of the screen, and questions about the content of those documents are shown in the other. On one page the researcher can read the methodology for a metric, investigate whether the answer can be found in the available sources, and add the result of their investigation as a metric value. A series of related metrics can be strung together as part of a Project, so that researchers answer sets of questions whose answers are usually found within the same document.

WikiRate is best described as a Peer Production [2] platform, all contributors are equal in principle and have permission to perform the same actions. This is a departure from many crowdsourcing initiatives, where "the crowd" is invited to perform certain tasks (like data capture or classification) but other tasks (like analysis of the resulting data) are reserved for the organisers of the initiative. User status on WikiRate is a social construct, established informally by one's contributions and how those are received by the community. The system is designed to function without designated gatekeepers or moderators.

One of the more difficult questions to answer when designing a peer production platform is "why will people be motivated to contribute?". Failing to find an answer to this question is the death of many such endeavours, and the only evidence that one has found the answer is a thriving community. At the moment, several hypotheses are being tested.

NGOs with strong followings are increasingly sensing the potential of a closer engagement with their members. Amnesty International has a program dedicated to this – Alt-Clicktivism [16] seeks to "harness the power of the collective to shed light on human rights abuses worldwide". Where such NGOs have an interest in conducting company-level research into any aspect of ESG performance, WikiRate is well positioned as a platform where that research can be conducted openly. An NGO can design metrics that ask questions about companies' performance, offer guidance on finding the answers, and invite their members to participate directly in the research effort.

For students of sustainability or corporate behaviour, the task of interrogating a company's reporting output to find concrete answers to questions about their performance offers an interesting perspective on CSR. WikiRate's metric framing helps to highlight the information that is missing, and puts the information that's presented into context. WikiRate also offers an opportunity to complete an assignment that generates a public (by-)product which can be integrated into the research of others. Several metric research pilot projects with university course organisers have been established and are ongoing or due to commence soon.

Researched Metric Examples. There follow some examples of how Researched Metrics are currently being used on WikiRate.

Ranking_Digital_Rights+RDR_Total_Score[11] is a percentage rating produced by Ranking Digital Rights:, it covers 16 major internet and telecommunications companies on 31 indicators and awards 3 sub-scores and a total score. PayScale+CEO_to_Worker_pay is a similar example of a metric based on external research that has been created in collaboration with the producers of the data. CDP+Scope_1_Emissions is a metric with values showing the number of tonnes of carbon (equivalent) emitted by 500 major companies, imported from the Carbon Disclosure Project's (CDP) public data-sets. This kind of data will make good raw material for calculated metrics related to climate change.

Amnesty_International+CMR_Lists_Smelters_and_Refiners is an example of a community-assessed metric. This metric was designed by *Amnesty International* as part of an ongoing edit-a-thon event pilot project. People are invited to attend and spend some time in groups researching companies to answer a set of questions about their conflict minerals reporting, and discussing what they find. Several other pilot projects experimenting with ways of generating different types of data are also ongoing. A set of metrics is being created to capture information related to GRI's G4 indicators in a way which is consistent with their XBRL taxonomy[12]. These are part of a pilot project inviting students of corporate sustainability to participate in liberating CSR reporting data from PDF files and making it available for analysis.

3.2   Calculated Metrics

Researched Metrics are containers for data about companies' behaviour, Calculated Metrics are designed to allow that data to be analysed transparently in public and in a modular fashion. There are three types of Calculated Metrics – Formula Metrics, Score Metrics and WikiRatings.

Formula metrics will allow mathematical and logical operations to be performed on metric data. These metrics will make it easy to produce ratios or sums of metric values, but will also allow for complex calculations involving many steps. The Centre for Sustainable Organisations' context-based carbon metric[13] has been identified as a challenging but achievable test case for a calculated metric. This is an open source metric that is currently available as a spreadsheet and used by some companies internally it calculates whether a company is emitting within its "allowance" of global emissions based on the RCP 2.6 scenario [25], using the company's economic output and greenhouse gas emissions as variables. The implementation of this metric on WikiRate will allow one to assess

---

[11] Metric names are of the format Designer+Title, the URL for this metric is http://www.wikirate.org/Ranking_Digital_Rights+RDR_Total_Score.

[12] Examples: Global_Reporting_Initiative+Environmental_fines_G4_EN29_a and Global_Reporting_Initiative+Collective_Bargaining_G4_11.

[13] http://www.sustainableorganizations.org/context-based-metrics-in-public-domain.html.

whether a company was emitting within its "allowance" of global emissions, once the input metric values have been added for a company this will be calculated automatically.

Score metrics will be used to add value judgments to a metric by mapping its range of possible values onto a 0–10 scale. The creator of a score metric imposes their opinions about what constitutes terrible/excellent performance by defining how values are mapped onto a 0–10 scale. The scoring approach will be slightly different for categorical and numerical metrics, but mapping different value types onto the same scale is a necessary step if we are to make all of this information available for inclusion in WikiRatings.

WikiRatings will calculate a weighted average for a number of Score metrics and produce a 0–10 score that measures how well companies perform on a theme defined by the designer. To design a WikiRating, a user selects a set of metrics that they want to include, selects or creates a Score metric for each (that maps it onto a 0–10 scale), and then specifies the weight that each input metric should have. As all of the input metrics are on the same scale, the weight can be used as a direct indicator of the importance the metric designer places on the answer to each question.

WikiRatings will offer an easy entry point to start comparing companies, because they embody an existing set of analyses that produce easily interpreted output. The primary consideration in the design of WikiRatings has been presenting them in the most transparent and understandable way possible. WikiRate wants to draw attention to the raw data that is being used in the calculations and show how each answer for a company contributes to a WikiRating.

3.3   Metrics Marketplace and the WikiRate Index of Transparency (WRIT)

With researched metrics to contain data, calculated metrics representing chunks of analysis, and every user being able to create metrics of any type, WikiRate's approach is likely to result in a large number of metrics[14]. To aid with the navigation of these proliferating metrics, WikiRate has the concept of a Metrics Marketplace powered by user preferences/votes. Any user can nominate metrics that are important to them (and these metrics will be displayed prominently for that user) or that they see no value in (and these metrics will be hidden from their view). These preferences also double as votes, the collective preferences for a Metric are used to produce a score that determines how visibly it is displayed. The metrics marketplace is designed to avoid a sprawling mass of undifferentiated metrics, it is a mechanism of determining what the most important metrics are and focusing the attention of readers (and companies) on those metrics. This is vital to facilitating better CSR reporting, because one of the complaints that companies make is that they are being asked to report too many different things by too many different entities [26]. The metric marketplace serves as a way of

---

[14] There are 290 metrics already available on the platform.

establishing what we care about collectively, which questions about companies' behaviour we most want to know the answers to.

The metrics marketplace will allow for the automated calculation of a "Wiki-Rate Index of Transparency" (WRIT) score for each company, this will be the only Metric that WikiRate designs and endorses directly. WRIT will work by producing an importance score for every researched metric that takes into account both the direct importance votes on that metric, but also the importance votes of every calculated metric that uses it to perform some kind of calculation. Every metric will have a certain number of "transparency points" associated with it, determined by how heavily it is used within the system. A company's WRIT score will be determined by *whether they have disclosed the answers to relevant researched metrics' questions*. This will allow WikiRate to present companies with a list of questions for which we do not yet have their answers, and see how much their WRIT score could be improved by answering each question. WikiRate is developing a system of gamification to incentivise user (and metric designer) participation [17] – the metrics marketplace and WRIT score can be thought of as an attempt to gamify CSR for companies.

4   Conclusion and Future Development

WikiRate is designed to make the task of researching companies' ESG performance one that can be tackled collectively – breaking the process of defining indicators, collecting data and analysing that data down into granular tasks that can be completed by a range of actors in a collaborative space.

The broad goals of this collective research project are to:

1. collect the available information about companies ESG performance in one public place
2. see how much insight we can gain into companies' behaviour using this data
3. identify gaps or weaknesses with the available data, important questions that we cannot currently answer about companies
4. lobby for greater disclosure of that information

The major questions to address moving forward are whether people will be motivated to contribute to WikiRate (considered above) and whether their contributions will result in the high-quality information resource that is required to illuminate corporate impacts. WikiRate's approach to data quality is modelled on wiki principles, most content types can be edited directly by users, and everything has a full revision history showing who has edited it and what they have changed. Through discussion and direct editing, contributors to WikiRate are empowered to define and enforce standards for the evidence they want to collect. Testing whether these peer-produced researched metric values are accurate, and finding ways to improve both their quality and quantity, are some of the next priorities for research on the project.

It may be possible to find shortcuts to assuring the quality of data through the reputation of contributors. The creators and editors of every piece of content on WikiRate are prominently credited, and a user's profile page gives a detailed history of their participation on the platform. WikiRate's applies the principle of transparency to user activity in the same way that it is applied to the behaviour of companies. This extends to users' voting histories, which are a matter of public record on WikiRate – a departure from how up-down voting is usually deployed, and likely to influence the manner in which people vote [18].

This level of user transparency[15] is important for WikiRate because of the nature of the subject matter. WikiRate is designed as a platform that can ultimately exert power over the behaviour of companies, this makes it a likely target for actors who want to distort how certain companies or themes are portrayed. Allowing users to see exactly what their peers are doing on the platform is necessary if the community is to self-moderate effectively. This level of transparency should also establish WikiRate as an interesting venue for research on how a collective awareness community behaves.

Acknowledgement. This work is supported by the WikiRate FP7 project, partially funded by the EC under contract number 609897.

References

1. Aguinis, H., Glavas, A.: What we know and dont know about corporate social responsibility a review and research agenda. J. Manag. **38**(4), 932–968 (2012)
2. Benkler, Y.: The Wealth of Networks: How Social Production Transforms Markets and Freedom. Yale University Press, New Haven (2006)
3. Bloomberg: Customers using esg data increased 76 % in 2014. http://www.bloomberg.com/bcause/customers-using-esg-data-increased-76-in-2014. Accessed 11 Mar 2015
4. Brammer, S.J., Pavelin, S.: Corporate reputation and social performance: the importance of fit. J. Manage. Stud. **43**(3), 435–455 (2006)
5. Chatterji, A.K., Levine, D.I., Toffel, M.W.: How well do social ratings actually measure corporate social responsibility? J. Econ. Manag. Strategy **18**(1), 125–169 (2009)
6. Chatterji, A.K., Toffel, M.W.: How firms respond to being rated. Strateg. Manag. J. **31**(9), 917–945 (2010)
7. Christmann, P., Taylor, G.: Firm self-regulation through international certifiable standards: determinants of symbolic versus substantive implementation. J. Int. Bus. Stud. **37**(6), 863–878 (2006)
8. Corporate Knights Capital: Measuring sustainability disclosure: ranking the world's stock exchanges. http://www.corporateknights.com/wp-content/reports/2014_World_Stock_Exchange.pdf. Accessed 11 Mar 2015
9. Doolin, B., Troshani, I.: Organizational adoption of xbrl. Electron. Markets **17**(3), 199–209 (2007)
10. European Commission: non-financial reporting. http://ec.europa.eu/finance/company-reporting/non-financial_reporting/index_en.htm. Accessed 11 Mar 2015

---

[15] It is up to users whether they identify themselves or how they describe themselves, with the exception of users who have the authority to speak on behalf of companies or other registered organisations.

11. Fonseca, A.: How credible are mining corporations' sustainability reports? A critical analysis of external assurance under the requirements of the international council on mining and metals. Corp. Soc. Responsib. Environ. Manag. **17**(6), 355–370 (2010)
12. Gatziaki, V., Papadopoulos, S., Tsampoulatidis, Y., Diplaris, S., Mills, R., Kompatsiaris, Y.: Scalable analytics techniques for user contributions v2. Technical report D5.5.2, WikiRate Project, September 2015
13. Global Witness and Amnesty International: Digging for transparency. https://www.globalwitness.org/en/campaigns/conflict-minerals/digging-transparency/. Accessed 11 Mar 2015
14. Laufer, W.S.: Social accountability and corporate greenwashing. J. Bus. Ethics **43**(3), 253–261 (2003)
15. legislation.gov.uk. Modern slavery act 2015t (2015). http://www.legislation.gov.uk/ukpga/2015/30/section/54/enacted
16. Marin, M.: How the wisdom of crowds can help defend human rights. https://www.amnesty.org/en/latest/campaigns/2015/10/wisdom-of-crowds-defend-human-rights/. Accessed 11 Mar 2015
17. Mills, R., Paoli, S.D.: Interim report on user and community dynamics. Technical report D3.3.3, WikiRate Project, September 2015
18. Mills, R., Fish, A.: A computational study of how and why reddit.com was an effective platform in the campaign against SOPA. In: Meiselwitz, G. (ed.) SCSM 2015. LNCS, vol. 9182, pp. 229–241. Springer, Heidelberg (2015)
19. Prasad, A.: India's new csr law sparks debate among ngos and businesses 2014. http://www.theguardian.com/sustainable-business/india-csr-law-debate-business-ngo. Accessed 11 Mar 2015
20. ProPublica: Free the files. https://projects.propublica.org/free-the-files/sessions/new. Accessed 11 Mar 2015
21. Quartz: This is how much a bloomberg terminal costs. http://qz.com/84961/this-is-how-much-a-bloomberg-terminal-costs/. Accessed 11 Mar 2015
22. SEC: Dodd-frank conflict minerals disclosure fact sheet. https://www.sec.gov/News/Article/Detail/Article/1365171562058. Accessed 11 Mar 2015
23. Sestini, F.: Collective awareness platforms: engines for sustainability and ethics. IEEE Technol. Soc. Mag. **31**(4), 54–62 (2012)
24. Sharkey, A.J., Bromley, P.: Can ratings have indirect effects? Evidence from the organizational response to peers environmental ratings. Am. Sociol. Rev. (2014)
25. Stockholm Environment Institute: A guide to representation concentration pathways. https://www.sei-international.org/mediamanager/documents/A-guide-to-RCPs.pdf. Accessed 11 Mar 2015
26. United Nations Environment Programme. Raising the bar
27. Verrecchia, R.E.: Discretionary disclosure. J. Account. Econ. **5**, 179–194 (1983)
28. Sutantoputra, A.W.: Social disclosure rating system for assessing firms' csr reports. Corp. Commun. Int. J. **14**(1), 34–48 (2009)

SOCRATIC, the Place Where Social Innovation 'Happens'

Inés Romero, Yolanda Rueda$^{(\boxtimes)}$, and Antonio Fumero

Fundación Cibervoluntarios, Project Management Office, Madrid, Spain
{ines.romero,yolanda.rueda,
antonio.fumero}@cibervoluntarios.org

Abstract. For many years, we've been growing the number and variety of inno-
vation-related buzzwords by simply attaching different adjectives/attributes to
such a keyword within Economics, and Engineering. Social Innovation (SI) and
its associated conceptual framework are in its infancy. There are a lot of on-going
efforts focused on its theoretical development, and at the same time a growing
number and variety of empirical experiments aimed at extracting its characteristics. SOCRATIC is proposing its own SI methodology to be built on top of
different test-bed scenarios, and a consistent technological platform. The experience from two of such scenarios will be mapped against the state-of-the-art
conceptual frameworks for briefly presenting the baseline for SOCRATIC methodology and platform in this position paper.

Keywords: Social innovation · Innovation · Entrepreneurship · Sustainability ·
Citizenship

1 Introduction

According to Schumpeter [1], the economic development is a historical process of
structural changes caused largely by Innovation; a process with four basic dimensions:
invention, innovation, dissemination and imitation.

The rhetoric of Innovation has led us to prevail at all times a certain aspect, one of
its particular dimensions. Far from recognizing it as a situation of complexity in which
intervene organizational, technological, individual and processual-elements, we have
decided to particularize such situations as Technological Innovation, Social Innovation
[2], Open Innovation [3], etc., developing management methodologies and conceptual
tools for each of them.

We approach Innovation as a process, with a considerable inner complexity: it
involves organizational, individual, and technological dimensions; and the three of them
have to be tuned according to the process specific requirements, coming from a variety
of contextual or environmental conditions.

When dealing with specific Innovation projects, we are not so worry about the
definition of Social Innovation, but mainly focused on how to manage a complex Social
Innovation Process (SIP); and here comes the growing variety of scenarios where quite
different organizations are embracing their own methodological approaches.

© Springer International Publishing AG 2016
F. Bagnoli et al. (Eds.): INSCI 2016, LNCS 9934, pp. 89–96, 2016.
DOI: 10.1007/978-3-319-45982-0_8

Aiming at developing our own methodology, we are adopting a systemic approach based on an Universal Framework for Modelling (UFM) [11] where we, as (Human) 'observateurs' (H) are using a handful of conceptual tools (innovation life-cycle management models) as our Interface (I) for visualizing our Object of analysis (O) i.e. the Social Innovation Process (SIP) supporting some of our partners' operations (EiT & AppLabs at NTNU, and cybervolunteers missions at CIB), defining our very own Image of the Object (IO), i.e. our own model.

That's the dynamics of the UFM expressed by the synthetic formula H x I x O = IO (see Fig. 1) that is supporting our rationale within this brief position paper. Hence, the sections below will briefly present:

- Social Innovation term as a moving target.
- Managing the whole life cycle of Social Innovation Process (SIP) as an organizational ability supported by the right methodology.
- NTNU, and CIB as a way of 'exposing' such a methodology to the Innovation reality, for extracting the baseline of our SOCRATIC concept.
- Our architectural view for implementing SOCRATIC platform on top of such a concept.

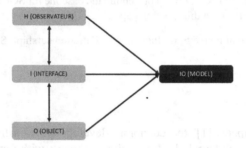

Fig. 1. Visual representation of a universal framework for modeling (Source: [11])

2 A Moving Target

The simple exercise of searching for the term "Social Innovation" in any scientific publications database permits us going through a variety of well-established definitions of the same term coming from well-known sources and institutions. Let's summarized a few of them that are supporting our own approach within SOCRATIC.

From the European Commission Guide to Social Innovation [2], we can highlight the following one: "*Social innovation can be defined as the development and implementation of new ideas (products, services and models) to meet social needs and create new social relationships or collaborations. It represents new, which affect the process of social interactions. It is aimed at improving human well-being. Social innovations are innovations that are social in both their ends and their means. They are innovations that are not only good for society but also enhance individual's capacity to act.*"

Reading the Stanford Social Innovation Review [4] we come across another well-known and accepted definition regarding SI: "*a novel solution to a social problem that is more effective, efficient, sustainable, or just than existing solutions and for which the value created accrues primarily to society as a whole rather than private individuals"*.

A detailed reading of [5] should point us to the definition from [6]: *"Social innovation as opposed to other narrower notions of innovation, is characterized by the following features: It contributes to satisfy human needs that would otherwise be ignored; It contributes to empower individuals and groups; It contributes to change social relations"*.

Out from our own state-of-the-art review of literature, we have to conclude that there is no universally accepted definition of Social Innovation (SI) [7] beyond its meaning as **an innovation creating value primarily to society, making social impact**.

3   Social Innovation, Coming to an Organization Near You

Digging into the Open Book of Social Innovation [8] we can find a reference model we've found quite useful for clearly identifying the key stages within the innovation projects lifecycle (Fig. 2).

Fig. 2.   Social innovation process (Source: Open Book of Social Innovation [8])

According to [8] these six different phases can be summarized as follows:

- **Prompts:** this step occurs before the SI process itself. In short it corresponds to identify and understand the social need(s) to be met by the social innovation. This identification serves as the base for the formalization of challenges to be addressed.
- **Ideation:** this stage is covered by many SI support process. It is the stage which would come after a societal problem has been observed, but a solution has not yet been found. It corresponds to more precisely identifying challenges based in the diagnose of the context of actions, choosing a challenge and generating and shaping an idea that can solve it.
- **Prototyping:** this stage is common to all SI methodologies, and in all of them it is described that the prototyping should be done fast and developed through multiple iterations, similarly to the 'Lean' philosophy. The rationale is that an innovation will rarely be fully formed from its first idea and that it needs to be validated and tested early, so that it is mature when it reaches the market.
- **Sustaining:** this stage corresponds to bring the innovation to the market and being adopted by the end-users. It may require much iteration to get it right and it also requires the innovators to organize themselves appropriately.
- **Scaling:** this is the stage which allows the innovation to spread, to reach new markets, regions or levels of implementation. It may be done through the expansion of the organization behind the innovation or through licensing and other mechanisms to allow other organizations to explore it as well. It deals with increasing the supply and finding the demand for the innovation artifact.

- **Systematic Change:** this one maps to a long-term effect of change in the public or private sector triggering a change of social relationships and powers.

Once the basic stages we need to have in place for effectively manage a generic SIP, we retrieve our SOCRATIC heritage from Extreme Factories [9] an EU-funded project (FP7, GA 285164); and here comes the Agile Innovation Process we defined partly inspired by the Agile Development Methodologies that are placed in the core of our Software Engineering Capabilities.

Following this methodology, the SIP is an iterative process aiming at a social impact by means of introducing an innovative artifact. In the terms of Fig. 3, the Inception stage could directly match Ideation in [8] while Implementation could be the Prototyping stage. Differently from [8], our Agile Innovation Process requires a Prioritization stage to be splinted from the Ideation one. The sustainability, scale, and even systematic change capacity of our process are finally gathered in terms of following up our implementations; that is intended to be fed back into the process.

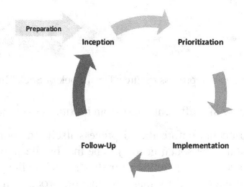

Fig. 3. Visual representation of the agile innovation process (Source: Extreme Factories)

4   The Reality Check

4.1   Southern Exposure

CIBervoluntarios (CIB) Foundation is a non-for-profit Spanish organization created and composed by 'social entrepreneurs', i.e. individuals passionate enough on using IT for volunteering in solving social challenges. The members of CIB, 1,500+ 'cybervolunteers' mostly active in Spain and Latin America, work on a daily basis with the mission of using IT to boost Social Innovation enabling citizens' empowerment. CIB's vision is to increase everyone's rights, opportunities and capacities within their social context, by means of tools and technological applications.

These cybervolunteers play an active role achieving a true societal change by developing volunteer work, promoting the usage of technological tools among the population with low access to IT and training. These agents are a crucial link between a local demand from different target groups and global solutions in Information Society. They are continuously detecting existing needs and demands from such target groups, and proactively proposing innovative, creative solutions.

CIB has been managing their activity through 'boots-on-the-field' missions (i.e. training sessions, workshops, seminars/webinars, and awareness actions) that are carried out mostly by self-organised teams supported by a quite lean infrastructure (CIB management team) offering a handful of on-demand services, resources and capabilities (mainly logistics, and documents/collaterals provision and delivery).

Hence, we've been translating CIB's ad-hoc, bottom-up, Social Innovation Process into the conceptual framework and modelling coming from [8, 9].

4.2 Northern Exposure

Norwegian University of Science and Technology (NTNU) will be from 2016 the largest university in Norway with 30,000 students. NTNU currently runs two innovative programs for their students: Experts in Team (EiT) and AppLabs.

The EiT Project is a disruptive study program at the NTNU that runs in the Spring semester over 14 weeks. EiT is taken by 2,000 students every year, divided in approximately 70 classes (called "villages") of 30 students each, who are composed into 6 teams of 5 students from last year courses of different disciplines/studies. Each village is supervised by a professor, who has described a fairly open ended challenge for that village. The students in that village have to provide specific ideas for that challenge that will also implement in teams.

AppLabs purpose is to stimulate innovation through inspiration, collaboration, new knowledge and relationship building. The program is intended for especially motivated students with knowledge in programming, app development and innovation who are impatient and want to do something "for real". The program runs for six months with several mandatory objectives. At an end, a Beta version of the app launched on the stores. Along the participants will get close monitoring and professional input of AppLabs team, which consists of selected business actors, professors, etc.

These programs offer us a systematic, top-down, case study for managing the Social Innovation Process; a quite different approach than the one from CIB. SOCRATIC platform will support the combination of both programs, EiT and AppLabs, aiming at covering the whole life-cycle of social innovation according to our own SOCRATIC methodology, from ideation and proof of concept (carried out within the EiT program) to implementation and exploitation (carried out within the AppLabs program).

4.3 The SOCRATIC Concept

Roughly mapping our experience against the previously presented conceptual models, we can identify the following stages (see Fig. 4, and Table 1):

- **The Challenge/Prompts:** A challenge is an invitation to solve a social need. In SOCRATIC, a challenge addresses a need entering in the themes supported by the following three UN goals selected by the project:

- – "Ensuring healthy life and promote well-being for all at all ages" (UN's Goals 3);
- – "Ensuring inclusive and equitable quality education and promote lifelong learning opportunities for all" (UN's Goals 4); and,
- – "Promoting sustained, inclusive and sustainable economic growth, full and productive employment and decent work for all" (UN's Goals 8).

- **The idea:** The idea is the first step towards a solution to a particular identified challenge.
- **The project:** A project corresponds to the formalization of the uptake of the idea by the project team, which is based on those who were involved in the ideation. Within the project, the team elaborates a plan for bringing the idea towards a prototype, solution, scalable solution and systematic change; in other words, to follow the Social Innovation Process.
- **Prototyping:** As presented in the previous phase, the prototyping of the solution takes place through different activities, in search for the most adequate solution to the problem exposed by the challenge.
- **Solution and Sustaining:** Irrespective of the adopted development methodology, the social innovation is iterative where the development process periodically releases a prototype, collates feedback from beneficiaries and plan the subsequent prototype based on the data provided. In this phase, the purpose is reach a solution consisting of either a product or service that is deployed in the desired environment.
- **Scaling:** At this phase, the focus moves beyond sustainability and towards scale. There are many different ways to facilitate scalability of the social innovation. However, irrespective of the adopted methodology, the SOCRATIC process relies on the use of KPIs to evaluate the how the social innovation is growing in terms of number of beneficiaries or communities addressed.

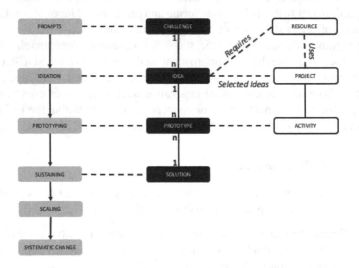

Fig. 4. Mapping SOCRATIC scenarios against a generic social innovation process

- **Systematic Change:** In this phase of the SOCRATIC process, the solutions are considered sustainable and have scaled in dimension such that it attracts stakeholders with societal influence, thus changes to the underlying systems underpinning society are subject to change.

Table 1. Mapping EiT and CIB scenarios against SOCRATIC social innovation process

SOCRATIC	CIB	EiT
Challenge	Project (Coming from a larger challenge)	Challenge
Idea	Mission	Idea
Prototype	Development and organization of the material, courses, training, …	Prototype, mock up
Solution	The material courses, training, …	Solution

5 Technology Is not Enough, IT's a Must

SOCRATIC is a research project funded under the Collaborative Awareness Platforms for Sustainability and Social Innovation (CAPS) [10] program of Horizon 2020. The initiative was first started under the EU FP7 ICT Work Program. CAPS initiative aims at designing and piloting online platforms creating awareness of sustainability problems and offering collaborative solutions based on networks (of people, of ideas, of sensors), enabling new forms of social innovation. SOCRATIC will make use of existing Service Oriented Architecture Implementation Frameworks (SOAIF), specifically the one used for the implementation of Extreme Factories [9].

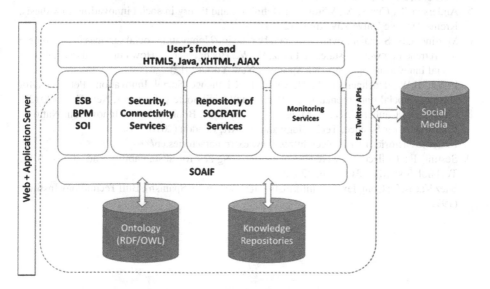

Fig. 5. SOCRATIC Architecture (as it presented in the Description of Action)

These frameworks implement all the necessary components in a service architecture (see Fig. 5 below), such as the Enterprise Service Bus paradigm (ESB, communication channel for enterprise and external applications), Business Process Model (BPM, services implementing business processes), Service Oriented Integration (SOI, to guarantee interoperability inter and intra applications), standard services for security (LDAP, TLS), service connectivity (J2EE,.Net, Web Services), communication through Java Messaging System (JMS), etc.

The architecture already integrates a service/component to search, raise and make available the knowledge bases, including the SOCRATIC ontology, modelled with RDF/OWL notation via Web Protégé. Regarding the user's front end, accepted standards will be used, such as HTML5 artefacts to ensure the validity of the portal in any type of device.

Acknowledgment. This position paper is supported by the on-going work of the whole SOCRATIC consortium, that is funded under the EC H2020 CAPS project, Grant Agreement 688228.

References

1. Becker, M.C., Knudsen, T.: Schumpeter 1911: farsighted visions on economic development. Am. J. Econ. Sociol. **61**(2), 387–403 (2002)
2. European Commission Guide to Social Innovation. http://s3platform.jrc.ec.europa.eu/documents/20182/84453/Guide_to_Social_Innovation.pdf
3. Chesbrough, H.W.: Open Innovation: The New Imperative for Creating and Profiting from Technology. Harvard Business School Press, Boston (2003)
4. Phills, J.A., Deiglmeier, K., Miller, D.T.: Rediscovering social innovation. Stanf. Soc. Innov. Rev. **6**(4), 34–43 (2008)
5. Anderson, T., Curtis, A., Wittig, C.: Definition and theory in social innovation. MA thesis, Krems, Danube University Krems (2014)
6. Martinelli, F.: Social innovation or social exclusion? Innovating social services in the context of a retrenching welfare state. In: Franz, H.-W., Hochgerner, J., Howaldt, J. (eds.) Challenge Social Innovation, pp. 169–180. Springer, New York (2012)
7. Franz, H.W., Hochgerner, J., Howaldt, J.: Challenge Social Innovation: Potentials for Business, Social Entrepreneurship, Welfare and Civil Society. Springer, New York (2012)
8. Murray, R., Caulier-Grice, J., Mulgan, G.: The Open Book of Social Innovation. National Endowment for Science, Technology and the Art, London (2010)
9. Extreme Factories FP7 Project. http://www.extremefactories.eu/
10. Sestini, F.: Collective awareness platforms: engines for sustainability and ethics. IEEE Technol. Soc. Mag. **31**, 54–62 (2012)
11. Sáez Vacas, F.: Complexity & information technology (in Spanish). Bull Technology Institute (1992)

Application Design and Engagement Strategy of a Game with a Purpose for Climate Change Awareness

Arno Scharl[1(✉)], Michael Föls[1], David Herring[2], Lara Piccolo[3], Miriam Fernandez[3], and Harith Alani[3]

[1] Department of New Media Technology, MODUL University Vienna, Vienna, Austria
{arno.scharl,michael.foels}@modul.ac.at
[2] National Oceanic and Atmospheric Administration (NOAA), Climate Program Office, Silver Spring, USA
david.herring@noaa.gov
[3] Knowledge Media Institute, The Open University, Milton Keynes, UK
{lara.piccolo,miriam.fernandez,harith.alani}@open.ac.uk

Abstract. The *Climate Challenge* is an online application in the tradition of games with a purpose that combines practical steps to reduce carbon footprint with predictive tasks to estimate future climate-related conditions. As part of the *Collective Awareness Platform*, the application aims to increase environmental literacy and motivate users to adopt more sustainable lifestyles. It has been deployed in conjunction with the *Media Watch on Climate Change*, a publicly available knowledge aggregator and visual analytics system for exploring environmental content from multiple online sources. This paper presents the motivation and goals of the *Climate Challenge* from an interdisciplinary perspective, outlines the application design including the types of tasks built into the application, discusses incentive mechanisms, and analyses the pursued user engagement strategies.

1 Introduction

Mitigating the impact of climate change is among the most important and complex contemporary issues, requiring an interdisciplinary response including, but not limited to, technical innovations, economic plans, global political agreements, and societal engagement. Although the problem is widely recognized, changing attitudes and citizens' lifestyle choices has proven to be a societal challenge from educational, social and psychological perspectives (Marshall 2014).

The *Climate Challenge* (www.ecoresearch.net/climate-challenge) is part of a collective awareness platform conceived to contribute to this societal challenge. Going beyond informing citizens and focusing on triggering environmental action and behavioural change, the *Climate Challenge* as a platform-independent social media application that engages citizens with a competition in the tradition of games with a purpose (Ahn and Dabbish 2008; Rafelsberger and Scharl 2009). It provides different strategies to help people learn more about Earth's climate, assess climate knowledge, and promote the adoption of sustainable lifestyle choices.

© Springer International Publishing AG 2016
F. Bagnoli et al. (Eds.): INSCI 2016, LNCS 9934, pp. 97–104, 2016.
DOI: 10.1007/978-3-319-45982-0_9

The application motivates participants through a gamification strategy, in which individuals are immersed in a context that favours play and healthy rivalry within a growing online community. Measuring the distribution of opinions among citizens in a monthly prediction task as shown in Fig. 1, for example, represents a first step in harnessing the wisdom of the crowd in ways that benefit society – e.g. decision making in the face of a high degree of uncertainty.

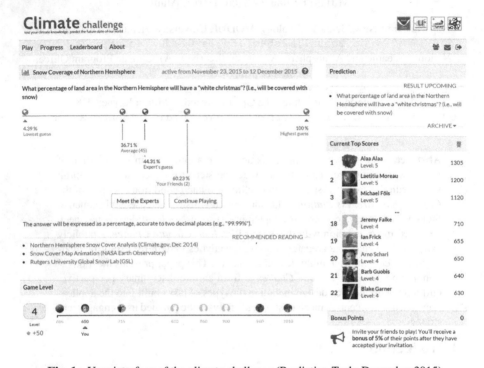

Fig. 1. User interface of the climate challenge (Prediction Task; December 2015)

The competition-based approach is intended to overcome the perceived lack of personal efficacy among individuals. As part of DecarboNet (www.decarbonet.eu), the *Climate Challenge* has been continuously been updated with new content for 18 months. It engages a diverse population and measures not only changes in energy conservation habits, but also capturing the evolution of environmental knowledge and attitudes, which are at the foundation of sustainable changes in behaviour.

Advantages of using social networking platforms to engage citizens include a large number of potential participants, intrinsic motivation in an environmental context, and effective mechanisms to detect and combat attempts of cheating or manipulating results. Viral mechanisms will trigger behavioural change, track the pursuit of common goals and induce competitive behaviour. Using real-time updates whenever possible, the strategy to engage Climate Challenge participants and sustain the competition among them includes regular content updates and the unlocking of new task types.

2   Motivation and Goals

The DecarboNet project develops a *Collective Awareness Platform* to empower citizens, help translate awareness into behavioural change, and provide visual analytics tools to understand processes that underlie this behavioural change. The platform engages environmental stakeholders with a focus on carbon footprint reductions.

Climate Challenge is designed to appeal to citizens of various backgrounds, leveraging their interest in the domain as a motivational factor together with the application's entertainment value. Users learn about changes in the Earth climate system, and how to adopt more sustainable lifestyles.

To harness the player's intrinsic motivation, to keep them interested in the game and to encourage them to invite their friends, a variety of tasks is being offered – avoiding repetition and resulting in a richer dataset to analyse. Built-in notification systems and real-time progress statistics help engage users and leverage the wisdom of the crowds for scientific purposes. A differentiating feature of *Climate Challenge* compared to other knowledge acquisition games is its pronounced educational goal, a feature resembling virtual citizen science projects.

3   Earth Hour 2016 Competition

In collaboration with the organizers of the WWF Earth Hour, the world's largest grassroots movement for the environment that took place 19 March 2016, a special edition of the Climate Challenge was announced. Individuals interested in the Earth Hour were invited to join the online competition and win one of the monthly prizes.

Earth Hour 2016 represented an ideal opportunity to engage users with an interest climate change, as it came at a moment when the world stood at a climate crossroads – emerging from a year that was marked by a universal climate deal, but at the same time learning that 2015 had been the hottest year on record.

The provided content consisted of a set of multiple choice questions about the history and impact of Earth Hour, polarity assessments of keywords related to the event, and a prediction question asking users to guess at how many people would use their "social power" for Earth Hour 2016 (including all Facebook users who either changed their profile picture or allowed the Earth Hour application to post on their behalf).

The *Media Watch on Climate Change* (Scharl et al. 2016; Scharl et al. 2013) is a content aggregator on climate change and related environmental issues, publicly available at www.ecoresearch.net/climate. It not only provided the keywords for the polarity assessment task mentioned above, but also was used by the *Earth Hour* team to monitor the online coverage before, during and after the event. The screenshot in Fig. 2 exemplifies the system's analytic capabilities by showing a query on "earth hour" resulting in a total of 106,000 documents from a wide range of online sources – including news media, social media platforms, as well as the Web sites of environmental organizations and Fortune 1000 companies.

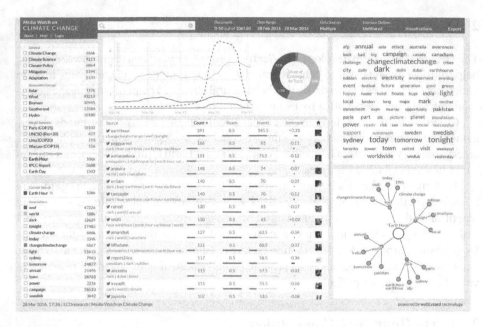

Fig. 2. Screenshot of the media watch on climate change (www.ecoresearch.net/climate)

4   Application Design and Task Types

The *Climate Challenge* was launched in March 2015 and offers 12 monthly game rounds per year, where players accumulate points by solving various game tasks. Each round combines one prediction question about future climate conditions with a range of additional tasks to earn game points throughout the month. Currently, there are four general tasks built into the game:

- **Awareness** | Test your climate change knowledge
- **Prediction** | Correctly guess the future state of our planet, in terms of both global and regional indicators
- **Change** | Reduce your carbon footprint and adopt a more sustainable lifestyle
- **Sentiment** | Assess keywords in news media coverage about climate change

A flexible task management and prioritization system, together with the ability to directly link to specific task types, enables the system to personalise content. The bar chart visualization shown in Fig. 3 is available via the "Progress" menu, increasing transparency by presenting an overview of the game structure. It lets users track their progress by task type, and informs them about the total number of available questions.

Introducing new questions and game elements is central to the engagement strategy of the Climate Challenge (see Sect. 6) to motivate players and achieve a critical mass of interactions for analytic purposes. In addition to the generic task types described in the previous section, the Climate Challenge can also be used to address specific domains or communities, and serve as a supporting mechanism for

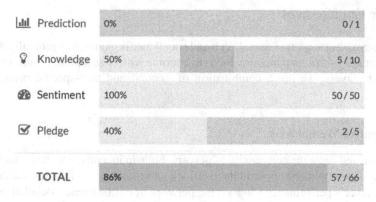

Fig. 3. Progress bar to track task completion by task type

environmental campaigns – aiming to strengthen player identification with a particular cause, and to amplify dissemination activities via social networking platforms.

5   Application Development

The *Climate Challenge* has been developed using the *uComp Human Computation Engine* (www.ucomp.eu). It is based on HTML5 to address desktop and mobile users alike. A custom login framework is used to authenticate users, based on the OAuth2.0 protocol (www.oauth.net). To increase the flexibility of the system, a custom module supports logins through popular third-party social media platforms including Twitter, Google+ and Facebook. The module allows users to connect via these services and match their profiles to a unified user account that distinguishes users based on the provided e-mail address. The framework enforces the user to be authenticated in applications that are not written in PHP. The stored user profiles include selected attributes together with application-specific details such as the number of invitations and solved tasks in the *Climate Challenge*, or the access rights and ownership status of documents created with the context-sensitive document editor to support knowledge co-creation processes (Scharl et al. 2013).

Selected profile attributes (acquired with the explicit consent of the user) in conjunction with game statistics reflect engagement levels, shed light on the behavioural impact of certain tasks, and help create a richer and more personal experience for the user – e.g., by showing the scores of the user's friends.

Climate Challenge uses a Model-View-Controller (MVC) design pattern to allow easy maintenance and extensibility. The different task types within the game are handled through a game class, which decides which type of task is given to the user, which specific task will be used, and it is also used to generate the basic HTML outline for the given task. For interactive game elements, JavaScript events trigger a communication with certain PHP hooks, which further process the request and are used to store answers and handle the navigation on the site.

6   User Engagement

In the fast-growing market of casual Web and social media games, it is generally difficult for applications with a sustainability focus to become widely accepted. To attract a large number of players, we use a combination of general and task-specific promotional activities.

6.1   Incentive Mechanisms

To maintain and grow the community of players, built-in incentive mechanisms include a levelling system with the opportunity to unlock additional games features, the comparison of a player's performance vis-à-vis the network of online friends, detailed progress statistics for each of the tasks, and a leaderboard with aggregate monthly scores. Ongoing evaluation and targeted promotion per task type leverages the existing communities of the DecarboNet core and associate partners – e.g., social media activities around the presented carbon reduction strategies by employees of *WWF Switzerland* or monthly promotion of the prediction tasks by the *Climate Program Office* of the *National Oceanic and Atmospheric Administration* (NOAA).

6.2   Analysis of User Data

To evaluate the potential impact of the *Climate Challenge* on behaviour beyond the direct interaction within the game, we follow the *Five-Doors Theory of Behaviour Change* that integrates formal theories from psychology and social sciences to enable relationships between people and modify the technological and social contexts (Robinson 2012; Robinson 2015). Figure 4 outlines five factors that must be present in the actors' lives to trigger behavioural change, which have guided the development of the *Climate Challenge*.

Fig. 4. Enabling factors according to the *Five Doors Theory of Behaviour Change*

The study includes the logged data of 645 users registered between 25 March 2015 and 16 December 2015, only considering players who provided answers to all task types ($n = 288$). Players' actions were extracted as numerical features, which can be automatically processed by applying unsupervised pattern mining algorithms. Pledges and association with social media platforms were key elements to assess user participation, while tasks with specific targets were not considered for the analysis:

- **NP:** Number of pledges answered (at least five to be considered in the analysis).
- **%NPA:** Percentage of accepted pledges.
- **%NPR:** Percentage of rejected pledges.
- **%NPD:** Percentage of pledges that the user is already doing.
- **NPo/NL:** Number of points per visit.
- **SUP** (Social User Profile): Whether the user connected a social media account with the Climate Challenge; i.e. either Twitter, Facebook, or Google+.

Users were then clustered based on the proposed features that represent the Five-Doors' conditions. Results of the cluster analysis revealed that:

- 24 people (8.3 % of the users) were in the *"Desirability"* stage – players with the lowest level of knowledge and also the second lowest level of pledge participation.
- *"Enabling Context"* has the majority of the users ($n = 111$, 38.5 %). They have adequate environmental knowledge (5.4 points per visit), and show the lowest participation in pledges (56 %), but the highest will of participation (35 %).
- 13.2 % of the users are in the *"Can Do"* stage, characterised by the second highest participation level in pledges (64 %), but relatively low number of points per visit.
- *"Buzz"* refers to 35.1 % of the users with a relatively high participation in pledges (64 %) and a good environmental knowledge (8.5 points per visit).
- The last stage *"Invitation"* contains only 4.9 % of the users. They already do 70 % of the pledges presented to them, achieve the highest number of points per visit (13), and access the Climate Challenge using their social media accounts.

Current users are mostly concentrated either on *Enabling Context* stage (38.5 %), where awareness is transformed into behaviour change, or *Buzz* (35.1 %). Both stages have a high number of points per visit, evidencing the potential of the game to raise awareness and build knowledge around climate change. Pursuing users' progress along the behaviour change process towards the *Invitation stage*, the Climate Challenge should strengthen practical information around changes in behaviour, and enhance the connection with social media, creating more incentives for people to share, cite and invite other people within their social network.

7   Summary and Outlook

Climate Challenge (www.ecoresearch.net/climate-challenge) is a social media application in the tradition of games with a purpose that provides an engaging way to help people learn about Earth's climate, assess the level of climate knowledge among citizens, create collective awareness, and promote the adoption of sustainable lifestyle choices.

In conjunction with data streams from the *Media Watch on Climate Change* (www.ecor-esearch.net/climate), a news aggregation and visual analytics platform, the *Climate Challenge* provides a rich dataset for longitudinal engagement monitoring. This paper presents the motivation and goals that guided the development of the application, outlines the range of task types offered (generic as well as target-group specific), and groups the participating users in terms of engagement levels.

Future work will provide new content elements such as new task types that can only be solved in collaboration with other users, more complex energy consumption monitoring scenarios, or language-specific tasks to assess not only the sentiment of keywords associated with current events, but also other emotional categories.

In terms of dissemination and user engagement, a combination of general and task-specific activities will help to ensure an active user base, connecting and mobilizing different online communities around energy and climate issues.

Acknowledgement. The Climate Challenge has been launched as part of the DecarboNet project, which receives funding by the EU 7th Framework Program for Research, Technology Development & Demonstration under Grant Agreement No 610829 (www.decarbonet.eu). It builds upon the crowdsourcing engine developed within the uComp project, which receives funding by EPSRC EP/K017896/1, FWF 1097-N23, and ANR-12-CHRI-0003-03, in the framework of CHIST-ERA ERA-NET (www.ucomp.eu).

References

Ahn, L.V., Dabbish, L.: Designing games with a purpose. Commun. ACM **51**(8), 58–67 (2008)

Marshall, G.: Don't Even Think About It: Why Our Brains are Wired to Ignore Climate Change. Bloomsbury, New York (2014)

Rafelsberger, W. and Scharl, A.: Games with a purpose for social networking platforms. In: Cattuto, C., et al. (eds.) 20th ACM Conference on Hypertext and Hypermedia, Torino, Italy, pp. 193–197. Association for Computing Machinery (2009)

Robinson, L.: Changeology: How to Enable Groups, Communities and Societies to do Things They've Never Done Before. Green Books, Cambridge (2012)

Robinson, L.: Five doors – an integrated theory of behaviour change (2015). http://www.enablingchange.com.au/enabling_change_theory.pdf

Scharl, A., Herring, D., et al.: Semantic systems and visual tools to support environmental communication. IEEE Syst. J. **PP**, 1–10 (2016)

Scharl, A., Hubmann-Haidvogel, A., et al.: From web intelligence to knowledge co-creation – a platform to analyze and support stakeholder communication. IEEE Internet Comput. **17**(5), 21–29 (2013)

Collective Intelligence or Collecting Intelligence?

Richard Absalom[1(✉)], Dap Hartmann[2], and Aelita Skaržauskienė[3]

[1] Den Haag, The Netherlands
durhamzoo@ymail.com
[2] Delft University of Technology, Delft, The Netherlands
l.hartmann@tudelft.nl
[3] Mykolas Romeris University, Vilnius, Lithuania
aelita@mruni.eu

Abstract. The 'Open Data', 'Open Knowledge' and 'Open Access' movements promote the dissemination of information for societal benefit. Sharing information can benefit experts in a particular endeavour, and facilitate discovery and enhance value through data mining. On-going advances in Artificial Intelligence (AI) are accelerating the development of invention machines to which few individual information donors have access. Is the movement toward open information further empowering the few? Does open information promote collective intelligence, or does the collection of information both from and about many individuals present a collection of intelligence that can be leveraged by a very few? We propose the Durham Zoo project to develop a search-and-innovation engine built upon crowd-sourced knowledge. It is hoped that this will eventually contribute to the sharing of AI–powered innovation whilst funding academic research.

Keywords: Collective intelligence · Artificial Intelligence · Classification · Search engine · Fuzzy · Innovation · Problem solving · Crowd sourcing

1 Introduction

Durham Zoo (DZ) is a research project in progress. Earlier work has focused on the development of a knowledge representation to facilitate the crowdsourcing of classification of the kind used in patent classification. The knowledge representation enables experts from across science and engineering to both contribute to, and benefit from, a knowledge base as a whole. The system can find use in searching the prior art and in problem solving. A not-for-profit foundation, eventually funding research and academia, has been proposed as best suited to the 'by the people, for the people' operating model.

Data mining is increasingly able to extract information from data, knowledge from information, and invention from knowledge. Will engineers and scientists at the coalface of innovation and discovery find their efforts increasingly leveraged by the information scientist? This paper proposes a *quid pro quo* in terms of rewarding the input of information and ideas into DZ, with time-limited confidentiality of the searched idea and search results, as well as access to data mining.

© Springer International Publishing AG 2016
F. Bagnoli et al. (Eds.): INSCI 2016, LNCS 9934, pp. 105–111, 2016.
DOI: 10.1007/978-3-319-45982-0_10

2   Collective Intelligence and Collective Benefit

Wikipedia defines Collective Intelligence as 'shared or group intelligence that emerges from the collaboration, collective efforts, and competition of many individuals' [1]. Wikipedia itself is often the first-cited example of Collective Intelligence.

Contributors to Wikipedia share their knowledge freely, and this knowledge sharing is frequently linked to the democratisation of knowledge. Open Knowledge describes itself as 'a worldwide non-profit network of people passionate about openness, using advocacy, technology and training to unlock information and enable people to work with it to create and share knowledge' [2]. In relation to research, science and culture, Open Knowledge believes that 'free access to the sum of human knowledge enables everyone to fully understand their lives and our world, to make informed choices, and to build a better future together'. In the medical field and elsewhere, a shared interest means that knowledge sharing makes good sense.

IBM's collective intelligence is the 'cognitive computing' that runs on Watson. IBM has assimilated the knowledge of many human experts. The Artificial Intelligence (AI) built on top of such knowledge in Watson has 'created an architecture of discovery' [3]. Until relatively recently, humans had the monopoly on discovery. However, AI is a real game changer because IBM's Watson needs no sleep and does not forget anything it 'reads' – and it can read a great deal at a very high speed.

And what of contribution and reward? What of collective intelligence and collective benefit? While the discovery of the structure of DNA was a collective effort, Rosalind Franklin's contribution is contested. The Rosalind Franklin entry on Wikipedia states that her X-ray diffraction images were shown to James Watson 'without her approval or knowledge' [4], while the Wikipedia page on Maurice Wilkins (her colleague at King's College), states that Wilkins 'having checked he was free to use the photograph to confirm his earlier results showed it to Watson' [5]. While perceptions may differ, it is documented that Francis Crick attested to Franklin's contribution as having been key, and a matter of fact that only Crick, Watson and Wilkins became Nobel Laureates. There is still on-going debate as to whether Rosalind Franklin would have become a Nobel Laureate had she not died from ovarian cancer at the age of 37, four years prior to the 1962 Nobel Prize in Physiology or Medicine.

If scientists discover and publish many parts of an almost complete puzzle relating to a new drug, and a commercial company puts in the final piece, should the scientists who solved most of the puzzle benefit in any other way than as mere consumers? With progressively more powerful AI, will academia and large parts of industry increasingly find themselves the Rosalind Franklins rather than the Cricks or Watsons?

That an invention has been developed using AI is no grounds for not granting an associated patent. Invention in the pharmaceutical industry, which in the obscure art of chemistry is often best served by a massive screening process, promoted reform in the United States 1952 Patent Act [6], replacing the requirement for a 'flash of creative genius' with a 'non-obviousness test' of the skilled person. This non-obviousness relates to the skilled person and not the skilled computer.

Whilst Genetic Algorithms that mimic natural selection have been used with success [7], much recent research effort in AI has been in Machine Learning. Knowledge acts as an amplifier in Machine Learning. Those with the means to process the masses of open knowledge can make use of it. Are calls for open knowledge unwittingly feeding the commercial aspirations of a relative few?

Whether or when AI will culminate with Superintelligence, touted as both the saviour and nemesis of humanity, is fortunately still a subject of debate. That AI will become *the* key enabling technology is less contentious.

With AI and the world's knowledge at their disposal, cannot the relative few stand on the shoulders of all the giants and all the not-so-giants that ever lived? Could this eventually have a negative effect on society? [8].

3   Durham Zoo System Design

DZ is a research project that seeks to combine the sharing of knowledge with the sharing of the benefits. The goal of DZ is to support innovation, both in improving search within what already exists (prior art search) as well as finding new solutions to unsolved problems (solution search). Solution search may adapt known solutions in one particular field to similar problems from non-obvious areas of technology in other fields, or look for solutions in the natural world.

3.1   Concepts and Classification

Searching a concept is rendered complex by the fuzzy nature of a concept, the many possible implementations of a concept, and the various ways that the many implementations can be expressed in natural language. Classification can address these problems by harnessing the intellect and knowledge of human experts to extract and encode concepts in a form that facilitates search [9]. Using classification as the basis for concept search can facilitate searching across foreign language collections and behind copyright-protecting paywalls. Improved concept search would reduce duplicitous research and avoid 'reinventing the wheel.' It would also likely improve the quality of granted patents, given that inventions, born with a particular application in mind, are commonly drafted as a more general concept in order to broaden the scope of legal protection: classification can be effective in addressing such abstraction. Classification can also promote innovation by helping to find inspiration or non-obvious solutions to similar problems in non-related fields: so called cross-industry innovation. For example, Ampelmann, a company founded by former students from Delft University of Technology, took the concept of moving a platform with respect to a stationary base from an aircraft simulator, 'stood the concept on its head', and invented a stationary platform on a moving base, where the stationary platform provides safe passage along a walkway above a heaving sea [10].

DZ Classification. The DZ knowledge representation facilitates the classification of concepts in science and technology [9, 11]. In DZ a concept is defined via five criteria: the Solution to a Problem in a context described by the Application, the Technology and the Operating Mode. A classification code from each of the five criteria, or facets as they are known in classification terminology, can be combined in a search query and matched with similarly classified concepts in the database. In traditional Boolean classification schemes, a match can be on all, or n-out-of-N of the facets in the query.

DZ implements fuzzy classification to be able to search fuzzy concepts [12]. Each classification code, called a Zootag, has an associated ontology of similar Zootags. Each Zootag for each facet of the query returns a ranked list of those documents in the literature that match the Zootag perfectly, followed by those documents that have similar Zootags. The necessary degree of similarity can be tailored in the different facets. In such a way a search can be restricted to a particular Problem or a particular Solution. Alternatively, the similarity in the different facets can be combined to rank the query with the literature according to their holistic similarity. For example, the search for a solution to the problem of stent thrombosis could be inspired by a disclosure where a non-smooth surface of a catheter is used to prevent bacterial build-up. The catheter, a tube used in medicine to deliver or drain a fluid, is similar to a stent, a tube used to keep a lumen open. The problem of stent thrombosis is similar to bacterial build-up given that bacterial build-up can be the cause of stent thrombosis. Thus whilst neither catheter nor bacterial build-up appears in the search query, a potential solution to the problem of stent-thrombosis would be high in the ranked results.

3.2 Classification and Crowdsourcing

In 1895, Paul Otlet and Henri La Fontaine, began the project that grew into the Mundaneum, an attempt at collecting and classifying the world's knowledge [13, 14]. Their card catalogue, initially called the Universal Bibliographic Repertory, compiled links to books, newspaper and magazine articles, pictures and other documents from libraries and archives around the world. People were able to submit queries via mail or telegraph. In essence, it was the Internet (complete with hyperlinks) long before its time. The project, which is today known as the 'paper Internet', led to the Universal Decimal Classification (UDC) that is still in use today [15]. But how to create a Mundaneum 2.0?

DZ Crowdsourcing. Classification schemes such as the UDC typically require centralised management to restrict the ambiguity of created classes. This overhead restricts the capacity of conventional schemes to be crowdsourced: both the necessary training to use the system and the management of the evolving system would be prohibitive. The new classification system of DZ separates the 'what is what' from the 'what goes where' of traditional classification.

As regards the 'what is what', a central repository, managed by Registrar volunteers, ensures that any newly created Zootags are different with respect to existing Zootags, and that their definition is clear and unambiguous. Ambiguity is the poison of classification and it is at this level that the problem is managed.

As regards the 'what goes where', competent domain experts build the ontology of similar Zootags for each Zootag. Experts from different fields are free to build ontologies in their areas of expertise. They can only use the Zootags and definitions from the central repository, however they are free to decide on what and how similar other Zootags are.

The collection of the work of the different experts, generated independently, can be processed algorithmically, linking the whole. The algorithm that combines similarity with Occam's Razor reduces central control to managing the central repository.

Importantly it is the use of similarly as the unique relationship across all the ontologies that enables a simple algorithmic processing of related Zootags and related literature from potentially unrelated technical fields. Similarity, in the guise of fuzziness, is also key to the power of the classification scheme and search capability, and is also simple and intuitive for both users and contributors to understand.

The third and final group in DZ are the searchers and classifiers. We foresee a system where everyone is free to both search and classify, adding Zootags to documents in the knowledge base as they see fit. We understand that poor classification represents noise and have proposed solutions for its elimination.

3.3 Search and Classification and Competitive Advantage

Feeding the System During Search. The system can be fed with explicit classifications of the literature with the Zootags, however there is much information to be gained from the use of the system. Results from a preliminary search retrieving a ranked list of pertinent documents would be scrutinised more closely. Interesting documents may be retained in 'drawers' relating to different aspects of the concept being searched. This is implicit classification given that similar documents are grouped together. The creation and attribution of (explicit) Zootags during the search process is encouraged. Furthermore, individual document passages may be highlighted, and further annotation added to define the content of a document. The database is thus enriched 'on the fly', with the annotation legacy of multiple searches and multiple classifications of a same document combined. Natural language processing and other 'semantic computing' methods are foreseen to be integrated to improve the overall power of the search. The profiles of individual classifiers can also be included to promote particular classification information.

Providing the Motivation. The best incentive to use the DZ system would be provided if the system were to work as intended. We believe that our design will work given both the simplicity of use and the power of the system. However, it would only work if fed with information from the crowd, and so the main challenge is to gain the necessary critical mass. We foresee the use of legacy classification information for this first stage.

More generally, the incentive to classify is provided by a feedback loop to the search engine. Information from classified documents can be used to further refine the search in progress, a technique known as relevancy feedback [16]. However, it is also possible to reward a conscientious contributor: the more a searcher classifies, the more processing power their search will be offered in return. The knowledge representation of DZ has a fractal-like nature. One possible incentive is to provide access to deeper levels of classification for active classifiers.

A working system could provide further motivation through targeted advertising that could relate to the concept in hand. Revenues would be managed by the non-profit organisation (NPO) to fund the development of the system, to support academia, and to fund research into societally important subjects such as climate change mitigation technologies.

Building a Competitive Advantage. A researcher or innovator may prefer to distribute their idea or their work freely. They may decide to develop the idea with others via a collaborative innovation company such as Quirky [17], or they may decide that their interests are best served by developing their ideas confidentially.

In the operating method as described above, the searcher effectively discloses the subject of their search in the search queries and the selection of pertinent literature. The European Patent Office's Espacenet software is a free service provided for inventors and patent professionals [18]. DZ seeks to provide a similarly free and confidential service. In the patent world, a patent application is typically not disclosed to the public for a certain period after filing. This provides an applicant with time in which to develop their ideas, and potentially to build a competitive lead over the ensuing competition. Our proposal is to do something similar.

Imagine an inventor has an idea to develop a camera system for a driverless car that resembles the arrangement of multiple pairs of eyes on a spider. The inventor can search the idea on the DZ search engine. It may be that the idea is known, or alternatively that no document exists that links the two elements. Either way, the inventor can classify the literature and add annotations to documents to suit their purpose. As a default, the information relating to the search (including the queries) and the documents selected and classified are not initially used to enhance the search engine. However, perhaps 36 months later, when the inventor has enjoyed three years to develop their ideas without them being disclosed in any way, the information from the original search can be fed into the search engine. Even if no document about the concept exists in the database, the search engine will be able to make the link between a camera system for driverless cars and the eyes of a spider. For example, someone searching for biomimetic image systems for driverless cars could retrieve an image of the eyes of a spider. As such the idea is not lost.

The act of searching with an NPO would not risk disclosing an invention to a commercial information provider that may be developing similar technology in an associated science and engineering company.

4   Conclusions and Outlook

We have proposed solutions to providing the necessary user motivation to develop a system that will democratise AI whilst earning revenue to fund research to address societal issues. We hope to create a citizen search engine, powered by AI and fuelled by collective human intelligence, democratizing invention whilst funding good causes. The next stage is a proof of concept.

References

1. Wikipedia definition of Collective Intelligence. https://en.wikipedia.org/wiki/Collective_intelligence. Accessed 20 June 2016
2. Open Knowledge. https://okfn.org/. Accessed 20 June 2016

3. Kelly III, J.E., Hamm, S.: Smart Machines: IBM's Watson and the Era of Cognitive Computing. Columbia Business School Publishing, New York (2013)
4. Wikipedia entry on Rosalind Franklin. https://en.wikipedia.org/wiki/Rosalind_Franklin. Accessed 20 June 2016
5. Wikipedia entry on Maurice Wilkins. https://en.wikipedia.org/wiki/Maurice_Wilkins. Accessed 20 June 2016
6. U.S. Patent Act. http://www.uspto.gov/web/offices/pac/mpep/consolidated_laws.pdf. Accessed 20 June 2016
7. Wongsarnpigoon, A., Grill, W.M.: Energy-efficient waveform shapes for neural stimulation revealed with a genetic algorithm. J. Neural Eng. **7**, 046009 (2010)
8. Vaidhyanathan, S.: The Googlization of Everything: (And Why We Should Worry). University of California Press, Berkeley (2011)
9. Absalom, R., Hartmann, D., Luczak-Rösch, M., Plaat, P.: Crowd-sourcing fuzzy and faceted classification for concept search. In: Proceedings of Collective Intelligence, Boston (2014)
10. Ampelmann Operations B.V. http://www.ampelmann.nl/. Accessed 20 June 2016
11. Absalom, R., Hartmann, D.: Durham Zoo: powering a search-&-innovation engine with collective intelligence. Soc. Technol. **4**(2), 245–267 (2014)
12. Baruchelli, B., Succi, G.A.: Fuzzy Approach to Faceted Classification and Retrieval of Reusable Software Components. Disa, Trento (2013)
13. Wright, A.: Cataloging the World: Paul Otlet and the Birth of the Information Age. Oxford University Press, Oxford (2014). pp. 8–15
14. Rayward, W.B.: Visions of Xanadu: Paul Otlet (1868–1944) and hypertext. JASIS **45**, 235–250 (1994)
15. McIlwaine, I.C.: Universal Decimal Classification (UDC). In: Bates, M.J. (ed.) Encyclopedia of Library and Information Sciences, 3rd edn, pp. 5432–5439. Taylor & Francis, New York (2010)
16. Büttcher, S., Clarke, C.L.A., Cormack, G.V.: Information Retrieval: Implementing and Evaluating Search Engines. MIT Press, Cambridge (2010)
17. Quirky. https://www.quirky.com/. Accessed 20 June 2016
18. Espacenet Patent Search. http://www.epo.org/searching-for-patents/technical/espacenet.html. Accessed 20 June 2016

Collaboration, Privacy and Conformity in Virtual/Social Environments

Non-trivial Reputation Effects on Social Decision Making in Virtual Environment

Mirko Duradoni[3]([✉]), Franco Bagnoli[1,2], and Andrea Guazzini[2,3]

[1] Department of Physics and Astronomy, University of Florence and INFN, Florence, Italy
[2] Center for the Study of Complex Dynamics, University of Florence, Florence, Italy
[3] Department of Science of Education and Psychology, University of Florence, Florence, Italy
mirko.duradoni@gmail.com

Abstract. Reputation systems are currently used, often with success, to ensure the functioning of online services as well as of e-commerce sites. Despite the relationship between reputation and material cooperative behaviours is quite supported, less obvious appears the relationship with informative behaviours, which are crucial for the transmission of reputational information and therefore for the maintenance of cooperation among individuals. The purpose of this study was to verify how reputation affects cooperation dynamics in virtual environment, within a social dilemma situation (i.e., where there are incentives to act selfishly). The results confirm that reputation can activate prosocial conducts, however it highlights also the limitations and distortions that reputation can create.

Keywords: Reputation · Cooperation · Social dilemma · Social decision making · Social heuristics

1 Introduction

The cooperative behaviours have been and are still the object of study of many disciplines, including evolutionary biology, antropology, sociology, social psychology and sociophysics. This interest arises from the fact that this type of conduct apparently seems to fall outside the natural selection theory. In addition, social behaviours appear in contrast, especially when these entail costs for the one who puts them in place, with the view of humans as self-interested agents (*i.e., homo oeconomicus*). Indeed humans appear to regulate their own behaviour in accordance with rules and standards different from rationality [1]. In particular, most of our actions are determined by the social environment, or, better, by what we perceive as our social environment. Laboratory [2] and field studies [3,4] have shown that subtle cues of the presence of other people, such as a photograph or a synthetic eyes on a computer screen, alter, in an almost unconscious way, our prosocial behaviour, as well as our performance and physiological activation.

© Springer International Publishing AG 2016
F. Bagnoli et al. (Eds.): INSCI 2016, LNCS 9934, pp. 115–122, 2016.
DOI: 10.1007/978-3-319-45982-0_11

Human beings are therefore deeply influenced by the social environment in which they live. This influence according to the *social heuristics hypothesis* [5] leads to the development of intuitive and economic models of decision-making (*i.e.,* heuristics) adaptive in the social context. Rand and his colleagues have shown that humans have a generalized tendency to cooperate, which however can be overridden by deliberation. It is therefore essential to identify the factors that promote cooperation. One of the mechanisms that seem to be able to maintain cooperation among humans is reputation [6]. Both computational models [7] and laboratory studies [8,9], emphasize the role of reputation in supporting the evolution and the maintenance of cooperation. Experimental studies mentioned confirm that humans are able to maintain high levels of cooperation through the indirect reciprocity mechanisms offered by it. Reputation indeed allows to identify the cooperators and, at the same time, to exclude non-cooperators through social control [10,11]. This information through communication can flow freely. Indeed we can obtain informations about the reliability (*i.e.,* the reputation) of a partner even without previous interactions with that individual and adjust our behaviour consequently. Reputation and communication are therefore intimately related. The language and the exchange of socially relevant information appears essential to ensure the functioning of the human cooperation model [12]. We know that reputation can improve material prosocial behaviour (*e.g.,* a fairer allocation in a dictator game), however, is less clear the relationship between reputation and informative behaviours, which are fundamental for the transmission and propagation of the reputational information. Feinberg and colleagues [13] have shown that individuals are ready to share information on non-cooperative individuals, even in situations where this action results expensive. Humans seem to have a strong tendency to transmit social evaluations, however the studies conducted so far have examined this behaviour only in contexts where there was no particularly reason to omit information. Indeed, participants were not competing with each other. Today, more than ever before, information and communication technologies connect people around the globe allowing them to exchange information and to work together, overcoming the physical separation constraints. These new opportunities for large-scale interaction, as well as the chance to make accessible our opinion to the community of Internet users (*i.e.,* bi-directionality), allowed the development of systems based on online feedback mechanisms. Actually we are witnessing the proliferation of services that rely on reputation systems [14]. The exchange of information and the presence of "reputation systems" ensure the functioning of e-commerce sites, such as Amazon and e-Bay, as well as services like Tripadvisor. Given the expansion of communication possibilities introduced by ICTs, it become necessary to understand how reputation and gossip affect cooperation and competition dynamics in virtual environment, as well as to verify how cooperative informative behaviours change within a competitive frame.

2   Overview of Present Studies

It was settled a 2×2 design in which reputation and cost of gossip could be present or absent. Four conditions were identified:

- Condition 1 (Reputation system ON, Cost of gossip OFF)
- Condition 2 (Reputation system OFF, Cost of gossip OFF)
- Condition 3 (Reputation system ON, Cost of gossip ON)
- Condition 4 (Reputation system OFF, Cost of gossip ON)

According to this, three studies were conducted. The first saw the activation of the conditions 1 and 2, the second one considered the condition 3 and 4, and the third compared the conditions in which the cost of gossip was present (*i.e.*, Condition 3 and 4) and those where it was absent (*i.e.*, Condition 1 and 2).

3   Participants

The first study (condition 1 and 2) involved 72 volunteers (38 females), with an average age of 22 (s.d. 3, 7). Instead, a total of 174 participants (129 females), with an average age of 22 (s.d. 4,7), were engaged in the study 2 (condition 3 and 4). A sample consisting of 246 individuals (167 females), derived from previous studies, was selected for the analyses concerning the study 3 (condition 1 and 2 vs. condition 3 and 4). For each study the participants were recruited through complete voluntary census.

4   Methods Ad Procedures

Trustee Game. Taking inspiration from the most famous social dilemma games (e.g., Ultimatum Game, Trust Game) we have developed a multiplayer virtual game called Trustee Game. The game was realized through *Google Apps*, using the *Google Script* programming language. Within Trustee Game, groups formed by 6 players interacted anonymously with the instruction to win the game for a total time of 30 min and 45 turns. Each player has had three types of resources. One was equal to 50 units (i.e., maximum resource) and the other two constituted the minimal resources (i.e., 5 units for each). The maximum resource was always different for each player and it was assigned at random. Each participant covered all the roles in the game for the same number of times (i.e., 15) and leaded for each role 3 interactions with each other member within the group. Moreover, the interaction sequence was random. The player with the highest minimum resource after 45 rounds achieved the victory.

Roles.
Donor: has the task to make an offer and a request to the receiver. The donor offers his greatest resource, among the three at his disposal, and asks in return his minimum resource to the receiver. Using the sliders the donor chooses how

much resource to offer and the amount to ask in return. Once established the terms of trade the donor should click on the "Go" button within 10 s.

Observer: has to judge the donor's action. The observer has a clear vision of the exchange proposed. Indeed, the observer is able to display both the amount and the type of resources involved in the deal. In addition, the observer can provide a hint to the receiver, clicking on the button "suggest to accept", "no hint" or "suggest to refuse". The time available for the observer to make his choice is 10 s.

Receiver: can only see the amount and the type of the resource offered by donor. Indeed, he is unaware of what the donor asked in return. The receiver may decide to "accept" or "reject" the donor deal without other information, or may require the observer suggestion (by clicking on "ask suggestion" button). When reputation system is active, the receiver can also see the rating (i.e., the number of like and dislike accumulated) of the observer with which he interacts. After his decision about the exchange offer, the receiver becomes completely aware of the donor request. If the receiver asked the suggestion and the like system is active, he has the opportunity to reward or punish (i.e., give a like or a dislike) the observer. The receiver has 18 s to make his own decisions.

Setting. The experiments took place within the computer lab of the School of Psychology of Florence. Each computer station was isolated from the others through separators in order to avoid interactions outside the game. Once arrived, participants were instructed about the rules of the game and were invited to clear up their doubts about the game mechanics. After that the game was launched on the machines.

Data Analysis. The preconditions necessary to inferential analyzes were verified on the data produced by the experiments. For all the continuous variables that were under investigation, the normality of the distribution was assessed through the analysis of asymmetry and kurtosis values. Also, were verified the presence of an adequate sample size in order to obtain robust statistics. On continuous variables that do not respect the preconditions a discretization were made, using the median as a reference, and thus defining two levels for each variable. The analyzes on these parameters were conducted using the Pearson's chi-square test.

Table 1. Reputation effect on material prosocial behaviour

Variables	Reputation Off	Reputation On	
(Study 1) Difference Donation-Request	34.5 % (+)	43.2 % (+)	29.25**
(Study 2) Difference Donation-Request	46.6 % (+)	48.2 % (+)	ns

$**p < .01$

5   Results

Within the environment characterized by a free information transmission, reputation seems to induce donors to propose fairer deals (i.e., more positive difference between amount offered and asked). However, when the communication involves a cost, reputation fails to influence a prosocial allocation behavior (Table 1).

Table 2. Reputation effect on informative prosocial behaviour

Variables	Reputation Off	Reputation On	χ^2
(Study 1) Suggestion Coherence	54.9 % (+)	54.4 % (+)	ns
(Study 2) Suggestion Coherence	29.7 % (+)	39.1 % (+)	35.66**

$**p < .01$

The presence of reputation mechanisms does not seem to influence the goodness of the suggestion provided in first study (i.e., where information flow freely). Whereas, the reputational influence on informative conducts occurs in the second study (i.e., where the suggestion involves a cost) (Table 2).

Table 3. Capability of reputation to identify those who behave cooperatively

Variables	Bad Reputation	Good Reputation	χ^2
(Study 1) Suggestion Coherence	47.6 % (+)	55.3 % (+)	9.60**
(Study 2) Suggestion Coherence	27.5 % (+)	49.8 % (+)	77.09**
(Study 2) Suggestion provided	55.7 % (0)	33.7 % (0)	71.86**

$**p < .01$

In both studies, reputational mechanisms show themselves able to identify who acts in a prosocial manner as observer. Indeed, those who provide coherent suggestions more frequently earn a good reputation. This effect is more pronounced in the study 2, where the communication cost drastically reduces the number of suggestions in the system. This reduction of information occurs both in situations with ($\chi^2 = 107.05, p < .01$) and without ($\chi^2 = 225.28, p < .01$) reputation mechanisms. Reputation also identifies those who provide an expensive suggestion to the receiver. Indeed, individuals with good reputation abstain themselves less frequently from providing a costly suggestion than those who have obtained a bad one (Table 3).

Both in Study 1 and in Study 2 those with a good reputation (i.e., those who mainly act in a prosocial manner in the observer role), use less time to take a decision. Within the Study 3, the same trend (i.e., minor decision time) is recorded among participants who provide an expensive suggestion and those

Table 4. Prosocial informative behaviour response time

Variables	Bad Reputation	Good Reputation	χ^2
(Study 1) Time to take a decision	11.7 % (+)	8.6 % (+)	3.98*
(Study 2) Time to take a decision	8.5 % (+)	5.7 % (+)	3.83*
	Costly suggestion not provided	**Costly suggestion provided**	
(Study 3) Time to take a decision	13.4 % (+)	1.4 % (+)	131.32**

$^*p < .05; \, ^{**}p < .01$

Table 5. Relationship between reputation and feedback behaviour in condition 1

	Coherence on like	Reputation		χ^2
		Good	Bad	
Suggestion Coherent	Coherent	56	10	44.11**
	No like	34	13	
	Not Coherent	6	23	
Suggestion Not Coherent	Coherent	8	11	10.40*
	No like	19	7	
	Not Coherent	17	4	

$^*p < .05; \, ^{**}p < .01$

Table 6. Relationship between reputation and feedback behaviour in condition 3

	Coherence on like	Reputation		χ^2
		Good	Bad	
Suggestion Coherent	Coherent	131	28	27.09**
	No like	183	130	
	Not Coherent	26	14	
Suggestion Not Coherent	Coherent	21	37	13.04**
	No like	73	63	
	Not Coherent	19	5	

$^{**}p < .01$

who does not perform such action. As we can see from the Table 4, those who decide to pay a cost in order to transmit an evaluation to the receiver, use a significantly shorter time.

Another interesting result emerges from the relationship between the like/dislike action (i.e., the social feedback) of the receiver and the reputation of the observer. In all the conditions in which the reputation mechanisms were active, an observer with a good reputation (i.e., more likes than dislikes) attracted more frequently a reward (i.e., likes) apart from the fact that the suggestion provided was good or bad. Specularly, an observer with a bad reputation received more frequently a dislike, even when he provided a coherent suggestion (Tables 5 and 6).

6 Discussion

The experiments showed that in a virtual social dilemma situation, reputation could trigger cooperative behaviours. The type of prosocial conduct that is elicited by reputation (i.e., material or informative) depends on the amount of available information. When the suggestion does not involve any penalty, and therefore there is a greater possibility that the observer provides an assessment of the deal, the reputation seems to be able to influence the donor behaviour. Instead, when the amount of information decreases due to the introduction of a cost to provide a suggestion, donors no longer appear to be influenced by reputation. However, this cost seems to act as a filter for those that do not provide coherent suggestions. Indeed, in an environment characterized by an expensive transmission of information, reputation appears to push the observers to provide more coherent advice.

Another result highlighted by this work concerns the reputation ability to efficaciously identify those who behave in a prosocial manner. Indeed, those who earn a good reputation act in a more prosocial way (i.e., provide good suggestion, pay a cost to trasmit information). The fact that those who perform prosocial actions employ a smaller amount of time, seems to confirm Rand and colleagues results [5]. Indeed, cooperative decision making seems to rely on social heuristics (i.e., fast and prosocial decision rules) which are made salient by reputation.

However, reputation also influences people's decision making in a non-trivial manner. As we have seen, reputation once acquired tends to perpetuate itself apart from the goodness of the suggestions. Through gossip the social prototype (good partner or bad one) is transmitted but at the same time, this prototype ends up influencing the decision making of individuals in the direction of its confirmation. Reputational systems are frequently used in virtual environment to ensure the functionality of online services. Nevertheless, their use is often unaware of the limits with which reputation can be applied. The future perspective is therefore to create adaptive automatic systems that will correct the distortion made by reputation, allowing a true estimate of people online behaviours as well as to be aware of the real value of a good or of a company. Indeed, the information conveyed by the reputation could not be connected to the real behaviour and so induce people to make mistakes (e.g., trust the wrong social partner). In addition, through the study of reputation dynamics the present work aims to contribute to the efficiency and effectiveness of virtual platforms, by means of an accurate modeling of their psychosocial ergonomy. Finally, allowing to design virtual envrionments able to elicit prosocial (e.g., energy consumption reduction) and health (e.g., support for addictions) behaviours from users.

A limit of the present work lies in the low power (i.e., too few participants), and in the high homogeneity of the results. Therefore, future research should deal with this aspect.

Acknowledgement. This work was supported by EU Commission (FP7-ICT-2013-10) Proposal No. 611299 SciCafe 2.0.

References

1. Lewin, K.: Psychological ecology. In: Cartwright, D. (ed.) Field theory in social science. Social Science Paperbacks, London (1943)
2. Haley, K.J., Fessler, D.M.: Nobody's watching?: subtle cues affect generosity in an anonymous economic game. Evol. Hum. Behav. **26**(3), 245–256 (2005)
3. Bateson, M., Nettle, D., Roberts, G.: Cues of being watched enhance cooperation in a real-world setting. Biol. Lett. **2**(3), 412–414 (2006)
4. Ernest-Jones, M., Nettle, D., Bateson, M.: Effects of eye images on everyday cooperative behavior: a field experiment. Evol. Hum. Behav. **32**(3), 172–178 (2011)
5. Rand, D.G., Peysakhovich, A., Kraft-Todd, G.T., Newman, G.E., Wurzbacher, O., Nowak, M.A., Greene, J.D.: Social heuristics shape intuitive cooperation. Nat. Commun. **5**, 411–419 (2014)
6. Alexander, R.D.: The biology of moral systems. New York: Aldine deGruyter. In: Dana, J., Cain, D.M., Dawes, R. (2006). What you dont know wonthurt me: Costly (but quiet) exit in a dictator game. Organ. Behav. Hum. Decis. Process. 100(2), 193–201 (1987)
7. Nowak, M.A., Sigmund, K.: Evolution of indirect reciprocity by image scoring. Nature **393**(6685), 573–577 (1998b)
8. Milinski, M., Semmann, D., Krambeck, H.J.: Reputation helps solve the tragedy of the commons. Nature **415**(6870), 424–426 (2002)
9. Piazza, J., Bering, J.M.: Concerns about reputation via gossip promote generous allocations in an economic game. Evol. Hum. Behav. **29**, 172–178 (2008)
10. Ohtsuki, H., Iwasa, Y.: How should we define goodness? reputation dynamics in indirect reciprocity. J. Theor. Biol. **231**(1), 107–120 (2004)
11. Giardini, F., Conte, R.: Gossip for social control in natural and artificial societies. Simulation **88**(1), 18–32 (2012)
12. Dunbar, R.I.: Gossip in evolutionary perspective. Rev. Gen. Psychol. **8**(2), 100 (2004)
13. Feinberg, M., Willer, R., Stellar, J., Keltner, D.: The virtues of gossip: reputational information sharing as prosocial behavior. J. Pers. Soc. Psychol. **102**(5), 1015 (2012)
14. Dellarocas, C.: The digitization of word of mouth: Promise and challenges of online feedback mechanisms. Manage. Sci. **49**(10), 1407–1424 (2003)

Small Group Processes on Computer Supported Collaborative Learning

Andrea Guazzini[1], Cristina Cecchini[2(✉)], and Elisa Guidi[2]

[1] Department of Education and Psychology and CSDC,
Università di Firenze, Florence, Italy
andrea.guazzini@unifi.it
[2] Department of Information Engineering and CSDC,
Università di Firenze, Florence, Italy
{cristina.cecchini,elisa.guidi}@unifi.it

Abstract. Today, information and communication technologies (ICTs) are often applied to assist learning processes. Peculiar objectives of ICT use in this topic are to facilitate collaboration and to increase learning through sharing and distributing knowledge. This study aimed to investigate the effects that a small group has on the individual and collaborative learning. A virtual environment was used to study the dynamics of social behaviors in collaborative and non-collaborative experimental conditions. Our results seem to support the hypothesis that social scripts are started, even when people are in non-interactive situations, and this is shown in virtual environments, too. Such outcomes, and the virtual interactions content analysis may suggest useful advices about collective reasoning and e-learning dynamics, which are very relevant topics in the study of web communities and educational communities.

Keywords: Virtual dynamics · Group dynamics · Gender difference · DRM paradigm · Content analysis

1 Introduction

In the field of learning processes, the information and communication technologies (ICTs) are usually applied in order to increase learning through the sharing of knowledge, and computer supported collaborative learning (CSCL) is an educational paradigm that tries to pursue this goal [1]. CSCL may create benefits for the members of a collaborative group, increasing the sharing of culturally different knowledge [2]. To estimate the "costs and benefits" of remembering in a group, the collaborative recall paradigm was designed (for review, see [3]). In this paradigm, the influence of recalling with someone else is often assessed by comparing the result of collaborative groups (a group of people learning together) with the result of nominal groups (a group of people tested individually) or individuals alone [4]. Research on CSCL has shown mixed results: may occur the phenomenon of collaborative facilitation, that is groups outperform individuals, or the process of collaborative inhibition, namely groups perform the same as or even worse than individuals [5].

© Springer International Publishing AG 2016
F. Bagnoli et al. (Eds.): INSCI 2016, LNCS 9934, pp. 123–132, 2016.
DOI: 10.1007/978-3-319-45982-0_12

1.1 Data from the DRM Paradigm

Some collaborative recall experiments have employed the Deese-Roediger-McDermott (DRM) paradigm [6]. The DRM paradigm contains lists with semantically related words (e.g., bed, rest, wake, tired, dream, etc.), that converge on the most common words (i.e., "critical lures"; e.g., sleep). These critical lures are removed from the lists. By means of DRM paradigm, researchers may measure the false recall of both critical lure and other words mentioned in error during collaboration (e.g., non-studied words). Some studies have evaluated collaborative recall performance in a group context and these studies showed mixed results. Through DRM list, one study demonstrated that participants within collaborative groups had a worse performance compared to those of equivalent sized nominal groups ([7] Experiment 1 and 2). In contrast, another study showed that collaborative groups recalled more studied words than nominal groups [8]. This inconsistency of results may be explained by the size of the groups: more members in a group could generate more disruption during recall, increasing the possibility of collaborative inhibition [9]. Moreover, some researchers assumed the importance of the encoding strategies of the lists. In particular, a study found that an imagery strategy could decrease false memories more than a word-whispering strategy [10]. Few studies have evaluated false learning in subsequent individual recall, in order to understand if collaboration has effects on recall. A study found that prior collaboration, if characterized by group pressure, retained later individual critical lure [9]. Moreover, another research showed that there was an increased individual recall after collaboration [4].

1.2 Gender Composition Group Impact on CSCL

A remarkable topic about collaborative learning and recall is whether all members of a group earn a similar profit by working in such environments [11]. Some studies not concerning the DRM paradigm have analyzed the gender effect in groups performing CSCL. Literature is pretty discordant, since studies supported same-gender groups because they work more purposefully than mixed-gender groups [12]. On the contrary, other studies revealed that mixed-gender groups perform better in CSCL, with respect to same-gender groups [13]. Recently, another research highlighted that female groups and balanced-gender groups obtain a better outcome in CSCL [14]. Analyzing the individual performance within the CSCL, a research discovered that female students in same-gender groups had a better performance than those in mixed-gender groups [13]. Instead, other researchers showed that male participants in mixed-gender groups significantly outperformed compared to males in single-gender groups [14] or to female participants in mixed-gender groups [15]. Some individual tendencies may clarify such diverse findings: females appear to be more comfortable in same-gender [14,16], while males appear to be more comfortable in mixed-gender, such as gender-balanced and gender-majority groups [14]. By contrast, a research revealed that both males and females perform better in same-gender groups, as they may better understand the style of communication applied [17].

1.3   The Effect of Social Interaction on CSCL

To understand what variables affect the CSCL, it is also necessary to analyze the interactions of the group, because the success or failure of one performance may be attributed to the content of the interaction that precede it [18]. This analysis seems particularly significant in the CSCL, where social interaction is the instrument through which participants, verbalizing their opinions, can develop a collective knowledge [19]. Moreover, participants' interaction is one of the most important predictors of success in online environments [20]. More specifically, the interaction style was positively correlated with the performance of virtual teams [21].

1.4   Aims of the Study

The general aim of this study was to compare costs and benefits for recall in three different experimental conditions (i.e., individual, nominal and collaborative), using two lists of the DRM paradigm [6] by means of a virtual environment recently implemented by our lab [22,23]. This study sought to verify the impact of different variables: (1) gender composition groups (i.e. same-gender, mixed-gender and gender majority group), (2) group size (i.e. individual, dyad, triad and quartet), (3) stimulus materials (i.e. list of concrete versus abstract words), and (4) experimental conditions order (i.e. individual-nominal vs nominal-individual or nominal-collaborative vs collaborative-nominal) on the global performance. Moreover, we analyzed the social interactions among the members of collaborative conditions.

2   Method

2.1   Participants

The participants were 144 (50 % female). All participants had reached the age of majority (age: M = 29.28, SD = 10.698), they were volunteers and unknown to each other. The average educational level of the sample was 14.66 years (SD = 3.946).

2.2   Procedures and Experimental Design

The research was conducted in accordance with the guidelines for the ethical treatment of human participants of the Italian Psychological Association (i.e., AIP). All participants were recruited with the snowball sampling strategy, and signed an informed consent. Participants were randomly assigned to one of three conditions: individual, nominal and collaborative condition, and sex and number of members in groups were balanced. The laboratory consists of two rooms: a larger room provided from two to four laptops, where the nominal and collaborative conditions were carried out, while the smaller room has been used for the individual condition. The study was composed of two protocols divided into two sub-protocols. In the first protocol, 72 participants were involved, and half of them was first tested individually and subsequently in the condition of nominal group (pairs, triplets or quartets) (Protocol 1a: individual-nominal). By contrast, the other half of the participants were first in nominal groups (pairs, triplets or quartets) and later in the individual condition (Protocol 1b: nominal-individual). In the second protocol, other 72 participants were involved, and half of them was first in nominal groups (couples, triplets or quartets) and later in collaborative groups (pairs, triplets or quartets) (Protocol 2a: nominal-collaborative). Vice versa, the other half of the participants were in collaborative groups (pairs, triplets or quartets) and then in nominal groups (pairs, triplets or quartets) (Protocol 2b: collaborative-nominal). The experiment consisted of two successive sessions. In the first session, the participant had to remember a first list of words (presented for thirty seconds). Then, the participant completed a 3-minute mathematical filler task (balanced across the experiment) to prevent rehearsal in short-term memory [24]. Finally, the subject recalled the studied words by marking on a list with both studied and non-studied words, and a critical lure [6]. In the second session, all participants were together (nominal condition) and they used the virtual environment to interact with each other anonymously for 3 min. Then, they were asked to remember a second list of words, they completed another 3-minute mathematical filler task and, unlike the first session, each participant completed the recall task at a separate computer with the presence of all participants. In the protocol 1b, the two successive sessions were inverted. In the protocol 2a, the first session (nominal condition) was identical to that of Protocol 1a and 1b. In the second session, after the presentation of the second list of words and the 3-minute mathematical filler task, the participants recalled the list in collaboration through the virtual environment, with no special instructions on how to coordinate recall, manage speaking turns, or resolve disagreements (free-for-all collaboration). In the protocol 2b, the two successive sessions were inverted. The order of the two lists within each protocol was balanced.

2.3   Measures

Deese-Roediger-McDermott (DRM) paradigm. Four lists were developed from the materials of Roediger and McDermott's article [6]. More specifically, (1) the first list was the Anger 15-Word List, composed by abstract words (i.e., words about feelings); (2) the second list was the Music list, composed by concrete words (i.e., words about musical instruments); (3) the third and fourth lists were composed by a "critical lure" (i.e., Anger for the first list and Music for the second one), 10 "real words" (i.e., already read in the original list of 15 words), and 9 "false words" (i.e., not presented in the original list). The lists produced five dependent variables: non-studied words (i.e., the sum of the critical lure and false words); studied words (i.e., real words); true negatives (i.e., false words not filled by the participants); false negatives (i.e., real words not filled by the participants); number of answers (i.e., total number of words filled by the participants). The score of each participant was added with the score of the other members of the same group, to obtain an average score of the group performance.

2.4   Data Analysis

We calculated the descriptive statistics, assessing the pre-conditions required by the inferential analysis, checking the Gaussian distribution of the continuous variables (i.e. skewness and kurtosis), and the balancing and size of the sub-samples of interest (i.e., gender, experimental condition, and list type). Then, we conducted the inferential analyses (i.e., Pearson's r correlation and Student's t-tests) to verify our aims (i.e., gender effect, group size, experimental condition order), while a MANCOVA analysis has been adopted to evaluate the connected role of experimental condition and list type on the performances. The independent variables were the experimental condition and the type of list, the dependent variables were studied words and non-studied words, and the education variable was introduced as covariate. Finally, we analyzed the linguistic content of group chat in the collaborative condition through the Linguistic Inquiry Word Count computer-program (LIWC) [25,26]. A Pearson's r correlation has been carried out to assess the relation between the LIWC dimensions and the performance scores (i.e., studied words and non-studied words).

3   Results

The analysis on the size of the group, the performance (i.e., the number of studied words, the non-studied words and the total number of answers) and the experimental condition shows that the number of members does not affect the group performance, regardless the experimental condition. Instead, the analysis on the experimental conditions order reveals that the collaborative condition shows a higher number of studied words ($t = -2.5 p < .05$; $M = 7.64\,VS\ M = 8.44$), and a lower number of false negatives ($t = 2.5 p < .05$; $M = 2.36\,VS\ M = 1.56$) whether the nominal condition comes first, while any other condition is significant. Regarding the gender difference analysis on the performance (Table 1), females perform better than males, achieving a higher number of studied words in the total sample and, particularly, in the nominal condition. Regarding the same-gender groups and the gender majority groups, both of them show a higher number of studied words in the individual and nominal conditions, but only the same-gender groups also show a lower number of non-studied words in the collaborative condition. The same-gender groups display a higher number of studied words than the mixed-gender groups in the collaborative condition, and a higher number of total answers in both total sample and collaborative condition. Finally, the same-gender groups show a higher number of non-studied words in the nominal condition. The MAN-COVA analysis (Table 2) highlights an effect of the experimental condition and the type of list on the performance, for both studied words and non-studied words, as well as the interaction effect between the two factors. The performance of collaborative groups shows a greater number of studied words ($Collaborative : M = 8.29$; $Nominal : M = 7.61$; $Individual : M = 7.56$), while any significant difference is found between individual and nominal groups' number of studied words. The analysis on the type of list shows that the list A about abstract words displays more non-studied words ($A : M = 2.60$; $B : M = 1.84$) and less studied words

$(A : M = 7.57; B : M = 8.07)$ than list B about concrete words. The analysis of the interaction between the experimental condition and the type of list shows that the experimental condition reduces the effect of the type of list on the non-studied words, while the difficulty difference between lists increases from nominal $(A : M = 2.25; B : M = 2.21)$, to individual $(A : M = 2.54; B : M = 1.60)$, to collaborative condition $(A : M = 2.99; B : M = 1.69)$. Finally, we analyzed the groups' virtual interactions in the collaborative condition. As we can see in Table 3, a greater communication and interaction in the group (i.e., LIWC variables: number of words produced, word count) and addressing the messages to all members (i.e., LIWC variables: 2nd person plural; references to other people), rather than a unique member (i.e., LIWC variables: 2nd person singular), are related to a higher number of studied words within the collaborative groups. However, the use of the 2nd person plural and a higher number of non-studied words are also associated. Finally, the use of certain words (i.e., LIWC variable: certainty) and negations in the communication is related to a lower number of non-studied words, while a clear assent in the group and the use of swear words are associated with a lower number of studied words.

Table 1. In table, the gender differences affecting the group performance (Studied words, Non-studied words, Total responses) are reported. In particular, the t value is positive when the first mentioned group has a significantly higher mean, and viceversa. In the tabel Ho. and He. indicate respectively the homogeneous and heterogeneous case, while Pr. indicates the word 'prevalence of'.

Group	Variable	t value	Sig.
Males VS females	Studied words tot sample	−1.971	$p < .05$
	Studied words individual condition	−	ns
	Studied words nominal condition	−2.104	$p < .05$
	Non-studied words collaborative condition	−	ns
Ho. male VS Ho. females	Studied words tot sample	−2.496	$p < .05$
	Studied words individual condition	−2.053	$p < .05$
	Studied words nominal condition	−2.164	$p < .05$
	Non-studied words collaborative condition	1.983	$p < .05$
Pr. males VS Pr. females	Studied words tot sample	−2.759	$p < .01$
	Studied words individual condition	−2.634	$p < .01$
	Studied words nominal condition	−2.413	$p < .05$
	Collaborative condition	−	ns
Ho. group VS He. group	Total responses total sample	2.096	$p < .05$
	Individual condition	−	ns
	Non-studied words nominal condition	2.008	$p < .05$
	Studied words collaborative condition	2.946	$p < .001$
	Total responses collaborative condition	2.389	$p < .05$

Table 2. MANCOVA on the group performance

Effect	Wilks' Lambda	F	Sig
Education	0.92	9.37	p. < .001
Experimental condition	0.95	4.41	p. < .05
Type of list	0.91	10.60	p. < .001
Exp. cond. * list	0.95	2.67	p. < .05
Source	Dependent V	F	Sig
Education	Non-studied words	18.80	p. < .001
Experimental condition	Studied words ($C > N/I$)	4.12	p. < .05
Type of list	Non-studied words ($A > B$)	14.06	p. < .001
Exp. cond. * list	Non-studied words	2.59	p. < .05

In brackets the scores, from the higher to the lower. C indicates collaborative, N indicates nominal and I individual. A and B are the types of list.

Table 3. Correlations between LIWC dimensions and performance's scores. The dimensions labelled as Studied words and Non-studied words refer to the total performance of the group, while the Average Non-studied words values refer to the total performance divided by the size of the group

LIWC categories	Performance's scores	Pearson's r	Sig
Word count	Studied words	.48	p.<.01
2nd pers. sing.	Studied words	−.33	p.<.05
Negations	Non-studied words	−.28	p.<.05
	Average non-studied words	−.33	p.<.05
Assent	Average Studied words	−.35	p.<.01
Certainty	Non-studied words	−.30	p.<.05
	Average non-studied words	−.30	p.<.05
Other people	Studied words	.34	p.<.05
Swear words	Average studied words	−.41	p.<.01
2nd pers. plur.	Non-studied words	.52	p.<.01
	Studied words	.61	p.<.01

4   Discussion

Adapting DRM paradigm in a virtual environment, the present research shows that collaborative facilitation may emerge and the collaborative inhibition may disappear. Collaborative groups show a higher number of studied words, confirming an effect of collaborative facilitation, whereas the similar number of non-studied words in all conditions (i.e., individual, nominal, and collaborative) suggests an absence of collaborative inhibition. Such results partially confirm the past literature [8] and they disconfirm another study [7]. Moreover, peculiar features appear in collaborative small group dynamics in virtual environments. Contrasting past literature [27], the low number of non-studied words in the collaborative groups for the list B (Concrete list) may highlight that the group's advantage of collaboration could interact with the performance in more simple tasks. In our study, when nominal condition precedes collaborative condition, the second shows a better performance. This result contradicts past studies [4,9], and it suggests the need to better understand this dynamics with further studies.

A gender effect is also revealed in the study, as female groups exhibit a lower number of non-studied words and a higher number of studied words. Since mixed-gender groups display a lower number of non-studied words in the nominal condition, it appears that the females performance improves performance of males in this group, thanks to the female individual performance that increases the outcome of the all group [28]. Finally, the analysis of the virtual interactions in the collaborative groups displays that a higher participation and a communication addressed to all members of the virtual group rather than to one individual (e.g., *What do you think, guys?*) is related to a better performance of the group. Confirming past literature [20], these results also suggest that a virtual collaboration may increase the in-group perception among members, which might be crucial for a better performance. This study, combined with the virtual interactions content analysis, may propose useful advices about collective reasoning and e-learning dynamics, which are nowadays very relevant topics in the study of web communities and educational communities.

5 Conclusions

Future research might analyze the collaboration and social dynamics in virtual groups running a complex task, comparing such a task in real environments. Moreover, we could verify the development of a sense of virtual community (SOVC) in members of a virtual group, analyzing the development of the in-group membership perception and taking into account gender, age, education and type of task. Finally, the effect of the reputation of people and the social facilitation in collaborative groups might be detected in virtual environments, investigating whether the reputation might increase or inhibit such a phenomenon.

Acknowledgements. This work was partially funded by the European Commission, under the FP7 EINS Open Call Project FOCAL.

References

1. Barros-Castro, R.A., Midgley, G., Pinzón, L.: Systemic intervention for computer-supported collaborative learning. Syst. Res. Behav. Sci. **32**(1), 86–96 (2015)
2. Popov, V., Noroozi, O., Barrett, J.B., Biemans, H.J.A., Teasley, S.D., Slof, B., Mulder, M.: Perceptions and experiences of, and outcomes for, university students in culturally diversified dyads in a computer-supported collaborative learning environment. Comput. Hum. Behav. **32**, 186–200 (2014)

3. Harris, C.B., Paterson, H.M., Kemp, R.I.: Collaborative recall and collective memory: what happens when we remember together? Memory **16**(3), 213–230 (2008)
4. Harris, C.B., Barnier, A.J., Sutton, J.: Consensus collaboration enhances group and individual recall accuracy. Q. J. Exp. Psychol. **65**(1), 179–194 (2012)
5. Nokes-Malach, T.J., Richey, J.E., Gadgil, S.: When is it better to learn together? Insights from research on collaborative learning. Educ. Psychol. Rev. **27**, 645–656 (2015)
6. Roediger, H.L., McDermott, K.B.: Creating false memories: remembering words not presented in lists. J. Exp. Psychol. Learn. Memory Cognit. **21**, 803–814 (1995)
7. Basden, B.H., Basden, D.R., Bryner, S., Thomas III, R.L.: A comparison of group and individual remembering: does collaboration disrupt retrieval strategies? J. Exp. Psychol. Learn. Memory Cognit. **23**(5), 1176–1189 (1997)
8. Maki, R.H., Weigold, A., Arellano, A.: False memory for associated word lists in individuals and collaborating groups. Memory Cognit. **36**, 598–603 (2008)
9. Thorley, C., Dewhurst, S.A.: Collaborative false recall in the DRM procedure: effects of group size and group pressure. Eur. J. Cognit. Psychol. **19**(6), 867–881 (2007)
10. Olszewska, J., Ulatowska, J.: Encoding strategy affects false recall and recognition: evidence from categorical study material. Adv. Cognit. Psychol. **9**(1), 44–52 (2013)
11. Prinsen, F., Volman, M.L.L., Terwel, J.: The influence of learner characteristics on degree and type of participation in a CSCL environment. Br. J. Educ. Technol. **38**(6), 1037–1055 (2007)
12. Bennett, J., Hogarth, S., Lubben, F., Campbell, B., Robinson, A.: Talking science: the research evidence on the use of small group discussions in science teaching. Int. J. Sci. Educ. **32**(1), 69–955 (2010)
13. Ding, N., Bosker, R.J., Harskamp, E.G.: Exploring gender and gender pairing in the knowledge elaboration processes of students using computer-supported collaborative learning. Comput. Educ. **56**(2), 325–336 (2011)
14. Zhan, Z., Fong, P.S.W., Mei, H., Liang, T.: Effects of gender grouping on students' group performance, individual achievements and attitudes in computer-supported collaborative learning. Comput. Hum. Behav. **48**, 587–596 (2015)
15. Harskamp, E., Ding, N., Suhre, C.: Group composition and its effect on female and male problem-solving in science education. Educ. Res. **50**(4), 307–318 (2008)
16. Sopka, S., Biermann, H., Rossaint, R., Rex, S., Jäger, M., Skorning, M., Heussen, N., Beckers, S.K.: Resuscitation training in small-group setting-gender matters. Scand. J. Trauma resuscitation Emerg. Med. **21**(1), 1–10 (2013)
17. Liu, N., Lim, J., Zhong, Y.: Joint effects of gender composition, anonymity in communication and task type on collaborative learning. In: PACIS 2007 Proceedings, vol. 85 (2007)
18. Benbunan-Fich, R., Hiltz, S.R., Turoff, M.: A comparative content analysis of face-to-face vs. asynchronous group decision making. Decis. Support Syst. **50**(4), 457–469 (2003)
19. Kreijns, K., Kirschner, P.A., Jochems, W.: Identifying the pitfalls for social interaction in computer-supported collaborative learning environments: a review of the research. Comput. Hum. Behav. **19**(3), 335–353 (2003)
20. Arbaugh, J.B., Benbunan-Fich, R.: The importance of participant interaction in online environments. Decis. Support Syst. **43**(3), 853–865 (2007)
21. Potter, R.E., Balthazard, P.A.: Virtual team interaction styles: assessment and effects. Int. J. Hum. Comput. Stud. **56**(4), 423–443 (2002)
22. Guazzini, A., Lió, P., Bagnoli, F., Passarella, A., Conti, M.: Cognitive network dynamics in chatlines. Procedia Comput. Sci. **1**(1), 2355–2362 (2010)

23. Guazzini, A., Vilone, D., Bagnoli, F., Carletti, T., Grotto, R.L.: Cognitive network structure: an experimental study. Adv. Complex Syst. **15**(06), 1250084 (2012)
24. Meade, M.L., Gigone, D.: The effect of information distribution on collaborative inhibition. Memory **19**(5), 417–428 (2011)
25. Pennebaker, J.W., Chung, C.K., Ireland, M., Gonzales, A., Booth, R.J.: The Development and Psychometric Properties of LIWC2007. LIWC.net.Pennebaker, Austin (2007)
26. Tausczik, Y.R., Pennebaker, J.W.: The psychological meaning of words: LIWC and computerized text analysis methods. J. Lang. Soc. Psychol. **29**(1), 24–54 (2010)
27. Kirschner, F., Paas, F., Kirschner, P.A.: Task complexity as a driver for collaborative learning efficiency: the collective working memory effect. Appl. Cogn. Psychol. **25**(4), 615–624 (2011)
28. Michailidou, A., Economides, A.: Gender and diversity in collaborative virtual teams. In: Orvis, K.L., Lassiter, A.L.R. (eds.) Computer Supported Collaborative Learning: Best Practices and Principles for Instructors, pp. 199–224. IGI Global, Hershey (2008)

Perceived Versus Actual Predictability of Personal Information in Social Networks

Eleftherios Spyromitros-Xioufis[1]($\boxtimes$), Georgios Petkos[1], Symeon Papadopoulos[1], Rob Heyman[2], and Yiannis Kompatsiaris[1]

[1] CERTH-ITI, Thessaloniki, Greece
{espyromi,gpetkos,papadop,ikom}@iti.gr
[2] iMinds-SMIT, Vrije Universiteit Brussel, Brussels, Belgium
rob.heyman@vub.ac.be

Abstract. This paper looks at the problem of privacy in the context of Online Social Networks (OSNs). In particular, it examines the predictability of different types of personal information based on OSN data and compares it to the perceptions of users about the disclosure of their information. To this end, a real life dataset is composed. This consists of the Facebook data (images, posts and likes) of 170 people along with their replies to a survey that addresses both their personal information, as well as their perceptions about the sensitivity and the predictability of different types of information. Importantly, we evaluate several learning techniques for the prediction of user attributes based on their OSN data. Our analysis shows that the perceptions of users with respect to the disclosure of specific types of information are often incorrect. For instance, it appears that the predictability of their political beliefs and employment status is higher than they tend to believe. Interestingly, it also appears that information that is characterized by users as more sensitive, is actually more easily predictable than users think, and vice versa (i.e. information that is characterized as relatively less sensitive is less easily predictable than users might have thought).

Keywords: Privacy · Social networks · Personal attributes · Inference

1 Introduction

Online Social Networks (OSNs) have had transforming impact on the overall Internet landscape. OSNs have affected the way people communicate, are being informed or even make business online. An issue that is sometimes overlooked though is the exposure of personal information through the OSNs. Participation in an OSN means that a certain amount of data related to the user is accessible from a) other OSN users and b) the OSN service. The disclosure of specific types of information may pose serious threats to the users. For instance, in several cases, information about the gender, age, ethnicity, political or religious beliefs, sexual preferences, and financial status of a person have been used for

© Springer International Publishing AG 2016
F. Bagnoli et al. (Eds.): INSCI 2016, LNCS 9934, pp. 133–147, 2016.
DOI: 10.1007/978-3-319-45982-0_13

unjustified discrimination, for instance, in the context of personnel selection [2] and for loan approval and pricing based on social media profiles [24].

In this paper we look into the issue of privacy in the context of OSNs. In particular, we study the predictability of various types of personal information based on shared OSN data, and contrast it to the users' perceptions about the exposure of their personal information. To perform this analysis, we employ a real life dataset that was composed through a user study that involved 170 participants. Each participant was asked to answer a questionnaire that included questions about his/her personal information as well as questions about his/her perceptions with respect to the disclosure of different types of information. Moreover, all users granted us access to their Facebook data (posts, likes and images) via a specially designed Facebook application.

Utilizing the collected OSN data and user responses, we train and evaluate the accuracy of classifiers that predict various personal user attributes using the OSN data as input. Different classifiers and a number of meta-learning techniques are tested (such as fusion of different feature modalities and multi-label classification). Eventually, we obtain indications of the actual predictability of different types of personal information and compare them to users' perceptions about the predictability and sensitivity of the corresponding types of information. It appears that users' perceptions about the predictability of different types of information are sometimes correct and sometimes not. For instance, users tend to correctly believe that their demographics information, such as their age, gender and nationality can be predicted quite accurately, whereas they incorrectly believe that their political beliefs cannot be accurately predicted. Moreover, it appears that information that is characterized by users as more sensitive is actually more easily predictable than users might have thought, and vice versa. To the best of our knowledge, this is the first study to compare users' perceptions about the disclosure of their personal information through an OSN to the actual predictability of such information.

2   Background

2.1   Privacy in OSNs

The current social research on privacy in OSNs focuses on awareness, attitudes and practices [9] with regard to volunteered or observed personal information disclosure. Nevertheless, it neglects to explore awareness or attitudes with regard to *inferred* information, the third category of personal information identified by the World Economic Forum [11], which is the type of information we focus on in this paper. Several studies have investigated the attitudes of people towards information disclosure in OSNs. For instance, [17] identifies three main classes of users with respect to the level of information disclosure in OSNs: (a) privacy fundamentalists, (b) pragmatists, and (c) unconcerned. Other studies compare attitude with behavior; these studies map what users are willing to disclose and how this is reflected in settings and other proxies that reflect their behavior [6]. A related study [19] shows that OSN users have difficulties dealing with privacy within OSNs. In particular, among 65 users that were asked to look for sharing violations in their OSN profiles (i.e. cases in which they shared content with people that they really would not like to) every one found at least one such violation. This mismatch between intended and actual sharing policies has been attributed to incomplete information, bounded cognitive ability and cognitive and behavioral biases [1], which may be caused by the difficulty of managing privacy settings and opting-out defaults [15]. While few works have studied privacy with regard to observed data by first [31] and third parties [3], to the best of our knowledge this is the first work to investigate awareness, attitudes and behavior with regard to inferred information on OSNs.

This line of research is significant for two reasons. First, the existence and use of inferred data will increase. Secondly, on a theoretical level, little sociological or psychological models exist that take inferred information into account for privacy. Behavioural economics [1], Westin's [34] privacy definition, contextual integrity [20] and Communication Privacy Management (CPM) [23] are all limited to access control. This means that each privacy perspective presents privacy as a question of giving access or communicating personal information to a particular party. This is illustrated in Westin's definition of privacy: "The claim of individuals, groups, or institutions to determine for themselves when, how, and to what extent information about them is communicated to others." [34]. But for inferred information, this definition becomes: "The claim of individuals, groups, or institutions to determine for themselves when, how, and to what extent information about them is *inferred.*" However, such control is non-existent because users: (a) are unaware and (b) have no control over the logic of the inferences being made. Since this area is under-researched, our first aim is to understand if and how users intuitively grasp what can be inferred from their disclosed data.

2.2 Prediction of Personal Attributes

A major issue about privacy is the fact that information about a user may not only appear in an explicit manner, but it can also appear implicitly and may be obtained using appropriate inference mechanisms. For instance, one might easily guess that a user who is interested in university/educational issues is very likely to be a young adult. In the following, we briefly review some previous work on inferring personal information based on OSN data.

In the study of Kosinski et al. [18], 58,466 Facebook users provided their complete like history (170 likes/person on average), their profile information, as well as the results of several psychometric tests. Using likes data, and particularly a reduced (via Singular Value Decomposition) version of the user-like matrix as input, the authors trained linear and logistic regression models to predict numeric and binary variables respectively. The Area Under ROC Curve (AUC) scores for predicting the binary variables were: 95 % for ethnicity, 93 % for gender, 88 % for gays, 75 % for lesbians, 85 % for political affiliation, 82 % for religion, 73 % for cigarette smoking, 70 % for alcohol consumption, 67 % for relationship status, 65 % for drug use and 60 % for parents being together when the user was 21. The Pearson correlation coefficient for age was 0.75.

Schwartz et al. [28] studied a dataset of 15.4 million status updates from a total of 74,941 Facebook users, who also submitted their gender, age and Big-5 personality scores. They tested traditional techniques of linking language with personality, gender and age such as the Linguistic Inquiry and Word Count (LIWC), which uses a lexicon with pre-selected categories, but also developed a new approach, the Differential Language Analysis (DLA), which generates the lexicon categories based on the text being analyzed. The researchers first used Principal Component Analysis (PCA) to reduce the feature dimension, and then a linear Support Vector Machine (SVM) for classifying gender and ridge regression for predicting age and personality traits. Among other results, they were able to predict gender with 92 % accuracy.

Other approaches have looked at a variety of user attributes and techniques. For instance, Backstrom and Kleinberg [4] managed to predict whether a user is single or not with 68 % accuracy and whether he/she is single or married with 79 % accuracy. Jernigan et al. [16] looked at sexual orientation and achieved an accuracy of 78 %. Of particular interest is the study presented by Zheleva and Getoor [35], where different OSNs were considered; examined user attributes are the country, gender and political views. Rao et al. [25] evaluated the accuracy of predicting gender (72 %), age (74 %), regional origin (77 %) and political affiliation (83 %) from Twitter messages. Particularly good results (95 % accuracy) on political views were obtained by Conover et al. [8]. Very good results on political views (89 % accuracy) were also achieved by Penna et al. [21]. Interestingly, they utilized a set of network attributes as features, whereas they also consider two more attributes: for ethnicity they achieved an F-score of 70 % and for predicting whether a person is a Starbacks fan an F-score of 76 %. Finally, an interesting finding is that inferring personal information based on OSN data can be highly unreliable (close-to-random) for a significant number of users [32].

3   Data Collection and Experimental Setup

3.1   Data Collection

Our study is based on a set of 170 Facebook users who gave us their informed consent to collect their OSN data through a test Facebook application[1]. In particular we collected each user's likes, textual posts and images. In addition, all users answered a questionnaire that included questions about several personal attributes as well as questions related to their perceptions about the predictability and the sensitivity of different types of information. Feedback about the perceived predictability was provided by the users with a yes/no answer to the question: "Can this particular type of information be inferred based on your OSN data?", and feedback about the sensitivity of different types of information was provided in a scale from 1 to 7 with higher values denoting higher sensitivity. Personal user attributes were organized into 10 categories: 1. "Demographics", 2. "Employment status and income", 3. "Relationship status and living condition",

---

[1] https://databait.hwcomms.com.

4. "Religious views", 5. "Personality traits", 6. "Sexual orientation", 7. "Political attitude", 8. "Health factors", 9. "Location", 10. "Consumer profile", hereafter referred to as *disclosure dimensions* [22], to facilitate a more compact and intuitive presentation and handling of a user's personal information. For instance, the "Demographics" dimension includes the following personal attributes: "age", "gender", "education level", "language", "nationality" and "residence". Due to space limitations, the full organization of personal user attributes into disclosure dimensions along with some statistics about the collected data are provided in a supplementary document[2].

3.2 Experimental Setup

In the learning experiments we considered 96 questions from the questionnaire, corresponding to 9 of the 10 disclosure dimensions (location was not considered due to the high cardinality of possible responses, which would lead to a very sparse training set given the limited number of test users). Evaluation was performed using repeated random sub-sampling validation. In this procedure, the data is randomly split n times into training and test sets. For each split, a model is fit to the training set and its prediction accuracy is assessed on the test set. The final performance is calculated as the average over the n tests. For this study, 66 % of the data were used for training and the process was repeated 10 times. Since for many of the questions (user attributes), the distribution of responses is highly imbalanced we used AUC as the evaluation measure due to its better robustness with imbalanced classes compared to measures such as classification accuracy that tend to favor classifiers that frequently predict the majority class.

The features that we extract from the OSN data and use as input attributes for the classification models throughout the experiments are the following:

– likes: A binary vector where each variable indicates the presence or absence of a like in the set of likes of the user. The vocabulary consists of the 3,622 likes that appear in the sets of likes of at least two users.
– likesCats: Each like in Facebook is assigned to a general category, such as "Community" or "Music". This vector is a histogram of the frequencies of these categories in the set of likes of each user. The vocabulary consists of the 191 categories that appear in the sets of likes of at least two users.
– likesTerms: A Bag-of-Words (BoW) vector computed using the terms that appear in the description, title and about sections of all likes made by each user. We performed stop-word removal (using three language-specific lists of stop words for the three main languages that appear in the collected content: English, Dutch and Swedish) and kept only terms that appear in the profiles of at least two users. This resulted in a vocabulary of 62,547 terms.
– msgTerms: A BoW vector computed using the terms that appear in all posts of each user. The same pre-processing was applied as in the case of likesTerms, resulting in a vocabulary of 24,990 terms.

---

[2] http://usemp-mklab.iti.gr/usemp/prepilot_survey_data_statistics.pdf.

- **LDA-t:** The distribution of topics in the textual content of the user's posts and likes (description, title and about sections). Topics were extracted using Latent Dirichlet Allocation (LDA) [5] and different setups involving different numbers of topics were examined (t = 20, 30, 50 and 100 topics).
- **visual:** The concepts depicted in the images of the users. These concepts were detected using the Convolutional Neural Network (CNN) variant presented in [13]. For each image the 12 most dominant concepts were kept, which resulted in a vocabulary of 11,866 distinct concepts for the whole collection. We used three alternative representations:
  - **visual-bin:** a binary vector representing concept presence/absence.
  - **visual-freq:** a histogram vector representing concept frequencies.
  - **visual-conf:** a vector where each variable represents the sum of detection scores for each concept across all images of each user.

4   Experimental Results

Here, we first present a set of thorough learning experiments with the goal of assessing the predictability of different types of user attributes and then compare these results to the perceptions of users.

The first experiment explores the performance of various baseline and state-of-the-art classifiers using the features described in the previous section. In particular, the following classifiers were considered:

- **knn:** The k-nearest neighbors ($k = 10$) classifier using the Euclidean distance.
- **tree:** A simple decision tree classifier (Weka's REPTree class [14]).
- **nb:** The Naïve Bayes classifier.
- **adaboost:** The Adaboost M1 boosting meta-classifier with a decision stump (a one-level decision tree) as the base classifier [12].
- **rf:** The Random Forest classifier [7] using 100 random trees.
- **logistic:** An efficient implementation of L2-regularized logistic regression from LibLinear [10] with probabilistic estimates and tuning of the regularization parameter ($c \in \{0.1, 1, 10\}$).

Due to the high computational cost of evaluating all six classifiers using all types of features, instead of performing the evaluation on all 96 target attributes, we selected eight representative ones: 'BMI class', 'Income', 'Health', 'Use of cannabis', 'Smoking behavior', 'Employment status', 'Drinking behavior' and 'Sexual orientation'. For each classifier, Fig. 1 shows the best achieved AUC performance (across all types of features) on each target attribute. We see that logistic and rf are the two best-performing classifiers in most cases. Specifically, logistic achieves the best performance in five targets, rf in two targets and adaboost in one target. Given the good performance of logistic and rf and their better scalability (especially with respect to the number of features) compared to the competing classifiers, we opted for using these two classifiers in the rest of the experiments.

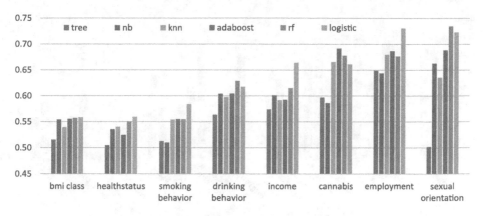

Fig. 1. Comparison of the performance of six different classifiers on eight target attributes. The best performance across all types of features is reported.

Our next experiment aims at evaluating the relative strength of the different types of features described in Sect. 3.2. Figure 2 shows the average AUC performance (across all 96 target attributes) using each type of feature by each classifier (logistic and rf). We observe that the best performance is obtained with likes, followed by LDA-t, likesTerms, msgTerms and likesCats. On the other hand, features based on visual concepts obtain lower performance scores, indicating that it is difficult to predict user attributes using this type of information alone. LDA-30 has a small edge over other LDA-based features, while visual-conf obtains the best performance among features based on visual concepts. With respect to the two classifiers, logistic is consistently better (on average) than rf with all feature types.

Since different features may capture different information about users, we also explore the possibility of increasing performance by combining features. To this end, we employ a simple late fusion scheme that consists of averaging the results produced by different single-feature classifiers. In this experiment, we use only the logistic classifier (as it was shown to significantly outperform rf in the previous experiment) and evaluate the performance of all possible two-classifier combinations. To avoid combining features that carry redundant information, we selected only the best performing variants of LDA-t (LDA-30) and visual (visual-conf). Thus, we ended up evaluating all 15 distinct pairs of the following features: likes, likesTerms, likesCats, msgTerms, LDA-30, and visual-conf. Figure 3 shows the average performance obtained by different late fusion schemes, along with the performance of models that are based on single features to facilitate a direct comparison. We see that the top four late fusion schemes include LDA-30 features and that two of them, LDA-30/likes and LDA-30/visual-conf obtain slightly better performance than the performance of the best single-feature model (the one based on likes). Another interesting observation is that although visual-conf and likesCats are the two worst performing features when used separately, their combination with LDA-30

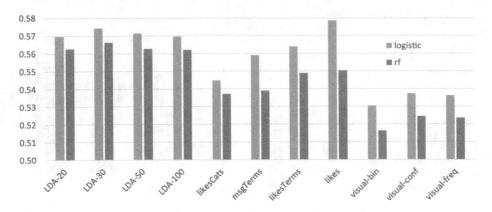

Fig. 2. Average AUC (across all 96 classification targets) for each type of feature using `logistic` and `rf`.

provides better results compared to e.g. the combination of LDA-30 with msgTerms. This is attributed to the fact that msgTerms are computed from the same data (terms appearing in a users likes) as the LDA-30 features and thus exhibit a lower degree of complementarity with them compared to likeCats and especially visual-conf features.

In classification problems where multiple target variables need to be predicted based on a common set of predictive variables, predictive performance can often be improved by taking target correlations into account [33]. Recognizing that different user attributes are likely correlated, we studied whether we could further improve predictive performance using *multi-label classification* methods. However, differently from typical multi-label classification problems where all target variables are binary, here we deal with a more general learning task since in addition to binary variables we also have to predict nominal variables with more than two levels. As a result, multi-label classification approaches that transform the problem into one or more multi-class classification problems where each class corresponds to a different combination of labels (e.g. [26]) are not directly applicable. On the other hand, approaches that build a separate model for each target can be easily adapted to handle different types of target variables by employing appropriate base models as shown in [30].

This category includes the baseline Single-Target (ST) approach that builds an independent model for each target variable and does not account for target dependencies (the approach that we have used so far), but also approaches that capture target dependencies by treating other target variables as additional feature attributes when predicting each target. A popular approach of this type is Classifier Chains (CC) [27]. CC constructs a chain of models, where each model involves the prediction of a single target and is built using a feature space that is augmented by the targets that appear earlier in the chain. During prediction, where the target values are unknown, CC uses estimated values obtained by sequentially applying the trained models. Here, we use this

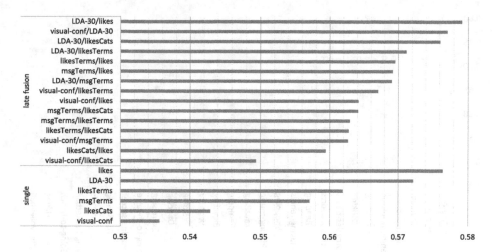

Fig. 3. Average AUC (over all 96 classification targets) of single-feature models and of models that combine two features with late fusion.

approach to predict a mixture of binary and nominal target variables by employing a multi-class instead of a binary classifier for nominal target variables with more than two levels. In addition to CC, we also use an ensemble version of the method called Ensemble of Classifier Chains (ECC) [27]. ECC builds multiple differently ordered random chains of classifiers and the final prediction for each target comes from majority voting.

We evaluated ST, CC and ECC (using 10 random chains) on each of the 96 targets using `likes` and `LDA-30` features (the two best performing features). All methods take the base single-target classifier as a parameter. Thus, we instantiated each method with `logistic` and `rf` and report, for each target, the best performance obtained using any combination of base classifier and feature. Figure 4 shows the results obtained by each method on the 28 targets related to the "Consumer profile" dimension (we do not show results on all 96 targets to improve the readability of the figure). We see that, although ST obtains the best performance in most targets (17 out of 28), it is outperformed by CC in 2 targets and by ECC in 9 targets. The picture is similar when all targets are considered. Again, despite the fact that ST obtains the best performance in most targets (50 out of 96), it is outperformed by CC in 17 targets and by ECC in 29 targets. As expected, ECC outperforms CC in most cases. A closer look at the results, reveals that CC tends to perform better than ST on targets that appear earlier in the chain but the performance starts deteriorating after a certain number of targets. This is due to the fact that prediction noise is accumulated along the chain, a known problem for CC on datasets with many targets such as this one.

Figure 5 shows the best AUC that we obtained for each target attribute, using any combination of features and classification approach. Target attributes are grouped by disclosure dimension and sorted in ascending AUC order within each dimension. The average best AUC achieved for all 96 attributes is 0.63

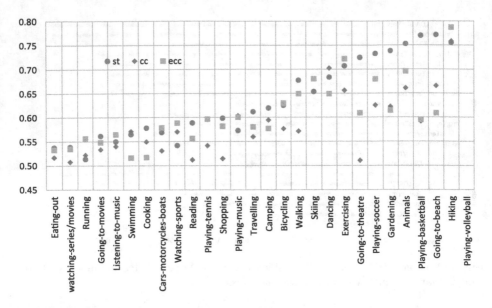

Fig. 4. Maximum AUC per target using ST, CC and ECC.

which represents a significant improvement over random performance (0.5) and is actually quite impressive if we take into account the limited number of training examples and the high cardinality of some classes.

Having performed a set of thorough experiments that measured the actual predictability of different types of personal information, we now proceed to a comparison with the perceived predictability and sensitivity of different types of information (according to users' responses in the survey). Table 1 presents the ranking of dimensions according to (a) their perceived predictability, (b) their actual predictability according to our experiments (obtained by averaging the performance over the attributes of each dimension) and, (c) their predictability according to [18] (for those dimensions for which data is available). It is noted that users perceive "Demographics" as the dimension that is most predictable (88.4%), and indeed it was found through our study that it is the dimension that can be predicted most accurately. Our conclusions also appear to mostly match those of [18]. In particular, "Demographics" and "Political views" are identified as the most predictable dimensions in both studies and the ranking of the remaining dimensions is quite similar (except for "Religious views").

Figure 6 presents an overall comparison between perceived and actual predictability of dimensions with respect to perceived sensitivity. Let us first focus on the relationship between perceived predictability and sensitivity. With the exception of the "Religious views" and "Relationships" dimensions, there appears to be a clear linear relationship between sensitivity and perceived predictability. That is, the more sensitive some dimension is perceived by users, the less predictable it is considered. For instance, "Demographics", the dimension that is perceived as the easiest to predict (and is actually the most predictable),

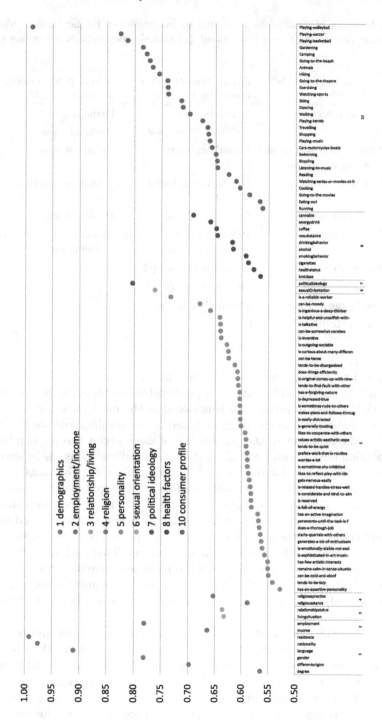

Fig. 5. Best AUC achieved on each target attribute using any combination of features, classifier and fusion approach.

Table 1. Ranking of dimensions according to (a) perceived predictability, (b) actual predictability (according to our study) and, (c) actual predictability according to [18].

Rank	Perceived predictability of dimension	Actual predictability according to our study	Actual predictability according to [18]
1	Demographics	Demographics	Demographics
2	Location	Political views (**+4**)	Political views
3	Relationship status and living condition	Sexual orientation	Religious views
4	Sexual orientation	Employment/Income (**+5**)	Sexual orientation
5	Consumer profile	Consumer profile	Health status
6	Political views	Relationship status and living condition	Relationship status and living condition
7	Personality traits	Religious views (**+1**)	
8	Religious views	Health status (**+1**)	
9	Employment/Income	Personality traits	
10	Health status		

is considered to be the least sensitive. At the same time, "Health status" the dimension that is perceived as the least predictable (and is actually among those that are the hardest to predict), is considered as the most sensitive.

Two more observations can be made based on the results shown on Fig. 6. The first is that the accuracy of the perceptions of users about the predictability of each dimension tends to vary considerably. For some dimensions, their perception is rather accurate, but for others it is far from accurate. For instance, users correctly believe that their demographics information is quite predictable (actual predictability is quite high) and also have a quite accurate perception about the predictability of their consumer profile information and factors related to their personality traits. On the other hand, their perception about the predictability of their health related information is rather incorrect. This leads us to the second observation: the actual predictability of the more sensitive dimensions is higher than the perceived predictability. Vice versa, perceived predictability is higher than actual predictability for the less sensitive dimensions (with the exception of "Religious views").

It is also worth looking at any conclusions that may be reached by looking at the perceptions of individual users and in particular, users that belong to potentially sensitive groups; for instance, people that have answered that their health is poor or people that are not heterosexuals. We examined whether the sensitivity of particular dimensions differs for users belonging to different classes. We formed a two-way table with one dimension representing the class of the user (e.g. poor/good health) and the other dimension representing the sensitivity of the information. A $\mathcal{X}^2$ test was performed to examine if the perceptions of

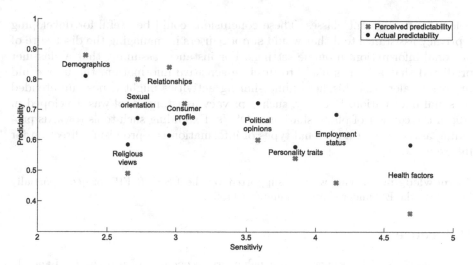

Fig. 6. Comparison of perceived and actual predictability of the disclosure dimensions with respect to sensitivity.

different classes of users about the sensitivity of some dimension differ. The test was positive (at the 0.05 level) for the following three dimensions: "Sexual orientation" (p-value: 0.000003), "Health factors" (p-value: 0.029) and "Religious beliefs" (p-value: 0.011). So, for instance, homosexual and bisexual users tend to view the disclosure of information about their sexual profile as more sensitive than heterosexual users. Also, users with good health tend to view the disclosure of information about their health as less sensitive than people with poor health.

5 Conclusions

The paper discussed the issue of privacy in the context of OSNs. In particular, it examined different mechanisms by which user attributes can be predicted based on content shared by users in an OSN. Importantly, the predictability of different types of personal information was compared against the perceptions of users about the predictability and sensitivity of each type. Experiments and analysis were carried out on a dataset collected for this purpose via a custom Facebook application. The dataset consisted of the posts, images and likes of 170 Facebook users along with their responses to a survey that considered both their personal information as well as their perceptions about privacy and disclosure of information in the OSN.

A number of insights were extracted with respect to the relationship between actual predictability, perceived predictability and sensitivity. In particular, it appears that users have both correct and incorrect perceptions about the predictability of specific types of information. Moreover, the more sensitive a type of information is, the more the users underestimate its predictability. Additionally, the sensitivity of particular types of information seems to be different for users belonging to different classes. These conclusions could be useful for developing a privacy assistance tool that would support users in managing the disclosure of personal information in online settings. For instance, assuming that a classifier predicted that a user is likely to disclose sensitive information, the user could receive an alert that his/her online sharing activities might expose unintended personal information. Recently, such a privacy assistance tool was developed in [29] in the context of photo sharing in OSNs. Extending such tools towards providing assistance for additional types of information is a promising direction for future work.

Acknowledgment. This work is supported by the USEMP FP7 project, partially funded by the EC under contract number 611596.

References

1. Acquisti, A.: The economics and behavioral economics of privacy. In: Lane, J., Stodden, V., Bender, S., Nissenbaum, H. (eds.) Privacy, Big Data, and the Public Good: Frameworks for Engagement, pp. 98–112. Cambridge University Press (2014)
2. Acquisti, A., Fong, C.M.: An experiment in hiring discrimination via online social networks. (2015). Available at SSRN 2031979
3. Agarwal, L., Shrivastava, N., Jaiswal, S., Panjwani, S.: Do not embarrass: re-examining user concerns for online tracking and advertising. In: Proceedings of the Ninth Symposium on Usable Privacy and Security (2013)
4. Backstrom, L., Kleinberg, J., Romantic partnerships, the dispersion of social ties: a network analysis of relationship status on facebook. In: Proceedings of CSCW 2014, pp. 831–841. ACM (2014)
5. Blei, D.M., Ng, A.Y., Jordan, M.I.: Latent dirichlet allocation. J. Mach. Learn. Res. **3**, 993–1022 (2003)
6. Brandimarte, L., Acquisti, A., Loewenstein, G.: Misplaced confidences: privacy and the control paradox. In: Ninth Annual Workshop on the Economics of Information-Security, p. 43, Cambridge (2010)
7. Breiman, L.: Random forests. Mach. Learn. **45**(1), 5–32 (2001)
8. Conover, M.D., Goncalves, B., Ratkiewicz, J., Flammini, A., Menczer, F.: Predicting the political alignment of twitter users. In: Privacy, Security, Risk and Trust (PASSAT) and SocialCom 2011, pp. 192–199 (2011)
9. Debatin, B., Lovejoy, J.P., Horn, A.-K., Hughes, B.N.: Facebook and online privacy: attitudes, behaviors, and unintended consequences. J. Comput. Mediated Commun. **15**(1), 83–108 (2009)
10. Fan, R.-E., Chang, K.-W., Hsieh, C.-J., Wang, X.-R., Lin, C.-J.: Liblinear: a library for large linear classification. J. Mach. Learn. Res. **9**, 1871–1874 (2008)
11. World Economic Forum. Rethinking personal data: strengthening trust. Technical report, May 2012
12. Freund, Y., Schapire, R.E., et al.: Experiments with a new boosting algorithm. ICML **96**, 148–156 (1996)
13. Ginsca, A.L., Popescu, A., Le Borgne, H., Ballas, N., Vo, P., Kanellos, I.: Large-scale image mining with flickr groups. In: He, X., Luo, S., Tao, D., Xu, C., Yang, J., Hasan, M.A. (eds.) MMM 2015, Part I. LNCS, vol. 8935, pp. 318–334. Springer, Heidelberg (2015)

14. Hall, M., Frank, E., Holmes, G., Pfahringer, B., Reutemann, P., Witten, I.H.: The weka data mining software: an update. ACM SIGKDD Explor. Newslett. **11**(1), 10–18 (2009)
15. Heyman, R., De Wolf, R., Pierson, J.: Evaluating social media privacy settings for personal, advertising purposes. Info **16**(4), 18–32 (2014)
16. Jernigan, C., Mistree, B.F., Gaydar: Facebook friendships expose sexual orientation. First Monday, **14**(10) (2009)
17. Knijnenburg, B.P., Kobsa, A., Jin, H.: Dimensionality of information disclosure behavior. Int. J. Hum. Comput. Stud. **71**(12), 1144–1162 (2013)
18. Kosinski, M., Stillwell, D., Graepel, T.: Private traits and attributes are predictable from digital records of human behavior. Proc. Nat. Acad. Sci. **110**(15), 5802–5805 (2013)
19. Madejski, M., Johnson, M., Bellovin, S.M.: A study of privacy settings errors in an online social network. In: PERCOM Workshops (2012)
20. Nissenbaum, H.: Privacy as contextual integrity. Wash. L. Rev. **79**, 101–139 (2004)
21. Pennacchiotti, M., Popescu, A.-M.: Democrats, republicans, starbucks afficionados: user classification in twitter. In: SIGKDD (2011)
22. Petkos, G., Papadopoulos, S., Kompatsiaris, Y.: PScore: A framework for enhancing privacy awareness in online social networks. In: Availability, Reliability and Security (ARES 2015), pp. 592–600. IEEE (2015)
23. Petronio, S.S.: Boundaries of Privacy: Dialectics of Disclosure. SUNY series in communication studies. State University of New York Press, Albany (2002)
24. Raman, A.S., Barloon, J.L., Welch, D.M.: Social media: emerging fair lending issues. Rev. Banking Financial Serv. **28**(7), 81–88 (2012)
25. Rao, D., Yarowsky, D., Shreevats, A., Gupta, M.: Classifying latent user attributes in twitter. In: Proceedings of the 2nd International Workshop on Search and Mining User-Generated Contents, pp. 37–44. ACM (2010)
26. Read, J., Pfahringer, B., Holmes, G.: Multi-label classification using ensembles of pruned sets. In: ICDM 2008, pp. 995–1000 (2008)
27. Read, J., Pfahringer, B., Holmes, G., Frank, E.: Classifier chains for multi-label classification. Mach. Learn. **85**(3), 333–359 (2011)
28. Andrew Schwartz, H., Eichstaedt, J.C., Kern, M.L., Dziurzynski, L., Ramones, S.M., Agrawal, M., Shah, A., Kosinski, M., Stillwell, D., Seligman, M.E.P., et al.: Personality, gender, and age in the language of social media: the open-vocabulary approach. PloS one **8**(9), e73791 (2013)
29. Spyromitros-Xioufis, E., Papadopoulos, S., Popescu, A., Kompatsiaris, Y.: Personalized privacy-aware image classification. In: Proceedings of the 6th ACM International Conference on Multimedia Retrieval, ICMR 2016 (2016)
30. Spyromitros-Xioufis, E., Tsoumakas, G., Groves, W., Vlahavas, I.: Multi-target regression via input space expansion: treating targets as inputs. Machine Learning, pp. 1–44 (2016)
31. Stutzman, F., Gross, R., Acquisti, A.: Silent listeners: the evolution of privacy and disclosure on Facebook. J. Privacy Confidentiality **4**(2), 7–41 (2012)
32. Theodoridis, T., Papadopoulos, S., Kompatsiaris, Y.: Assessing the reliability of facebook user profiling. In: WWW (2015)
33. Tsoumakas, G., Katakis, I., Vlahavas, I.: Mining multi-label data. In: Maimon, O., Rokach, L. (eds.) Data Mining and Knowledge Discovery Handbook, pp. 667–685. Springer, New York (2009)
34. Westin, A.: Privacy and Freedom. Bodley Head, London (1970)
35. Zheleva, E., Getoor, L.: To join or not to join: the illusion of privacy in social networks with mixed public and private user profiles. In: WWW (2009)

Conformity in Virtual Environments:
A Hybrid Neurophysiological
and Psychosocial Approach

Serena Coppolino Perfumi[1(✉)], Chiara Cardelli[1], Franco Bagnoli[3,4],
and Andrea Guazzini[2,3]

[1] Neurofarba Department (Neuroscience, Psychology,
Pharmacology and Children's Health), University of Florence, Florence, Italy
serena.c.perfumi@gmail.com
[2] Department of Educational Sciences and Psychology,
University of Florence, Florence, Italy
[3] Centre for the Study of Complex Dynamics,
University of Florence, Florence, Italy
[4] Department of Physics and Astronomy, University of Florence and INFN,
Florence, Italy

Abstract. The main aim of our study was to analyze the effects of a virtual environment on social conformity, with particular attention to the effects of different types of task and psychological variables on social influence, on one side, and to the neural correlates related to conformity, measured by means of an Emotiv EPOC device on the other. For our purpose, we replicated the famous Asch's visual task and created two new tasks of increasing ambiguity, assessed through the calculation of the item's entropy. We also administered five scales in order to assess different psychological traits. From the experiment, conducted on 181 university students, emerged that conformity grows according to the ambiguity of the task, but normative influence is significantly weaker in virtual environments, if compared to face-to-face experiments. The analyzed psychological traits, however, result not to be relatable to conformity, and they only affect the subjects' response times. From the ERP (Event-related potentials) analysis, we detected N200 and P300 components comparing the plots of conformist and non-conformist subjects, alongside with the detection of their Late Positive Potential, Readiness Potential, and Error-Related Negativity, which appear consistently different for the two typologies.

Keywords: Social conformity · Social influence · ERP · Group pressure · In-group dynamics

1 Introduction

Conformity has been widely analyzed by social psychology starting from the pioneeristic works of Sherif and Asch [1].

These experiments showed to what extent majority pressure can be powerful, even when the majority is giving a clearly incorrect answer.

© Springer International Publishing AG 2016
F. Bagnoli et al. (Eds.): INSCI 2016, LNCS 9934, pp. 148–157, 2016.
DOI: 10.1007/978-3-319-45982-0_14

However, from an evolutionary point of view, these results are not shocking, since conformity turns out to be an adaptive behavior that presents many benefits concerning human beings' fitness, reproduction and survival [2].

Recent cultural studies on conformity, analyzed its connection with protection and showed how the inhabitants of areas that historically had higher prevalence of disease tend to be more conformist, and this outcome is explained by the fact that conformity is a strong protective factor against the risk of contracting illness [3].

Among the different benefits, conformity can work as a protective shield against threats linked to group exclusion: infact human beings developed heuristics and neurally evolved with the ability to select similar individuals to bond with and to distinguish in-group members from out-group members. From this point of view, conformist behaviors like mimicry can be helpful in creating group membership [4].

The context plays a crucial role in fostering this type of behavior but the majority of the studies on conformity focused on face-to-face interaction.

However, considering the widespread of social networks and computer-mediated communication nowadays, it is necessary to shed light on how social influence works in a context characterized by anonymity.

Contrasting theoretical frameworks focused on the effects of anonymity on human interaction: on one side anonymity seems to be able to give individuals a feeling of protection that leads them to feel more free to speak their minds [5], but on the other side, the Social Identity Model of Deindividuation Effects (SIDE) perspective shows how anonymity can lead to deindividuation, and this factor, making less salient individual traits, can lead to a stronger tendency to conform to social norms [6].

From the very few studies on conformity in virtual environments emerged that social influence can occur also in virtual environments but with some differences according to the type of influence elicited.

A replication of Asch's experiment, showed no conformity in anonymous condition [7]. Asch's task, which consisted in confronting a reference bar with three options of different lengths, among which was present only one twin bar, is an example of normative influence, namely the tendency to conform in order not to appear as an outsider when confronting a group [8]. Asch's experimental organization consisted in a group of seven people among which only the person in sixth answer position was the experimental subject. In some trials, the majority was asked to provide unanimously the same incorrect answer, and the tendency to conform of the experimental subject was analyzed. Averagely, 32 % of the experimental subjects conformed to the majority [1]. In this case, since the task's ambiguity was low and detecting the correct answer was pretty easy, the reason that brought the subjects to conform is relatable to in-group dynamics, social norms and the desire not to break them [8].

The existing neurophysiological literature deepens the construct of normative influence, showing how Event-Related Potentials (ERP, measurable brain responses resulting from a specific cognitive, sensory or motor event) components such as N200 (that is a negativity associated with a variation in form or context of a predominant stimulus, typically evoked between 180 and 325 ms following the presentation of a particular visual or auditory stimulus) and P300 (which is a positivity typically emerging approximately between 300 and 400 ms following the stimulus presentation, and perhaps the most-studied ERP component in research concerning selective attention and information processing) respectively indicate the internal conflict experienced by the subject and the activation of inhibitory response mechanisms [9], as well as the awareness of the conformity of the response [10, 11].

Other experiments that used more ambiguous or difficult tasks, showed how informational influence, namely the tendency to conform when the subjects have lacking information on the task and for this reason reckon the group a reliable source, can occur also in virtual environments [12].

In this case, group dynamics are less relevant, since the goal is to give a correct answer [8].

The aim of the present study was to analyze how normative and informational influence could be affected by a virtual context, so the effect given by the type of task was taken into account. Besides replicating Asch's visual task, we created two more tasks whose items presented different levels of ambiguity, assessed with the measurement of the item's entropy.

The cultural items consisted in a target work associated with three adjectives with different levels of semantic relation with the target word, while the apperceptive items consisted in invented words associated with existing adjectives and vice-versa.

The experiment was also conducted in different conditions, the first one concerning different levels of anonymity, in order to see if a higher exposure could have an effect on conformity, and the second one making the subjects perform the experiment alone or with the physical presence of other subjects in the same room.

For our purpose, we created a virtual interface that simulated the responses of six non-existing people, with the experimental subject placed always in the sixth response position, in order to be able to see the responses of a majority of five subjects, inside a group of seven people.

Besides these variables, we also controlled the interaction with personality traits, in order to analyze whether it is possible to predict conformity from certain psychological features.

We performed the same experiment on subjects wearing an Emotiv EPOC device, a wireless EEG-based headset that enables the detection of electrical brain signals on the scalp's surface, in order to record and analyze the ERP components.

The ERP experiment focused on the differences between conformist and non-conformist subjects' cerebral activity within all the tasks.

Besides analyzing N200 and P300, ERP components such as LPP (Late Positive Potential), RP (Readiness Potential) and ERN (Error-Related Negativity) were reckoned to be potentially interesting for the phenomenon taken into consideration, because they respectively indicate emotional regulation [13], premotor planning of voluntary movement [14] and error awareness [15].

2 Participants

For our study we recruited 181 universitary students: 120 participated to the standard experiment and 60 participated to the ERP version of the study.

For the experimental typologies, we balanced them for the full and partial anonymity conditions and the group and single condition.

The only unbalanced condition is gender, with 139 females and 42 males, but in the data analysis phase, the factor has been controlled. The recruitment took place through voluntary census and the majority (80 %) came from the School of Psychology of the University of Florence.

3 Method and Procedure

To control the possible effect of psychological variables on conformity, the first experimental phase consisted in a preliminary survey composed of a battery of self-reported socio-demographic and psychological questionnaires and scales. After this phase, the subjects performed the experiment on a software that re-created a group condition. Finally they were asked to fill a questionnaire investigating their experience within the group.

Materials. The preliminary survey was composed by two sections. The first section consisted in socio-demographic items concerning age, gender, type of studies attended, educational level, marital status, presence of children and religious orientation. The second section consisted in a series of scales investigating psychological traits and status, in particular the scales administered were:

- The Fast Five Personality Questionnaire [16]
- The State-Trait Anxiety Inventory for Adults [17]
- The Multidimensional Sense of Community Scale [18]
- The Rosenberg's Self-Esteem Scale [19]
- The General Self-Efficacy Scale [20]

Software. To perform the experiment we created a software designed on Google Script, the functioning of which is similar to Crutchfield's [21] apparatus. Before starting, the experimental subjects were informed that six other subjects were about to log in and participate with them. After reading the instructions, they could log-in. The interface was organized in order to simulate the responses of six non-existing people, with randomized log-in and response times. The interface provided also the possibility to manipulate anonymity and to collect the subjects' response times.

On the left was placed a series of dots, vertically numbered from 1 to 7, associated with each group member: in the fully anonymous condition, the subjects could only see the numbers associated with the response order, while in the partially anonymous condition, they could see names and surnames. The experimental subjects were always placed in the sixth position, so that they could see the answers provided by five fake subjects before: when a subject answered, a number indicating their choice appeared beside their number or name, and when it was the experimental subject's turn, the stimulus appeared as long as three buttons (numbered 1, 2 and 3) in correspondence of the three alternatives. The first task was Asch's adaptation, with twenty items, the second the cultural and the third the apperceptive, each composed by forty-five items.

This second phase took averagely forty-five minutes to be completed.

Setting. The experiment was presented as a study on visual and semantic perception.

We collected the contacts and scheduled the appointments with the subjects via e-mail or text message, making sure that they fit the non-psychological disease condition.

The subjects were then randomly assigned to the experimental conditions.

The group-condition experiments took place in the computer science laboratory, with groups of six, seven or eight people.

They were equipped with headphones playing white noise and each workstation had a barrier isolating each subject.

The single-condition and ERP experiments took place in the social psychology laboratory, where the subjects performed the expriment alone with the presence of maximum three experimenters. Each experiment lasted approximately an hour and twenty minutes.

4 Data Analysis

The first step consisted in a pre-processing of the data, in which we verified the necessary pre-conditions for the inferential analysis.

For the t-Student analysis, we balanced the experimental conditions for each sub-group, made sure to have the necessary minimum numerosity, and verified a proper gaussian distribution for the dependent variables. Where necessary, we proceeded with a re-normalization of the dependent variables by means of a logarithmic function.

In order to calculate the entropy of each item in cultural and apperceptive tasks, we presented the items to two samples of people (71 subjects for the cultural and 79 for the apperceptive) without manipulation, collected their responses, and calculated the frequencies and percentages of the answers to each item.

On the basis of the percentages we calculated the entropy for each item "i" using the Eq. 1, with $p_j^k = (\Sigma_{i=1}^n r_i^k) / n$, "$n$" equal to the number of respondents, and "r_i^k" reporting the answer of the subject "i" to the item "k" (i.e., that can be "0" or "1").

$$E^k = \Sigma_{j=1}^3 - p_j^k \log p_j^k \tag{1}$$

When we collected all the entropies, we calculated the median for cultural and apperceptive tasks and according to that, we divided the items in high and low entropy. In order to assess the effect of entropy on the decision making we balanced the distribution of the entropy within the set of experimental stimuli.

We adopted a t-Student test for independent samples to analyze the relations between conformity, delay and type of anonymity.

To analyze the relations between type of task, conformity, delay and entropy, we proceeded with a χ^2 test. To analyze the psychological and socio-demographic variables effects we conducted a r Pearson's correlations and again a χ^2 analysis.

ERP data were properly filtered using Matlab in order to be analyzed and then t-Student tests were adopted.

5 Results

Starting from the anonymity effect, below are presented the significant values. The affected variables are the general delay and the delay in conformist answers in Asch's task, the general delay in the cultural task and conformity in the cultural task (Table 1).

Table 1. Anonymity effect on conformity and delay. $^* = p < .05$

Condition	Anonymity	Mean	St. Dev.	t
General delay (Asch)	No Anonymity	12438.60	41050.13	1.98*
	Anonymity	3781.48	1207.95	
Conformity-related delay (Asch)	No Anonymity	10576.88	30842.82	2.03*
	Anonymity	3911.59	1191.88	
General delay (Cultural)	No Anonymity	6500.87	3441.42	2.04*
	Anonymity	5649.43	1909.53	
Conformity (Apperceptive)	No Anonymity	0.27	0.14	−1.99*
	Anonymity	0.33	0.21	

For what concerns the type of task, significant relations appeared between all the investigated factors, which are conformity, entropy and delay (Tables 2 and 3).

Table 2. Type of task effect on conformity and delay. $^{***} = p < .001;\,^1 = 4805$ ms

		Asch	Task Cultural	Apperceptive	χ^2
Conformity	No	98.6 %	84.8 %	70.2 %	954.64***
	yes	1.4 %	15.2 %	29.8 %	
Delay	< median^1	59.8 %	60.2 %	56.9 %	19.15***
	> median^1	40.2 %	39.8 %	43.1 %	

Table 3. Entropy effect on conformity and delay. $^{***} = p < .001;\,^1 = 0.427;\,^2 = 4805$ ms

		Entropy		χ^2
		< median^1	> median^1	
Conformity	No	92.6 %	69.6 %	1065.396***
	Yes	7.4 %	30.4 %	
Conformity delay	< median^2	60.5 %	55.9 %	25.272***
	> median^2	39.5 %	44.1 %	

The correlations between conformity, delay and psychological traits did not provide significant results, only state and trait anxiety have an impact on the delays in the cultural task's responses. For what concerns the socio-demographic variables, no gender differences in conformist behavior appeared.

The ERP plots obtained by using Emotiv EPOC were divided according to two specific times of the test, namely considering two trigger-moments. In the patterns relating to the stimulus administration presented below, emerged how typical ERP components such as N200, P300 and LPP (Late Positive Potential) are significantly higher for the conformist subjects (Fig. 1) (Table 4).

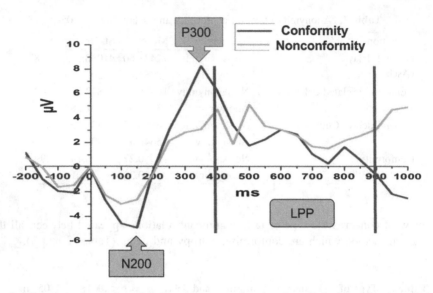

Fig. 1. Grand-averages of conformist and non-conformist plots elicited by stimulus administration. Left hemisphere electrodes (no occipital): E1, E2, E3, E4, E5. Y-axis reports the cerebral activation in microvolts, while x-axis reports the time in milliseconds from the stimulus administration. Blue line stands for conformist subjects, red line for non-conformist subjects. N200 indicates a negativity typically evoked 180 to 325 ms following the presentation of a stimulus (i.e. mismatch detection), P300 is a positivity that peaks approximately 300–400 ms post-stimulus (i.e. selective attention), finally LPP begins around 400 ms after the onset of a stimulus and lasts for a few hundred milliseconds (i.e. emotional activity). (Color figure online)

Table 4. Observable components' means, standard deviations and t-Students of conformist and non-conformist plots relative to the time of stimulus administration. *** = $p < .001$

Observable	Condition	Mean	St. Dev.	t
N200	Conformist	−4,06	5,18	38,80***
	Non-conformist	−2,10	5,17	
P300	Conformist	5,29	5,17	19,74***
	Non-conformist	2,94	5,18	
LPP	Conformist	0,64	5,18	−15,95***
	Non-conformist	2,97	5,19	

Conversely, in the plots related to the click time, which represented the final decision made by the subject, it clearly emerges how components such as RP (Readiness Potential) and ERN (Error-Related Negativity) are significantly higher for the non-conformist subjects (Fig. 2) (Table 5).

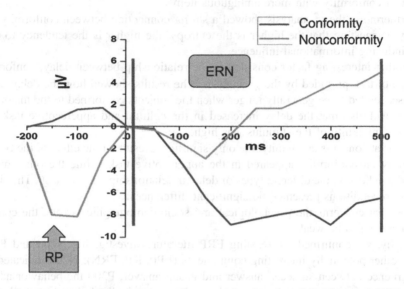

Fig. 2. Grand-averages of conformist and non-conformist plots following the click's time. Frontal electrodes: E1, E3, E12, E14. RP peaks around 250–0 ms before the time of movement execution (i.e. voluntary movement planning), and ERN is a negativity beginning around the first 50–100 ms after the click (i.e. error awareness).

Table 5. Observable components' means, standard deviations and t Students of conformist and non-conformist plots relative to the time of the click, *** $= p < .001$.

Observable	Condition	Mean	St. Dev.	t
RP	Conformist	0,97	5,19	31,54***
	Non-conformist	−5,08	5,52	
ERN	Conformist	− 6,42	5,23	−36,14***
	Non-conformist	1,38	5,53	

6   Discussion

This research, aimed at highlighting the differences that may be elicited by a virtual environment in social phenomena, specifically conformity and its interaction with other factors.

At first, we obtained different percentages of conformity according to the type of task: in Asch's task, only 1.4 % of the subjects conformed when the majority gave an unanimous incorrect answer, while the percentages grow in the cultural (15.2 %) and in the apperceptive (29.8 %). These preliminary results suggest that Asch's paradigm changes in a virtual, anonymous environment, and that normative influence might be less effective due to the characteristics of the setting and social norms might take longer to become effective.

The percentages emerged in the cultural and apperceptive tasks, however, showed a growth of conformity with more ambiguous items.

Furthermore, the χ^2 analysis showed a strong connection between conformity and entropy, suggesting that the higher is the entropy, the higher is the tendency to conform, inducing informational influence.

Another interesting factor consisted in the relationship between delay, conformity and type of task provided by the χ^2 analysis. The results showed how the delay in the responses tended to be generally longer when the subjects conformed to the majority's opinion, and also that the delay increased in the cultural and apperceptive tasks, so when the ambiguity of the stimulus was higher.

For what concerns anonymity, the only significant result on the effect of the type of anonymity on conformity appeared in the apperceptive task, while the other results showed an effect on the different types of delay in relationship with the task. The single and group conditions presented no significant differences.

For what concerns the psychological and socio-demographic factors, the correlations appeared to be weak.

Finally, we confirmed the existing ERP literature investigating N200 and P300, adding other potentially interesting components (LPP, RP, ERN): N200 indicated the incongruence between subjects' answer and group answer, P300 the behavior adjustment, LPP the consequential emotional regulation, and all of them resulted more pronounced in the conformist subjects. RP probably indicates the premotor click planning, ERN the awareness of the error committed, and these components are more evident in the case of the non-conformist subjects. Thanks to such identifications, we have been able to differentiate the conformist events from the nonconformist ones in the electroencephalographic patterns. The ERP analysis, for now, considered only the differences between conformists and non-conformists, further analysis will focus on the ERP components related to entropy and type of task.

Acknowledgements. This work was supported by EU Commission (FP7-ICT-2013-10) Proposal No. 611299 SciCafe 2.0.

References

1. Asch, S.E.: Effects of group pressure upon the modification and distortion of judgments. In: Guetzkow, H.S. (ed.) Groups, Leadership and Men, pp. 177–190. Carnegie Press, Pittsburgh (1951)
2. Morgan, T.J.H., Laland, K.N.: The biological bases of conformity. Front. Neurosci. **6**(87), 1–7 (2012). doi:10.3389/fnins.2012.00087
3. Murray, D.R., Trudeau, R., Schaller, M.: On the origins of cultural differences in conformity: four tests of the pathogen prevalence hypothesis. Pers. Soc. Psychol. Bull. **37**(3), 318–329 (2011)

4. Neuberg, S. L., Kenrick, D. T., Schaller, M.: Evolutionary Social Psychology. Handbook of Social Psychology (2010)
5. Kiesler, S., Siegel, J., McGuire, T.W.: Social psychological aspects of computer-mediated communication. Am. Psychol. **39**(10), 1123–1134 (1984)
6. Spears, R., Postmes, T., Lea, M., Wolbert, A.: When are net effects gross products? Commun. J. Soc. Issues **58**(1), 91–107 (2002)
7. Laporte, L., van Nimwegen, C., Uyttendaele, A.J.: Do people say what they think: Social conformity behavior in varying degrees of online social presence. In: Proceedings of the 6th Nordic Conference on Human-Computer Interaction: Extending Boundaries, pp. 305–314. ACM, October 2010
8. Deutsch, M., Gerard, H.B.: A study of normative and informational social influences upon individual judgment. J. Abnorm. Soc. Psychol. **51**(3), 629–636 (1955)
9. Chen, J., Wu, Y., Tong, G., Guan, X., Zhou, X.: ERP correlates of social conformity in a line judgment task. BMC Neurosci. **13**(1), 43–53 (2012)
10. Kim, B.R., Liss, A., Rao, M., Singer, Z., Compton, R.J.: Social deviance activates the brain's error-monitoring system. Cogn. Affect. Behav. Neurosci. **12**(1), 65–73 (2012)
11. Shestakova, A., Rieskamp, J., Tugin, S., Ossadtchi, A., Krutitskaya, J., Klucharev, V.: Electrophysiological precursors of social conformity. Soc. Cogn. Affect. Neurosci. **8**(7), 756–763 (2013)
12. Rosander, M., Eriksson, O.: Conformity on the internet-the role of task difficulty and gender differences. Comput. Hum. Behav. **28**(5), 1587–1595 (2012)
13. Dennis, T.A., Hajcak, G.: The late positive potential: a neurophysiological marker for emotion regulation in children. J. Child Psychol. Psychiatry **50**(11), 1373–1383 (2009)
14. Libet, B., Gleason, C.A., Wright, E.W., Pearl, D.K.: Time of conscious intention to act in relation to onset of cerebral activity (readiness-potential). Brain **106**(3), 623–642 (1983)
15. Klein, T.A., Endrass, T., Kathmann, N., Neumann, J., von Cramon, D.Y., Ullsperger, M.: Neural correlates of error awareness. Neuroimage **34**(4), 1774–1781 (2007)
16. Giannini, M., Pannocchia, L., Lauro-Grotto, R., Gori, A. (n.d.): A measure for counseling: the five factor adjective short test (5-fast). Unpublished manuscript, Department of Psychology, University of Florence, Florence, Italy
17. Spielberger, C.D., Gorsuch, R.L.: State-Trait Anxiety Inventory for Adults: Sampler Set: Manual, Tekst Booklet and Scoring Key. Consulting Psychologists Press, Palo Alto (1983)
18. Prezza, M., Pacilli, M.G., Barbaranelli, C., Zampatti, E.: The MTSOCS: A multidimensional sense of community scale for local communities. J. Community Psychol. **37**(3), 305–326 (2009)
19. Rosenberg, M.: Conceiving the self. In: Ciarrochi, J., Bilich, L.: Acceptance and Commitment Therapy. Measures Package. University of Wollongong, Wollongong, Australia (2007). (Unpublished manuscript)
20. Sibilia, L., Schwarzer, R., Jerusalem, M.: Italian adaptation of the general self-efficacy scale. Resource document. Ralf Schwarzer web site (1995)
21. Crutchfield, R.S.: Conformity and character. Am. Psychol. **10**(5), 191 (1955)

Internet Interoperability, Freedom and Data Analysis

Interoperable and Efficient: Linked Data for the Internet of Things

Eugene Siow[✉], Thanassis Tiropanis, and Wendy Hall

Electronics and Computer Science, University of Southampton, Southampton, UK
{eugene.siow,t.tiropanis,wh}@soton.ac.uk

Abstract. Two requirements to utilise the large source of time-series sensor data from the Internet of Things are interoperability and efficient access. We present a Linked Data solution that increases interoperability through the use and referencing of common identifiers and ontologies for integration. From our study of the shape of Internet of Things data, we show how we can improve access within the resource constraints of Lightweight Computers, compact machines deployed in close proximity to sensors, by storing time-series data succinctly as rows and producing Linked Data 'just-in-time'. We examine our approach within two scenarios: a distributed meteorological analytics system and a smart home hub. We show with established benchmarks that in comparison to storing the data in a traditional Linked Data store, our approach provides gains in both storage efficiency and query performance from over 3 times to over three orders of magnitude on Lightweight Computers. Finally, we reflect how pushing computing to edge networks with our infrastructure can affect privacy, data ownership and data locality.

Keywords: Interoperability · Internet of Things · Query translation · Linked Data

1 Introduction

The Internet of Things (IoT) envisions a world-wide, interconnected network of smart physical entities with the aim of providing technological and societal benefits [8]. However, as the W3C Web of Things Interest Group charter[1] points out, the IoT is currently beset by product silos and to unlock its potential, an open ecosystem based upon open standards for the interoperation of services is required. There is also a need for rich descriptions and shared data models, with close attention to security, privacy and scalability.

Linked Data is a set of best practices for publishing data on the Web so that distributed, structured data can be interconnected and made more useful by semantic queries [4]. A common representation is as a set of triples formed from a subject, predicate and object. For example, in the statement 'sensor1 has weatherObservation1', the subject is *sensor1*, the predicate is *has* and the object is

---

[1] https://www.w3.org/2014/12/wot-ig-charter.html.

© Springer International Publishing AG 2016
F. Bagnoli et al. (Eds.): INSCI 2016, LNCS 9934, pp. 161–175, 2016.
DOI: 10.1007/978-3-319-45982-0_15

weatherObservation1. 'weatherObservation1 hasValue 30knots' is another triple and the union of this set of triples forms a Linked Data graph. SPARQL is a language for querying Linked Data. Linked Data has demonstrated its feasibility as a means of connecting and integrating rich and heterogeneous web data using current infrastructure [7] and Barnaghi *et al.* [3] have supported the view that semantics can serve to facilitate interoperability, data abstraction, access and integration with other cyber, social or physical world data in the IoT.

In particular, Linked Data helps with interoperability in the IoT through:

1. The use and referencing of common identifiers (internationalised resource identifiers (IRIs)) and ontologies to help establish common data structures and types for integration e.g. the Semantic Sensor Ontology (SSN)[2].
2. The provision of machine-interpretable descriptions within Linked Data to describe what data represents, where it originates from, how it can be related to its surroundings, who is providing it, and what its attributes are e.g. a unit of measure of knots for each wind speed reading, the sensor that records it, its platform and its location.

The next question is whether Linked Data for the IoT can provide efficient access in terms of query performance and scalability. Buil-Aranda *et al.* [5] have examined traditional Linked Data stores on the web and shown that performance can be an issue. Performance for generic queries can vary by up to 3–4 orders of magnitude and stores generally limit or have worsened reliability when issued with a series of non-trivial queries.

IoT devices present even greater resource constraints, however, time-series sensor data from the IoT also presents a unique opportunity for optimisation. In this paper, we study the shape of IoT data in Sect. 2 and from that, design and implement a solution to optimise the storage and query performance of Linked Data using row storage and producing Linked Data 'just-in-time' in Sects. 3 and 4 on an IoT infrastructure across two varying scenarios described in Sect. 5. We show with established benchmarks how our approach compares to a traditional Linked Data store in terms of storage and query performance in Sect. 6 and reflect on the impact our infrastructure, which distributes computing and storage to edge networks, has on privacy and data ownership in Sect. 7. Finally, we look at the related work in the area in Sect. 8.

2 Shape of IoT Data

To investigate the shape of data produced by sensors in the Internet of Things, we collected a sample of the schema of over 20,000 unique IoT devices from public data streams on Dweet.io[3].

Dweet.io supports the publishing of data from IoT devices in JavaScript Object Notation (JSON). Since JSON is the data format, the schema for the data

---

[2] https://www.w3.org/2005/Incubator/ssn/ssnx/ssn.
[3] http://dweet.io/see.

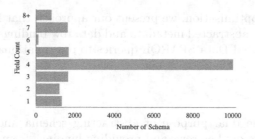

Fig. 1. Field counts from flat device schema

can be flat (row-like with a single level of data) or complex (tree-like/hierachical with multiple nested levels of data). We collected about 20,000 unique device schema from a one month period in January 2016 and analysed the structure of data. It was observed that out of 19,914 schema, 1542 (7.7 %) are empty. From the non-empty schema, 18,280 (99.5 %) are flat while 92 (0.5 %) are complex. Hence, non-empty Dweet.io schema was almost always, flat (99.5 %).

The field count of a schema refers to the number of values in a flat schema besides the timestamp. Figure 1 shows us the field counts of each flat device schema from Dweet.io. We found that 92.8 % of the devices sampled had a schema of 2 or more fields attached on top of the timestamp. The majority (54.7 %) had 4 fields attached to each timestamp. Hence, the schema indicates that sampled data on Dweet.io is largely wide. Our data is available on Github[4].

Therefore, through the study of public IoT device schema, we observe that our sample of over 20,000 unique IoT devices have data structures that are largely (1) flat and (2) wide (not just one, but multiple sensor values at a timestamp).

2.1 Optimising for Time-Series IoT Data

We hypothesise that flat and wide data, made up of a timestamp and multiple sensor values, can be succinctly represented as rows and the necessary Linked Data produced 'just-in-time' for interoperability. As compared to representation as traditional Linked Data triples:

1. Storage is efficient as each field in a row stores just the value without additional subject and predicate values.
2. Queries that retrieve two or more fields from a row require no joins.
3. Metadata triples produced 'just-in-time' (e.g. the location of a sensor or unit of measure of its value) can be kept in-memory and need not be retrieved as data and joined.
4. Intermediate nodes (e.g. observation identifier connecting time instant and actual value) might seldom be used and can be abstracted from data.

---

[4] https://github.com/eugenesiow/iotdevices/releases/download/data/dweet_release. zip.

To realise this optimisation, we present our approach that involves mapping (a representation of abstracted metadata and data row bindings) and translation of queries from Linked Data SPARQL queries to row/relational SQL queries.

3 Mapping

Mapping serves the dual purpose of abstracting schema and metadata from actual sensor data stored as rows and providing bindings from that row data to Linked Data, allowing the translation of queries from SPARQL to SQL.

We propose the SPARQL2SQL Mapping Language (S2SML), a simple and compact RDF-based language, designed with the structure of IoT data in mind, that is compatible with W3C recommendation, R2RML[5], for mapping.

3.1   Mapping as a Store of Abstracted Metadata

The difference and justification for S2SML over R2RML in the IoT is that S2SML mappings act like a Linked Data store for abstracted metadata (e.g. altitude of a sensor) and intermediate nodes (e.g. observation node connecting time instant, measurement data and sensor platform). It makes sense to abstract these to mappings as they are structurally different from the row data or are seldom projected from queries, however, they serve to connect and make Linked Data interoperable 'just-in-time'. R2RML on the other hand is designed just for binding relational datasets to Linked Data.

3.2   Formal Definition of Mappings

Mappings and elements unique to S2SML are defined in Definitions 1–4 and Table 1 gives descriptions and examples for each S2SML element.

Definition 1 (S2SML Mappings, M). *Given a set of all possible S2SML mappings, M and a mapping, $m \in M$, a triple pattern, $tp = (s, p, o)$ that is part of a mapping, $tp \in m$, has subject, s, predicate, p, and object, o where $s = \{I, I_{map}, B, F\}$, $p = \{I\}$ and $o = \{I, I_{map}, B, L, L_{map}, F\}$.*

Definition 2 (IRI Map, I_{map}). *An I_{map} is defined as a template that consists of the union of a set of IRI string parts, I_p and a set of table column binding strings, C, so $I_{map} = I_p \cup C$ and $|C| >= 1, |I_p| >= 1$. c is a string that consists of the table name and column separated by a '.' character, enclosed within braces and $c \in C$.*

Definition 3 (Literal Map, L_{map}). *An L_{map} is defined as a RDF literal that consists of a table column binding string, c, as its value and a specific IRI, $<s2s:literalMap>$ identifying it as an S2SML literal map as its datatype. c is a string that consists of the table name and column separated by a '.' character.*

---

[5] http://www.w3.org/TR/r2rml/.

Table 1. Examples of elements in (s, p, o) sets

Symbol	Name	Example
I	IRI	\<http://knoesis.wright.edu/ssw/ont/weather.owl#degrees\>
I_{map}	IRI Map	\<http://knoesis.wright.edu/ssw/{sensors.sensorName}\>
B	Blank Node	_:bNodeId
L	Literal	"-111.88222"^^\<xsd:float\>
L_{map}	Literal Map	"readings.temperature"^^ \<s2s:literalMap\>
F	Faux Node	\<http://knoesis.wright.edu/ssw/obs/{readings.uuid}\>

Definition 4 (Faux Node, F). *An F is defined as a template that consists of the union of a set of IRI string parts, I_p and a set of ID placeholders, U_{id}, so $F = I_p \cup U_{id}$ and $|U_{id}| >= 1, |I_p| >= 1$. $u = \{tablename.uuid\}$ and $u \in U_{id}$.*

3.3 Mapping Closure

Devices and hubs might have multiple sets of row data and their corresponding mappings. We define a mapping closure in Definition 5 that allows us to represent this collection of multiple mappings on a device.

Definition 5 (Mapping Closure, M_c). *Given the set of all mappings on a device, $M_d = \{m_d | m_d \in M\}$, where M is a set of all possible S2SML mappings. A mapping closure is the union of all elements in M_d, so $M_c = \bigcup_{m \in M_d} m$.*

3.4 Implicit Join Conditions

Sensor data that is represented across multiple tables within a mapping closure might need to be joined if matched by a SPARQL query. In the R2RML specification, one or more join conditions (rr:joinCondition) may be specified between triple maps of different logical tables.

In S2SML, these join conditions are automatically discovered as they are implicit within mapping closures from IRI template matching involving two or more tables.

Definition 6 (IRI Template Matching). *Following from Definition 2, I_{map_1} and I_{map_2} are matching if $\bigcup_{i_1 \in I_{p_1}} i_1 = \bigcup_{i_2 \in I_{p_2}} i_2$ and $\forall i_1 \in I_{p_1}, \forall i_2 \in I_{p_2} : pos(i_1) = pos(i_2)$ where $pos(x)$ is a function that returns the position of x within its I_{map}.*

From matching IRI templates in the mapping closure, join conditions can be inferred. Given Definitions 2 and 6, if $c_1 \neq c_2$, where $c_1 \in C_1, c_2 \in C_2$ and $pos(c_1) = pos(c_2)$, then, a join condition of $c_1 = c_2$ is required.

Figure 2 shows a mapping closure consisting of sensors and observations mappings. An IRI map, sen:system{sensors.name} and sen:system{readings.sensor}, in each of the mappings, fulfil a template matching. A join condition is inferred between the columns {sensors.name} and {readings.sensor} as a result.

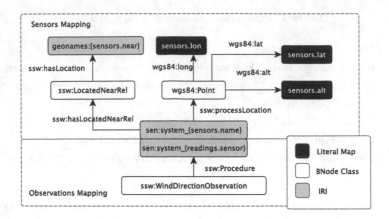

Fig. 2. Graph representation of an Implicit Join within a mapping closure

3.5 Cross Joins

Definition 7 (Table Selection, α). *Given a SPARQL query $q \in Q$, where Q is a set of all possible SPARQL queries, and a mapping closure, M_c, a table selection function $\alpha_{M_c}(q)$ returns a set of table names required by the query and referenced by elements in the mapping closure.*

The output of α is used in the FROM clause of the translated SQL query. If there are tables in the FROM where there are no corresponding join conditions, a cross join, resulting in the cartesian product of two tables, is performed. This is possible in a mapping, m, within the mapping closure, M_c, that refers to two or more logical tables within its collection of Literal or IRI Maps.

3.6 Compatibility with R2RML

Although S2SML is more compact in terms of verbosity and can be processed by any existing SPARQL engines (without needing any additional structures, translation or algorithms) it can be translated to and from R2RML without

Table 2. Other R2RML predicates and the corresponding S2SML construct

R2RML predicate	S2SML example
rr:language	"literal"@en
rr:datatype	"literal"^^ \<xsd:float\>
rr:inverseExpression	"{COL1} = SUBSTRING({{COL2}}, 3)"^^ \<s2s:inverse\>
rr:class	?s a \<ont:class\>.
rr:sqlQuery	\<context1\> {\<sen:sys_{table.col}\> ?p ?o.}
	\<context1\> s2s:sqlQuery "query".

losing expressiveness. Table 2 defines the other R2RML predicates and the corresponding S2SML construct.

In particular, {rr:inverseExpression} is encoded within a literal, L_{iv}, with a datatype of <s2s:inverse> and the {rr:column} denoted with double braces {{COL2}}. <rr:sqlQuery> is encoded by generating a named graph to group triples produced from that TripleMap and the query is stored in a literal object with context as the subject and <s2s:sqlQuery> as predicate as shown in Table 2.

4   Translation

The translation step is the process by which a Linked Data SPARQL query is applied to a mapping closure and translated to produce an SQL query that can be executed on the relational row store. Figure 3 describes the process whereby:

1. A SPARQL Query is parsed to SPARQL Algebra with Jena ARQ[6].
2. Each Basic Graph Pattern (BGP) in the algebra is visited so that:
   - A mapping closure, M_c, is built from the set of mappings (Sect. 4.1).
   - The BGP is expressed as a SPARQL select query and executed using any SPARQL engine in-memory on the mapping closure (Sect. 4.2).
   - The result set is processed to produce a map of variable bindings (e.g. ?var, table1.col1), table selection (α, Definition 7) and join list (Sect. 3.4).
3. Other operators like FILTER, GROUP, UNION and PROJECT are visited and referencing the map of variable bindings, table selection α and join list, an SQL query is generated by doing syntax translation (Sect. 4.3).

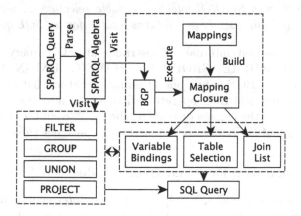

Fig. 3. Translation flow from SPARQL to SQL

---

6 https://jena.apache.org/documentation/query/.

4.1   Building a Mapping Closure

Following from Definition 5 of a Mapping Closure, M_c, a translation engine needs to perform, $\bigcup_{m \in M_d} m$, a union of all mappings on a device, M_d. Giving consideration to template matching as described in Definition 6, we replace all I_{map} within each mapping m with I_p, the union of IRI string parts, and extract C, the set of table column binding strings. C is then stored as a global map, m_{join}, with I_p as key and C as value. This map is used to produce the join list.

In the example in Fig. 2, `<sen:system_>` will replace `<sen:system_{sensors.name}>` and `<sen:system_{readings.sensor}>` and m_{join} will store under the key `<sen:system_>`, the set `{sensors.name, readings.sensor}`.

With this transformation, an M_c can be formed by adding all M_d to it. This can be done using any triple store that can be queried with a SPARQL engine.

4.2   BGP Resolution with Swappable SPARQL Engines

As mapping closures are standard Resource Description Format (RDF) triples, any SPARQL engine can perform BGP resolution. The BGP is expressed as a SPARQL select query (SELECT * WHERE {BGP}) and executed on a repository containing the mapping closure. We provide swappable Jena and OpenRDF Sesame engines in our implementation and a Java interface to extend to any other engine. Code is available on Github[7].

4.3   Operators and Syntax Translation

Definition 8 (Syntax Translation, *trans*). *trans*() *is a function that takes a set of operators from SPARQL algebra, a table selection α, map of variable bindings, v_{map}, and join list, J, and returns syntax for an SQL query.*

The *trans*() function internally constructs an SQL query with clauses SELECT, FROM, WHERE, GROUP BY, HAVING, UNION, ORDER BY, LIMIT and OFFSET. BGP is just one of many operators that are visited from the SPARQL algebra and each operator when input into the trans function, either modifies one of the clauses or adds to the v_{map}, α and J. Table 3 shows a sample of clauses and the operators & maps that construct them.

4.4   Compression with Faux and Blank Nodes

Blank nodes, B and faux nodes, F help to compress intermediate nodes unlikely to be accessed by abstracting them to the mapping and only if they are retrieved from a BGP and PROJECT are they generated 'just-in-time'. An ssn:Observation node in the SSN ontology can be connected to an ssn:SensorOutput, time:Instant and ssn:SensingDevice node. In turn an ssn:SensorOutput node is connected to a ssn:ObservationValue node which is connected to the actual value as a literal.

---

[7] https://github.com/eugenesiow/sparql2sql.

Table 3. Example of operators and clauses bindings in translation

Clause	Operators & Maps
SELECT	PROJECT, DISTINCT, UNION, v_{map}
FROM	α
WHERE	J, FILTER, UNION
HAVING	FILTER
GROUP BY	GROUP, v_{map}
LIMIT	SLICE

The intermediate ssn:Observation, ssn:SensorOutput, ssn:ObservationValue and time:Instant nodes are not required if we want to obtain the timestamp and the reading and can be compressed as B or F nodes.

If the intermediate nodes are required for some reason, F nodes can be used. When F is input to a PROJECT operator, an SQL update statement, `UPDATE table SET col=RANDOM_UUID()`, is run to generate a row of identifiers and the U_{id} part of F in the mapping is updated from {table.uuid} to {table.col}.

5 Linked Data Infrastructure for IoT Scenarios

We note from Sect. 2 that IoT time-series sensor data from our sample is flat and wide. In this section, we focus on two specific IoT scenarios with flat and wide IoT data: a distributed meteorological analytics system of weather sensors and a personal smart home hub.

In Fig. 4, we propose an inverse relationship between the level of distribution and compute and storage capability of components in a distributed architecture e.g. a cluster has high compute but cannot be distributed widely while sensors can be deployed widely but have minimal compute capability. Lightweight computers are compact and mobile machines that provide a balance of distribution and compute. We deploy our Linked Data Infrastructure on these lightweight computers, in close proximity to sensors and devices. The reference lightweight computer used is a Raspberry Pi 2 Model B+ with 1 GB RAM, a 900 MHz quad-core ARM Cortex-A7 CPU and a Class 10 SD Card.

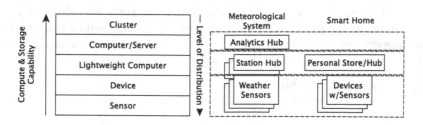

Fig. 4. Distributed infrastructure for IoT scenarios

5.1 Distributed Meteorological System and SRBench

This IoT scenario uses the established Linked Sensor Data [9] dataset that describes sensor data from about 20,000 weather stations across the United States with recorded observations from periods of hurricanes and blizzards. We used the Nevada Blizzard, with about 100k triples for storage and performance tests and the largest 300k triple Hurricane Ike dataset in storage tests.

At each station, there are a varying number of sensors (e.g. WindDirection, Rainfall) which produce observations at fixed intervals. This forms a stream of flat and wide rows of data. Each station and sensor also has metadata associated to it like the location, nearby stations or units of measure.

Figure 4 shows our design of the meteorological system. Lightweight computers serve as station hubs that store and make available for querying (as Linked Data) the stream of observations from weather sensors. An analytics hub on a server broadcasts queries to all station hubs and retrieves the results for visualisation.

We performed a benchmark with SRBench [15], an analytics benchmark for Linked Sensor Data. The benchmark uses streaming SPARQL queries but can be applied, with similar effect, to SPARQL queries constrained by time. Queries 1 to 10^8 were used as they involve time-series sensor data while the remaining queries involved integration or federation with DBpedia or Geonames which was not within the scope of the experiment. Queries are available on Github[9].

We transformed the Linked Sensor Data from Linked Data to row data[10]. Due to resource constraints, we ran the benchmarks for each station in series on a Pi, which is similar to parallel execution on a network with low latency, recording individual times and taking the maximum time among all stations.

5.2 Smart Home Hub and Analytics Benchmark

In this scenario, we used data from smart home sensors collected by Barker *et al.* [2] over 3 months in 2012. We utilised a variety of data including environment sensors, motion sensors in each room and energy meter readings to devise a set of queries that require space-time aggregations for descriptive and diagnostic analytics. Queries can be found on this wiki[11] and include (1) hourly aggregation of internal or external temperature (2) daily aggregation of temperature (3) hourly and room-based aggregation of energy usage (4) diagnose unattended energy usage with meter and motion, aggregating by hour and room.

Figure 4 shows our design of the smart home system with lightweight computers serving as the personal hub aggregating and storing sensor readings from energy meters, environment sensors, etc. Each device or sensor contributes a mapping in the mapping closure based on the SSN ontology[12].

---

[8] http://www.w3.org/wiki/SRBench.

[9] https://github.com/eugenesiow/sparql2sql/wiki.

[10] https://github.com/eugenesiow/lsd-ETL.

[11] https://github.com/eugenesiow/ldanalytics-PiSmartHome/wiki/.

[12] http://pi.webobservatory.me/info/datamodel.

5.3 Experiment

For both scenarios, we compared two Java-based database management systems, a **traditional Linked Data store, TDB**[13] and our approach with S2SML mapping and **S2S (SPARQL-to-SQL) translation** on a row-based store, H2[14]. Both stores were run in disk-based mode. Ethernet connections were used between the client and the Pis' to reduce network overhead for consistency. We took averages over 3 runs for each test. Running off the Java Virtual Machine on the Pis' gave a consistent platform for benchmarking with 512 mb the memory size allocated. We compared both storage efficiency and performance.

6 Results and Discussion

6.1 Storage Efficiency

Table 4 shows the difference in database storage sizes of different datasets for the S2S and TDB setups. As time-series sensor data benefits from the more succinct storage as rows, the S2S setup outperformed the Linked Data store, TDB, in terms of storage efficiency from one to two orders of magnitude. Furthermore, Linked Data stores rely on indexing all triples for performance [14] and TDB creates 3 triple indexes (OSP, POS, SPO) and 6 quad indexes to boost query performance. This increases storage size as observed.

6.2 Query Performance

The performance of the two setups for the SRBench queries from the Nevada Blizzard are shown in Fig. 5. Query performance for the S2S setup was from 3 times to 3 orders of magnitude better than the TDB setup.

The S2S setup performs consistently well for all the queries with similar execution times whereas the TDB setup differs significantly on different queries. The S2S setup does not have to perform joins between tables for all queries and hence the stable average run times.

The TDB setup performs much slower than the S2S setup on query 9 due to the join operation between two subtrees retrieved in the graph for two observations, WindSpeedObservation and WindDirectionObservation, being very time

Table 4. Database size by dataset

Dataset	S2S (mb)	TDB (mb)	$\%_{improve}$
Nevada Blizzard	90	6162	6847 %
Hurricane Ike	761	85274	11206 %
Smart Home	135	2103	1558 %

---

[13] https://jena.apache.org/documentation/tdb/.
[14] http://www.h2database.com/.

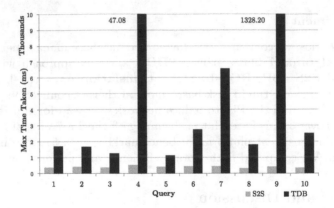

Fig. 5. SRBench query performance

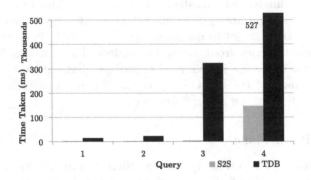

Fig. 6. Smart home query performance

consuming in the low-resource environment. An in-depth investigation showed the total query time was a 100 times more than the time to retrieve both subgraphs individually. Query 4 offers a similar situation with TemperatureObservation and WindSpeedObservation. The S2S setup, on the other hand, eliminates the need for this join as both observations belong to columns of the same row.

Figure 6 shows the Smart Home query performance for the S2S and TDB setups. Again the S2S Setup performed better for all queries, from 3 to 70 times faster. Both S2S and TDB performed much faster on queries 1 and 2 than 3 and 4 as they involved disk access (a limiting factor due to the SD card) on a much smaller portion of the database - environment sensor readings as compared to motion and meter sensor readings. The S2S setup still produced an order of magnitude better performance due to reducing joins e.g. between timestamp and the internal temperature values recorded in the same row.

Query 3 utilised smart meter data and query 4 involved both the smart meter and motion sensor data, a comparatively larger set of data and both did space and time aggregations on the data, hence, each took longer than the previous queries. Joins between tables (meter and motion) in Q4 affect both setups, as they belong to 2 different sensors, although the S2S setup still provides significant overall performance improvements in analytical queries. Table 5 summarises the results for both benchmarks.

Table 5. Average query run times of SRBench and smart home scenarios

SRBench	$T_{S2S}(ms)$	$T_{TDB}(ms)$	$\%_{improve}$
1	365	1679	460%
2	415	1651	398%
3	375	1258	335%
4	533	47084	8839%
5	415	1119	269%
6	457	2751	602%
7	455	6563	1444%
8	320	1785	558%
9	436	1328197	304865%
10	354	2514	709%

SmartHome	$T_{S2S}(ms)$	$T_{TDB}(ms)$	$\%_{improve}$
1	466	13709	2942%
2	2457	21898	891%
3	4685	322357	6881%
4	147649	527184	357%

6.3 Overall Efficient Access

Our approach, represented by the S2S setup, improves both storage efficiency and query performance. Most queries can be answered in sub-second times which means efficient access to time-series sensor data by IoT applications is possible while maintaining interoperability through the use of Linked Data.

7 Impact on Privacy, Data Ownership and Data Locality

The use of lightweight computers as distributed hubs in our proposed infrastructure means that data that is collected from sensors and devices are stored and processed locally. As Vaquero et al. [13] state, data ownership will be a cornerstone of distributed IoT networks, where some applications will be able to use the network to run applications and manage data without relying on centralised services. This approach has an advantage over storing encrypted data in traditional clouds as a means to maintain privacy because it is easier to perform processing (no need for crypto-processors or applying special encryption functions) over such data. In our smart home scenario, the use of a personal hub on a lightweight computer based on an open ecosystem helps to mitigate the fears proposed by Albrecht et al. [1] that a mega corporation owns our data (and the local supermarket) and has little incentive to value our privacy. Roman et al. [12] further emphasise with their study of centralised and decentralised

IoT infrastructures that when data is managed by the distributed entities, specific privacy policies and access control with additional trust and fault tolerance mechanisms can be created.

Data locality is beneficial in the sense that we no longer need to send all the data around the world all the time. In disaster management IoT scenarios, where last-mile connectivity is lost, having data locality and offline access is especially valuable. An example is the Nepal earthquake in 2015 where last-mile connectivity was lost though global connectivity was maintained.

Hence, our infrastructure that pushes both storage and compute to lightweight computers in edge networks within an open ecosystem, makes it more viable for end users to own their data. Specific privacy policies and technology can be built on top of this distributed infrastructure which has data locality as an added advantage. We show with our experiment that performance and storage efficiency for a variety of queries on data are sub-second and analytics-ready.

8   Related Work

SPARQL to SQL query translation has evolved with state-of-the-art engines like morph [10] and ontop [11] able to produce flatter & more efficient SQL queries. Both these engines, however, are designed for Ontology-Based Data Access (OBDA) or mapping relational stores to Linked Data with R2RML. Our work differs in that we build an R2RML-compatible mapping language that is additionally designed for the abstraction and storage of metadata within mappings. Secondly, we support the use of blank nodes (within the R2RML specification but not supported by other engines at the time of writing) and faux nodes to represent and compress intermediate nodes unlikely to be accessed. Lastly, we evaluate the performance of this approach on an IoT infrastructure with Pis'.

Previous work on SPARQL to SQL translation by Chebotko *et al.* [6] helped to establish formally that the full separation of translation from the relational database schema design was possible and that efficient queries significantly improved query performance. While work by Elliot *et al.* has the same aims of efficient SQL queries but covers a smaller subset of SPARQL 1.0 e.g. no support date functions required for time aggregation in analytical queries.

9   Conclusion

Our approach of storing time-series data from IoT sensors in rows on lightweight computers and allowing Linked Data SPARQL queries through translation via mappings is shown to increase the performance of both storage and compute in two IoT scenarios as compared to traditional Linked Data stores. The improvement in storage and query performance is significant, 3 times to three orders of magnitude. More essentially, it allows most benchmark queries and space-time aggregations for analytics to run in sub-second, providing a basis for IoT applications working on sensor data. With Linked Data produced 'just-in-time', the approach supports interoperability without exchanging efficient access. The proposed infrastructure also shows how compute and storage in the IoT can be distributed to edge networks with lightweight computers which is a boon for privacy, data ownership and situations where last-mile access breaks down.

References

1. Albrecht, K., Michael, K.: Connected: to everyone and everything. IEEE Technol. Soc. Mag. **32**, 31–34 (2013)
2. Barker, S., Mishra, A., Irwin, D., Cecchet, E.: Smart*: an open data set and tools for enabling research in sustainable homes. In: Proceedings of the Workshop on Data Mining Applications in Sustainability (2012)
3. Barnaghi, P., Wang, W.: Semantics for the Internet of Things: early progress and back to the future. Int. J. Semant. Web Inf. Syst. **8**(1), 1–21 (2012)
4. Bizer, C., Heath, T., Berners-Lee, T.: Linked data - the story so far. Int. J. Semant. Web Inf. Syst. **5**, 1–22 (2009)
5. Buil-Aranda, C., Hogan, A., Umbrich, J., Vandenbussche, P.-Y.: SPARQL Web-querying infrastructure: ready for action? In: Alani, H., et al. (eds.) ISWC 2013, Part II. LNCS, vol. 8219, pp. 277–293. Springer, Heidelberg (2013)
6. Chebotko, A., Lu, S., Fotouhi, F.: Semantics preserving SPARQL-to-SQL translation. Data Knowl. Eng. **68**(10), 973–1000 (2009)
7. Heath, T., Bizer, C.: Linked Data evolving the Web into a global data space. In: Synthesis Lectures on the Semantic Web: Theory and Technology (2011)
8. International Telecommunication Union: Overview of the Internet of things. Technical report (2012)
9. Patni, H., Henson, C., Sheth, A.: Linked sensor data. In: Proceedings of the International Symposium on Collaborative Technologies and Systems, pp. 362–370 (2010)
10. Priyatna, F., Corcho, O., Sequeda, J.: Formalisation and experiences of R2RML-based SPARQL to SQL Query Translation using Morph. In: Proceedings of the 23rd International Conference on World Wide Web, pp. 479–489 (2014)
11. Rodriguez-Muro, M., Rezk, M.: Efficient SPARQL-to-SQL with R2RML mappings. Web Semant. Sci. Serv. Agents World Wide Web **33**, 141–169 (2014)
12. Roman, R., Zhou, J., Lopez, J.: On the features and challenges of security and privacy in distributed internet of things. Comput. Netw. **57**(10), 2266–2279 (2013)
13. Vaquero, L.M., Rodero-Merino, L.: Finding your way in the fog: towards a comprehensive definition of fog computing. ACM SIGCOMM Comput. Commun. Rev. **44**(5), 27–32 (2014)
14. Weiss, C., Karras, P., Bernstein, A.: Hexastore: sextuple indexing for semantic web data management. Proc. VLDB Endowment **1**(1), 1008–1019 (2008)
15. Zhang, Y., Duc, P.M., Corcho, O., Calbimonte, J.-P.: SRBench: a streaming RDF/SPARQL benchmark. In: Cudré-Mauroux, P., et al. (eds.) ISWC 2012, Part I. LNCS, vol. 7649, pp. 641–657. Springer, Heidelberg (2012)

Stable Topic Modeling with Local Density Regularization

Sergei Koltcov[1], Sergey I. Nikolenko[1,2(✉)], Olessia Koltsova[1],
Vladimir Filippov[1], and Svetlana Bodrunova[1,3]

[1] National Research University Higher School of Economics, St. Petersburg, Russia
snikolenko@gmail.com
[2] Steklov Institute of Mathematics, St. Petersburg, Russia
[3] St. Petersburg State University, St. Petersburg, Russia

Abstract. Topic modeling has emerged over the last decade as a powerful tool for analyzing large text corpora, including Web-based user-generated texts. Topic stability, however, remains a concern: topic models have a very complex optimization landscape with many local maxima, and even different runs of the same model yield very different topics. Aiming to add stability to topic modeling, we propose an approach to topic modeling based on local density regularization, where words in a local context window of a given word have higher probabilities to get the same topic as that word. We compare several models with local density regularizers and show how they can improve topic stability while remaining on par with classical models in terms of quality metrics.

Keywords: Topic modeling · Latent Dirichlet allocation · Gibbs sampling

1 Introduction

Over the last decade, topic modeling has become one of the standard tools in text mining. In social sciences, topic models can be used to concisely describe a large corpus of documents, uncovering the actual topics covered in this corpus (via the word-topic distributions) and pointing to specific documents that deal with topics a researcher is interested in (via the topic-document distributions) [22,23]. Apart from exploratory analysis of large text corpora, topic modeling can also be used to mine latent variables from the documents such as [12,18]. These applications of topic modeling raise a number of problems regarding the evaluation of topic modeling results. First, it still remains an open problem to evaluate how "good" a topic is; the gold standard here is usually human interpretability, and the goal is to devise automated techniques that would come close to human estimates. Modern metrics include ones based on coherence [8,19] and its modifications [22], pointwise mutual information [6,19,21], and topics designed to match word intrusion and topic intrusion experiments [16].

However, apart from the actual quality of the resulting topics, *topic stability* is also a very important problem for real life applications of topic modeling,

© Springer International Publishing AG 2016
F. Bagnoli et al. (Eds.): INSCI 2016, LNCS 9934, pp. 176–188, 2016.
DOI: 10.1007/978-3-319-45982-0_16

especially in social sciences. The likelihood function of a topic model is usually very complex, with plenty of local maxima. If we considering inference in a topic model as stochastic matrix decomposition, representing the word-document matrix as a stochastic product of word-topic and topic-document matrices, we see that for every solution (Θ, Φ) there is an infinite number of equivalent solutions $(\Theta S, S^{-1}\Phi)$ for any invertible S; e.g., all permutations of the same topics are obviously equivalent. And there are plenty of substantially different solutions corresponding to different local maxima of the model posterior; the model may arrive to different local maxima depending on the randomness in initialization and sampling. For a practical application of topic models in social sciences, such as studies of Web content, it is highly desirable to have stable results: a social scientist is often interested in whether a topic is "there" in the dataset, and it would be hard to draw any conclusions if the topic was "blinking" in and out depending on purely random factors. Besides, it would be hard to rely on a study that cannot be reliably reproduced even in principle. Hence, it becomes especially important to develop topic models that produce stable, reproducible topic solutions, hopefully not at the cost of their quality (i.e., topic interpretability).

In this work, we introduce a new modification of the basic latent Dirichlet allocation (LDA) model called *granulated LDA* (GLDA) that assumes that topics cover relatively large contiguous subsets of a document and assigns the same topic with high probability to a window of words once the anchor word has been sampled in this window. We show that GLDA produces much more stable results while preserving approximately the same topic quality as classical topic models.

The paper is organized as follows. In Sect. 2, we introduce the topic models that we will consider below and the two approaches to inference in topic models. Section 3 contains a brief overview of regularization in topic models. Section 4 introduces our new approach to topic modeling, granulated LDA (GLDA). In Sect. 5 we show experimental results that prove that granulated LDA has solutions with similar quality or better than regular topic models but that are much more stable; we conclude with Sect. 6.

2   Topic Modeling

Let D be a collection of documents, and let W be the set of all words in them (vocabulary). Each document $d \in D$ is a sequence of terms $w_1, \ldots, w_{n_d}$ from the vocabulary W. The basic assumption of all probabilistic topic models is that there exists a finite set of topics T, and each occurrence of a word w in a document d is related to some topic $t \in T$, and the actual word depends only on the corresponding topic instance and not on the document itself or other words. Formally, we assume that the probability that a word w occurs in document d can be decomposed as

$$p(w \mid d) = \sum_{t \in T} p(w \mid t)p(t \mid d) = \sum_{t \in T} \phi_{wt}\theta_{td},$$

where $\phi_{wt} = p(w \mid t)$ is the distribution of words in a topic and $\theta_{td} = p(t \mid d)$ is the distribution of topics in a document. The problem of training a topic model on a collection of documents is, thus, the problem of finding the set of latent topics T, i.e., the set of multinomial distributions ϕ_{wt}, $t \in T$, and the set of multinomial distributions θ_{td}, $d \in D$, which we represent by the matrices $\Phi = (\phi_{wt})_{wt}$ and $\Theta = (\theta_{td})_{td}$ respectively.

There are two main approaches to solving this problem, i.e., reconstructing Φ and Θ. In the first approach, the total log-likelihood

$$L(\Phi, \Theta) = \sum_{d \in D} \sum_{w \in d} n_{wd} \ln \sum_{t \in T} \phi_{wt} \theta_{td} \to \max$$

is maximized with an expectation-maximization (EM) algorithm under constraints $\theta_{td} \geq 0$, $\phi_{w}t \geq 0$, $\sum_{t \in T} \theta_{td} = 1$, $d \in D$, and $\sum_{w \in W} \phi_{wt} = 1$, $t \in T$; n_{wd} denotes the number of times word w occurs in document d. This setting is the *probabilistic latent semantic analysis* (pLSA) model [13].

These ideas were further developed in the already classical *latent Dirichlet allocation* (LDA) model [4]. LDA is a Bayesian version of pLSA: it assumes that multinomial distributions θ_{td} and ϕ_{wt} are generated from prior Dirichlet distributions, one with parameter α (for the θ distributions) and one with parameter β (for the ϕ distributions). LDA inference can be done either with variational approximations or with Gibbs sampling, first proposed for LDA in [11]. Here the hidden variables z_i for every word occurrence are considered explicitly, and the inference algorithm produces estimates of model parameters as Monte Carlo estimates based on samples drawn for the latent variables. Gibbs sampling is a special case of Markov chain Monte Carlo methods where sampling is done coordinatewise, hidden variable by hidden variable. In the basic LDA model, Gibbs sampling with symmetric Dirichlet priors reduces to the so-called *collapsed Gibbs sampling*, where θ and ϕ variables are integrated out, and z_i are iteratively resampled according to the following distribution: $p(z_i = t \mid \boldsymbol{z}_{-i}, \boldsymbol{w}, \alpha, \beta) \propto$

$$q(z_i, t, \boldsymbol{z}_{-i}, \boldsymbol{w}, \alpha, \beta) = \frac{n_{-i,td} + \alpha}{\sum_{t' \in T} (n_{-i,t'd} + \alpha)} \frac{n_{-i,wt} + \beta}{\sum_{w' \in W} (n_{-i,w't} + \beta)},$$

where $n_{-i,td}$ is the number of words in document d chosen with topic t and $n_{-i,wt}$ is the number of times word w has been generated from topic t except the current occurrence z_i; both counters depend on the other variables $\boldsymbol{z}_{-w}$. Samples are then used to estimate model variables: $\theta_{td} = \frac{n_{-i,td} + \alpha}{\sum_{t' \in T} (n_{-i,t'd} + \alpha)}$, $\phi_{wt} = \frac{n_{-i,tw} + \beta}{\sum_{w' \in W} (n_{-i,w't} + \beta)}$, where ϕ_{wt} denotes the probability to draw word w in topic t and θ_{td} is the probability to draw topic t for a word in document d.

After it was introduced in [4], the basic LDA model has been subject to many extensions, each presenting either a variational or a Gibbs sampling algorithm for a model that builds upon LDA to incorporate some additional information or additional presumed dependencies. One large class of extensions deals with imposing new structure on the set of topics that are independent and uncorrelated in the base LDA model, including *correlated topic models* (CTM) [3], *Markov topic models* [17], *syntactic topic models* [7] and others. The other class of extensions takes into account additional information that may be available together with the documents and may reveal additional insights into the topical structure; this class includes models that account for timestamps of document creation [27,28], *semi-supervised LDA* that centers on specific topics [22], *DiscLDA* that uses document labels to solve a classification problem [15], and others. Finally, a lot of work has been done on nonparametric LDA variants based on Dirichlet processes, where the number of topics is also sampled automatically in the generative process; see [10] and references therein.

Additive Regularization of Topic Models (ARTM) [25,26] is a recently developed novel approach to topic models that avoids complications of LDA inference (it is no easy matter to develop a new LDA extension) while preserving the capabilities for extending and improving LDA. ARTM has several conceptual differences from the Bayesian approach [25]: in ARTM, regularizers are explicit, adding new regularizers is relatively easy, and inference is done via the regularized EM algorithm. We add regularizers $R(\Phi, \Theta) = \sum_i \tau_i R_i(\Phi, \Theta)$ to the basic pLSA model, where $R_i(\Phi, \Theta)$ is some regularizer with nonnegative regularization coefficient τ_i. Then the optimization problem is to maximize $L(\Phi, \Theta) + R(\Phi, \Theta)$, where $L(\Phi, \Theta)$ is the likelihood, and the regularized EM algorithm amounts to iterative recomputation of the model parameters as follows:

$$p_{dtw} = \frac{\phi_{wt}\theta_{td}}{\sum_{s \in T} \phi_{ws}\theta_{sd}}, \ \phi_{wt} \propto \left(n_{wt} + \phi_{wt}\frac{\partial R}{\partial \phi_{wt}}\right)_+, \ \theta_{td} \propto \left(n_{td} + \theta_{td}\frac{\partial R}{\partial \theta_{td}}\right)_+.$$

In this work, we use ARTM models with standard sparsity regularizers added to the Φ and Θ matrices.

3   Regularization in Topic Models

Whatever the inference method, the basic topic modeling problem is equivalent to stochastic matrix decomposition, where a large sparse matrix $F = (F_{dw})$ of size $|D| \times |W|$ that shows how words $w \in W$ occur in documents $d \in D$ is approximated by a product of two smaller matrices, Θ of size $|D| \times |T|$ and Φ of size $|T| \times |W|$. Note that almost by definition, the solution of this problem is not unique: if $F = \Theta\Phi$ is a solution of this problem then $F = (\Theta S)(S^{-1}\Phi)$ is also a solution for any nondegenerate $|T| \times |T|$ matrix S (for a simple example, note that we can permute topics freely, and nothing changes). In terms of the inference problem, this multitude of solutions means that an inference algorithm will converge to different solutions given different random factors in the algorithms and different starting points. In practice, by running the same algorithm on the same dataset we will get very different matrices Φ and Θ, which is obviously an undesirable property for applications.

In optimization theory, problems with non-unique and/or unstable solutions are called *ill-posed*, and a general approach to solving these problems is given by Tikhonov regularization [24]. In terms of the model definition, regularization can be viewed as extending the prior information which lets one reduce the set of solutions. Regularization is done either by introducing constraints on Φ and Θ matrices [20] or by modifying the sampling procedure [1].

We proceed with examples of regularizers relevant to the regularizer we propose in this work. First, the work [20] proposes to introduce a regularization procedure that uses external information on the relations between words. This information, possibly from an external dataset, is expressed as a $|W| \times |W|$ covariance matrix C; formally, this adds the prior $p(\phi_t \mid C) \propto (\phi_t^\top C \phi_t)^\nu$ for some regularization parameter ν, the total log posterior looks like

$$L = \sum_{i=1}^{W} N_{it} \log \phi_{it} + \nu \log \phi_t^\top C \phi_t,$$

and the ϕ matrix is now updated as

$$\phi_{wt} \propto \frac{1}{N_t + 2\nu}\left(N_{wt} + \frac{2\nu\phi_{wt}\sum_{i=1}^{W} C_{iw}\phi_{it}}{\phi_t^\top C \phi_t}\right).$$

Another regularizer proposed in [20] is based on the idea that ϕ_{wt} depends on some matrix C which, in turn, expresses the dependencies between pairs of unique words. In other words, now a topic is defined as a collection of related words with probability distribution ψ_t, but the probability distribution of their occurrences is $\phi_t \propto C\psi_t$. The total log posterior is now

$$L = \sum_{i=1}^{W} N_{it} \log \sum_{j=1}^{W} C_{ij}\psi_{jt} + \sum_{j=1}^{W}(\gamma - 1)\log \psi_{jt}$$

under the constraints that $\sum_{j=1}^{W} \psi_{jt} = 1$. One can update the Ψ matrix similar to the updates of Φ and Θ matrices:

$$\phi_{wt} \propto \sum_{i=1}^{W} \frac{N_{it}C_{iw}}{\sum_{j=1}^{W} C_{ij}\psi_{jt}} + \gamma.$$

However, in both cases one has to know the C matrix in advance; C is a very large matrix that should incorporate prior knowledge about every pair of words in the dataset, which represents a major obstacle to using these regularizers.

Another direction of LDA extensions that has been intended, at least in part, to improve the stability of topic solutions, is the direction of *semi-supervised LDA* (SLDA) and related extensions. Semi-supervised LDA is based on a special kind of regularizer; the idea is that in real life applications, especially in social science, it often happens that the entire text corpus deals with a large number of different unrelated topics while the researcher is actually interested only in a small subset of them. In this case, it is desirable to single out topics related to the subjects in question a make them more stable. If the subject are given as a set of seed words, the semi-supervised LDA model simply fixes the values of z for certain key words related to the topics in question; similar approaches have been considered in [1,2]. For words $w \in W_{\text{sup}}$ from a predefined set W_{sup}, the values of z are known and remain fixed to $\tilde{z}_w$ throughout Gibbs sampling:

$$p(z_w = t \mid \boldsymbol{z}_{-w}, \boldsymbol{w}, \alpha, \beta) \propto \begin{cases} [t = \tilde{z}_w], & w \in W_{\text{sup}}, \\ q(z_w, t, \boldsymbol{z}_{-w}, \boldsymbol{w}, \alpha, \beta) & \text{otherwise.} \end{cases}$$

Otherwise, the Gibbs sampler works as in the basic LDA model; this yields an efficient inference algorithm that does not incur additional computational costs.

In a straightforward extension, *interval semi-supervised LDA* (ISLDA), each key word $w \in W_{\text{sup}}$ is mapped to an interval of topics $[z_l^w, z_r^w]$, and the probability distribution is restricted to that interval. In the Gibbs sampling algorithm, we simply set the probabilities of all topics outside $[z_l^w, z_r^w]$ to zero and renormalize the distribution inside:

$$
p(z_w = t \mid \boldsymbol{z}_{-w}, \boldsymbol{w}, \alpha, \beta) \propto
\begin{cases}
I_{z_l^w}^{z_r^w}(z) \dfrac{q(z_w, t, \boldsymbol{z}_{-w}, \boldsymbol{w}, \alpha, \beta)}{\sum_{z_l^w \le t' \le z_r^w} q(z_w, t', \boldsymbol{z}_{-w}, \boldsymbol{w}, \alpha, \beta)}, & w \in W_{\text{sup}}, \\[4mm]
q(z_w, t, \boldsymbol{z}_{-w}, \boldsymbol{w}, \alpha, \beta) & \text{otherwise,}
\end{cases}
$$

where $I_{z_l^w}^{z_r^w}$ denotes the indicator function: $I_{z_l^w}^{z_r^w}(z) = 1$ iff $z \in [z_l^w, z_r^w]$. Interval semi-supervised LDA has been used in case studies related to social sciences in [5,22]; these works show that SLDA and ISLDA not only mine more relevant topics than regular LDA but also improve their stability, providing consistent results in the supervised subset of topics. In this work, we present a new LDA extension which provides even more stable results at no loss to their quality.

4   Granulated LDA

In this work, we introduce the *granulated sampling* approach which is based on two ideas. First, we recognize that there may be a dependency between a pair of unique words, but, unlike the convolved Dirichlet regularizer model, we do not express it as a predefined matrix. Rather, we assume that a topic consists of words that also often occur together; that is, we assume that words that are characteristic for the same topic are often colocated inside some relatively small window. The idea is to capture the intuition that words that are located close to each other in the document usually relate to the same topic; i.e., topics in a document are not distributed as independently sampled random variables but rather as relatively large contiguous streaks, or *granulas*, of words belonging to the same topic. Figure 1 illustrates the basic idea, showing a granulated surface as it is usually understood in physics (bottom right) and a sample partially granulated text that might result from the granulated LDA model (on the left).

Interestingly, the rather natural idea of granulas has not really been explored in topic models. The only similar approach known to us in prior work deals with using the additional information available in the text in the form of sentences and/or paragraphs. The work [9] adds a sentence layer to the basic LDA model; in sentence-layered LDA, each sentence is governed by its own topic distribution. Sentence and paragraph boundaries are also often used in LDA extensions dealing with sentiment analysis: it is often assumed that a single sentence or paragraph deals with only one aspect; see, e.g., the Aspect and Sentiment Unification Model (ASUM) [29] that extends the basic Sentence LDA (SLDA) model However, we are not aware of topic models that would use naturally arising granulas of fixed or variable size and assume that a granula is covered by

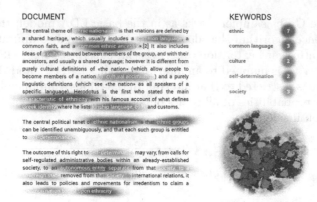

DOCUMENT

The central theme of ethnic nationalism is that «nations are defined by a shared heritage, which usually includes a common language, a common faith, and a common ethnic ancestry».[2] It also includes ideas of culture shared between members of the group, and with their ancestors, and usually a shared language; however it is different from purely cultural definitions of «the nation» (which allow people to become members of a nation by cultural assimilation) and a purely linguistic definitions (which see «the nation» as all speakers of a specific language). Herodotus is the first who stated the main characteristic of ethnicity with his famous account of what defines Greek identity, where he lists kinship language, cult and customs.

The central political tenet of ethnic nationalism is that ethnic groups can be identified unambiguously, and that each such group is entitled to self-determination.

The outcome of this right to self-determination may vary, from calls for self-regulated administrative bodies within an already-established society, to an autonomous entity separate from that society, to a sovereign state removed from that society in international relations. It also leads to policies and movements for irredentism to claim a common nation based upon ethnicity.

KEYWORDS

ethnic   7
common language   3
culture   2
self-determination   2
society   3

Fig. 1. Illustration for granulated LDA: granulated surface and granulated text.

the same topic. One could say that GLDA is in essence equivalent to a certain cooccurrence-based regularizer, but without the need to compute the entire cooccurrence matrix, everything is local.

Granulated Gibbs sampling is implemented as follows: we randomly sample anchor words in the document, sample their topics, but then set the topic of all words in a local context window with the use of the anchor word's sampling result. We sample as many anchor words as there are words in the document.

On the other hand, the topical distribution of words inside a window (granula) can have its own distribution, different from the distribution imposed by Dirichlet priors. By modifying the distribution function inside a window (local density) and changing the window size, we can influence the model's regularization. Thus, we regularize the topic model as follows: having sampled an anchor word $z_j = z$ in the middle of a window, we then set the topics of nearby words z_i, $|i - j| \leq l$, as $z_i = zK\left(\frac{|i-j|}{l}\right)$ for some kernel function K. The kernel function should satisfy $K(0) = 1$ and be monotone nonincreasing towards the ends of the window, modifying the distribution of topics inside a local window. We have compared three different kernels:

(1) step kernel $K(r) = 1$, when all topics in the window are set to z;
(2) Epanechnikov kernel $K(r) = 1 - r^2$;
(3) triangular Epanechnikov kernel $K(r) = 1 - |r|$.

Thus, formally speaking, after the initialization of Θ and Φ matrices as in regular Gibbs sampling, we run the following algorithm:

– for every document $d \in D$, repeat $|d|$ times:
   • sample a word instance $j \in d$ uniformly at random;
   • sample its topic $z_j = z$ as in Gibbs sampling;
   • set $z_i = zK\left(\frac{|i-j|}{l}\right)$ for all i such that $|i - j| \leq l$.

On the final inference stage, after sampling is over, we compute the Φ and Θ matrices as usual (see Sect. 2).

Note that unlike regular Gibbs sampling, we do not go over all words in the document but randomly sample anchor words. As a result of this process, words that are often found close together in different documents (inside a given window size) will be more likely to fall in the same topic.

5 Evaluation

In our experiments, we have used a dataset of 101481 blog posts from the *LiveJournal* blog platform with 172939 unique words in total; *LiveJournal* is a platform of choice for topic modeling experiments since the posts are both user-generated and much longer than a typical tweet or *facebook* post. We have trained six baseline models and several varieties of GLDA:

(1) the basic probabilistic latent semantic analysis model (pLSA);
(2) ARTM model with Φ sparsity regularizer;
(3) ARTM model with Θ sparsity regularizer;
(4) basic LDA model with inference based on Gibbs sampling [11];
(5) basic LDA model with inference based on the variational Bayes [4];
(6) supervised LDA model with a vocabulary consisting of ethnonyms; this vocabulary was developed in a previous case study of user-generated content designed to study ethnic-related topics [5,14,22];
(7) granulated LDA with three different windows: step, Epanechnikov, and triangular, and different window sizes, from $l = 1$ to $l = 3$;

In all cases, we have trained the models with $T = 200$ topics. Note that we train LDA with two different inference algorithms since they may have different stability properties. For SLDA, GLDA, and LDA with inference based on Gibbs sampling, we have set the Dirichlet prior parameters to be $\alpha = 0.1$ and $\beta = 0.5$, values that have been previously tuned for our datasets [14]. Regularization coefficients for the ARTM models were tuned to give the best possible topics.

In the experiments, we mostly strived for topic stability but we cannot afford to achieve stability at a significant loss of *topic quality*: useful topics have to be readily interpretable. For evaluation, we use the *coherence* and *tf-idf coherence* metrics. Coherence has been proposed as a topic quality metric in [8,19]. For a topic t characterized by its set of top words W_t, coherence is defined as $c(t, W_t) = \sum_{w_1,w_2 \in W_t} \log \frac{d(w_1,w_2)+\epsilon}{d(w_1)}$, where $d(w_i)$ is the number of documents that contain w_i, $d(w_i, w_j)$ is the number of documents where w_i and w_j cooccur, and ϵ is a smoothing count usually set to either 1 or 0.01. A recent work [22] proposed a modification of the coherence metric called *tf-idf coherence*:

$$c_{\text{tf-idf}}(t, W_t) = \sum_{w_1,w_2 \in W_t} \log \frac{\sum_{d:w_1,w_2 \in d} \text{tf-idf}(w_1, d)\text{tf-idf}(w_2, d) + \epsilon}{\sum_{d:w_1 \in d} \text{tf-idf}(w_1, d)},$$

Table 1. Overall metrics of topic quality and stability for granulated LDA and other models averaged over all runs of the corresponding model.

Topic model	Topic quality metrics		Topic stability metrics	
	Coherence	tf-idf Coherence	Stable topics	Jaccard
pLSA	−238.522	−126.934	54	0.47
pLSA + Φ sparsity reg	−231.639	−127.018	9	0.44
PLSA + Θ sparsity reg	−241.221	−125.979	87	0.47
LDA, Gibbs sampling	−208.548	−116.821	77	0.56
LDA, variational Bayes	−275.898	−112.544	111	0.53
SLDA	−208.508	−120.702	84	0.62
GLDA, step window, $l = 1$	−180.248	−123.231	195	0.64
GLDA, step window, $l = 2$	−171.038	−122.029	195	0.71
GLDA, step window, $l = 3$	−164.573	−121.582	197	0.73
GLDA, Epanechnikov window, $l = 1$	−226.394	−148.725	184	0.23
GLDA, Epanechnikov window, $l = 2$	−227.099	−174.475	192	0.33
GLDA, Epanechnikov window, $l = 3$	−206.347	−171.155	199	0.20
GLDA, triangular window, $l = 1$	−226.486	−148.147	162	0.16
GLDA, triangular window, $l = 2$	−234.096	−186.294	200	0.30
GLDA, triangular window, $l = 3$	−222.487	−184.187	200	0.68

where the tf-idf metric is computed with augmented frequency,

$$\text{tf-idf}(w, d) = \text{tf}(w, d) \times \text{idf}(w) = \left(\frac{1}{2} + \frac{f(w, d)}{\max_{w' \in d} f(w', d)} \right) \log \frac{|D|}{|\{d \in D : w \in d\}|},$$

where $f(w, d)$ is the number of occurrences of term w in document d. This skews the metric towards topics with high tf-idf scores in top words, since the numerator of the coherence fraction has quadratic dependence on the tf-idf scores and the denominator only linear. We have used both coherence and tf-idf coherence to evaluate topic quality in our solutions.

To evaluate topic stability, we have used the following approach. First, we introduce two natural similarity metrics for two topics [14]: symmetric Kullback–Leibler divergence between the probability distributions of two topics in a solution, defined as $\text{KL}(\phi^1, \phi^2) = \frac{1}{2} \sum_w \phi_w^1 \log \frac{\phi_w^1}{\phi_w^2} + \frac{1}{2} \sum_w \phi_w^2 \log \frac{\phi_w^2}{\phi_w^1}$, together with its normalized version [14] $\text{NKLS}(t_1, t_2) = 1 - \frac{\text{KL}(t_1, t_2)}{\max_{t'_1, t'_2} \text{KL}(t'_1, t'_2)}$, and Jaccard similarity of two sets of top words in two topics: for a given threshold T, we denote by Top_ϕ^T the set of T words with largest probabilities in a topic distribution ϕ and compute $J^T(\phi_1, \phi_2) = \frac{|\text{Top}_{\phi_1}^T \cap \text{Top}_{\phi_1}^T|}{|\text{Top}_{\phi_1}^T \cup \text{Top}_{\phi_1}^T|}$. We call two topics *matching* if their normalized Kullback-Leibler similarity is larger than 0.9 (a threshold chosen by

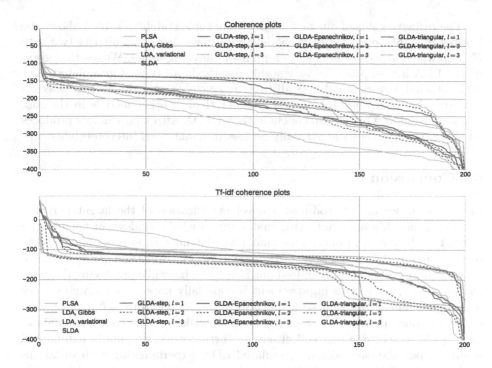

Fig. 2. Sorted topic quality metrics: coherence (top), tf-idf coherence (bottom).

hand so that the topics actually are similar), and we call a topic *stable* if there is a set of pairwise matching topics in every result across all runs [14].

Table 1 shows the results of our experimental evaluation, comparing the basic topic quality and topic stability metrics across several baseline topic models and granulated LDA with different window sizes. We have trained 200 topics for every model, averaging results over three runs. We see that granulated LDA with the step window produces topics that have quality matching that of baseline topic models or even exceeding it, but the other two windows, Epanechnikov and triangular, do not work nearly as well. One should be careful about using coherence to draw steadfast conclusions in this case, though, because granulated LDA naturally lends itself to optimizing coherence: it artificially sets words that cooccur in the same document (even in the same window) to the same topic. This effect is much less prominent for tf-idf coherence (many words in a window are likely to be common words with low tf-idf weights), and in tf-idf coherence we see GLDA with step window performing on par with other models. Figure 2 shows the distributions of coherence and tf-idf coherence metrics in more detail; namely, it shows the coherences (top) and tf-idf coherences (bottom) of all 200 topics for all models sorted in decreasing order, so a line higher on this plot means a better overall model. We can see that GLDA solutions, especially with the step window, hold up quite well compared with other models in our study.

The primary gains of our new model lie in topic stability. Table 1 shows the number of stable topics for every model and average Jaccard similarity (w.r.t. to 100 top words in each topic) between pairs of matching topics. We see that granulated LDA indeed produces very stable results: in all runs of granulated LDA with all window variants almost all topic were stable, and the average Jaccard similarity between them is also much higher than in other models in the case of a step window. Overall, we conclude that GLDA with step window produces much more stable topics at virtually no loss to quality and interpretability.

6   Conclusion

In this work, we have introduced a novel modification of the latent Dirichlet allocation model, granulated LDA, that samples whole windows of neighboring words in a document at once. This model was intended to improve the stability of the topic model results, and in the experimental evaluation we have shown that the results of GLDA are indeed much more stable while preserving the same overall topic quality. This improvement is especially important for web science and digital humanities that seek not only interpretable topics, but essentially entire solutions that could serve as a basis to make reliable conclusions about the topical structure of text collections. In further work, we plan to extend and improve upon the basic idea of granulated LDA, experimenting with variations of this model. We hope that designing topic models with an eye to topic stability will prove to be a promising new venue of research.

Acknowledgments. This work was supported by the Basic Research Program of the National Research University Higher School of Economics.

References

1. Andrzejewski, D., Zhu, X.: Latent Dirichlet allocation with topic-in-set knowledge. In: Proceedings of NAACL HLT 2009 Workshop on Semi-Supervised Learning for Natural Language Processing, SemiSupLearn 2009, pp. 43–48. Association for Computational Linguistics, Stroudsburg (2009)
2. Andrzejewski, D., Zhu, X., Craven, M.: Incorporating domain knowledge into topic modeling via Dirichlet forest priors. In: Proceedings of 26th Annual International Conference on Machine Learning, ICML 2009, pp. 25–32. ACM, New York (2009)
3. Blei, D.M., Lafferty, J.D.: Correlated topic models. In: Advances in Neural Information Processing Systems 18 (2006)
4. Blei, D.M., Ng, A.Y., Jordan, M.I.: Latent Dirichlet allocation. J. Mach. Learn. Res. **3**(4–5), 993–1022 (2003)
5. Bodrunova, S., Koltsov, S., Koltsova, O., Nikolenko, S., Shimorina, A.: Interval semi-supervised LDA: classifying needles in a haystack. In: Castro, F., Gelbukh, A., González, M. (eds.) MICAI 2013, Part I. LNCS, vol. 8265, pp. 265–274. Springer, Heidelberg (2013)
6. Bouma, G.: Normalized (pointwise) mutual information in collocation extraction. In: Proceedings of the Biennial GSCL Conference, pp. 31–40 (2013)

7. Boyd-Graber, J.L., Blei, D.M.: Syntactic topic models. In: Koller, D., Schuurmans, D., Bengio, Y., Bottou, L. (eds.) Advances in Neural Information Processing Systems, pp. 185–192. Curran Associates Inc. (2008)
8. Chang, J., Boyd-Graber, J., Gerrish, S., Wang, C., Blei, D.M.: Reading tea leaves: how humans interpret topic models. In: Advances in Neural Information Processing Systems 20 (2009)
9. Chen, R.-C., Swanson, R., Gordon, A.S.: An adaptation of topic modeling to sentences (2010). http://rueycheng.com/paper/adaptation.pdf
10. Chen, X., Zhou, M., Carin, L.: The contextual focused topic model. In: Proceedings of the 18th ACM SIGKDD International Conference on Knowledge Discovery and Data Mining, pp. 96–104. ACM, New York (2012)
11. Griffiths, T., Steyvers, M.: Finding scientific topics. Proc. Natl Acad. Sci. **101**(Suppl. 1), 5228–5335 (2004)
12. Grimmer, J., Stewart, B.M.: Text as data: the promise and pitfalls of automatic content analysis methods for political texts. Polit. Anal. **21**(3), 267–297 (2013)
13. Hoffmann, T.: Unsupervised learning by probabilistic latent semantic analysis. Mach. Learn. **42**(1), 177–196 (2001)
14. Koltcov, S., Koltsova, O., Nikolenko, S.I.: Latent Dirichlet allocation: stability and applications to studies of user-generated content. In: Proceedings of the 2014 ACM Conference on Web Science (WebSci 2014), pp. 161–165 (2014)
15. Lacoste-Julien, S., Sha, F., Jordan, M.I.: DiscLDA: discriminative learning for dimensionality reduction and classification. In: Advances in Neural Information Processing Systems 20 (2008)
16. Lau, J.H., Newman, D., Baldwin, T.: Machine reading tea leaves: automatically evaluating topic coherence and topic model quality. In: EACL, pp. 530–539 (2014)
17. Li, S.Z.: Markov Random Field Modeling in Image Analysis. Advances in Pattern Recognition. Springer, Heidelberg (2009)
18. McFarland, D.A., Ramage, D., Chuang, J., Heer, J., Manning, C.D., Jurafsky, D.: Differentiating language usage through topic models. Poetics **41**(6), 607–625 (2013)
19. Mimno, D., Wallach, H.M., Talley, E., Leenders, M., McCallum, A.: Optimizing semantic coherence in topic models. In: Proceedings of the Conference on Empirical Methods in Natural Language Processing, pp. 262–272. Association for Computational Linguistics, Stroudsburg (2011)
20. Newman, D., Bonilla, E.V., Buntine, W.: Improving topic coherence with regularized topic models. In: Advances in Neural Information Processing Systems 24, pp. 496–504. Curran Associates Inc. (2011)
21. Newman, D., Lau, J.H., Grieser, K., Baldwin, T.: Automatic evaluation of topic coherence. In: Human Language Technologies: The 2010 Annual Conference of the North American Chapter of the Association for Computational Linguistics, HLT 2010, pp. 100–108. Association for Computational Linguistics, Stroudsburg (2010)
22. Nikolenko, S.I., Koltsova, O., Koltsov, S.: Topic modelling for qualitative studies. J. Inf. Sci. (2015). doi:10.1177/0165551515617393
23. Ramage, D., Rosen, E., Chuang, J., Manning, C.D., McFarland, D.A.: Topic modeling for the social sciences. In: NIPS 2009 Workshop on Applications for Topic Models: Text and Beyond, Whistler, Canada, December 2009
24. Tikhonov, A.N., Arsenin, V.Y.: Solutions of Ill-posed problems. W.H. Winston, New York (1977)
25. Vorontsov, K.: Additive regularization for topic models of text collections. Doklady Math. **89**(3), 301–304 (2014)

26. Vorontsov, K.V., Potapenko, A.A.: Additive regularization of topic models. Mach. Learn. **101**(1), 303–323 (2015). Special Issue on Data Analysis and Intelligent Optimization with Applications
27. Wang, C., Blei, D.M., Heckerman, D.: Continuous time dynamic topic models. In: Proceedings of the 24th Conference on Uncertainty in Artificial Intelligence (2008)
28. Wang, X., McCallum, A.: Topics over time: a non-Markov continuous-time model of topical trends. In: Proceedings of the 12th ACM SIGKDD International Conference on Knowledge Discovery and Data Mining, pp. 424–433 (2006)
29. Yohan, J., O. A. H.: Aspect and sentiment unification model for online review analysis. In: Proceedings of the Fourth ACM International Conference on Web Search and Data Mining, WSDM 2011, New York, NY, USA, pp. 815–824 (2011)

An Empirically Informed Taxonomy
for the Maker Movement

Christian Voigt[1](✉), Calkin Suero Montero[2], and Massimo Menichinelli[3,4]

[1] Zentrum Für Soziale Innovation, Technology and Knowledge, Vienna, Austria
voigt@zsi.at
[2] University of Eastern Finland, Joensuu, Finland
calkin.montero@uef.fi
[3] IAAC | Fab Lab Barcelona, Barcelona, Spain
massimo@fablabbcn.org
[4] School of Art, Design and Architecture Media Lab Helsinki, Aalto University, Helsinki, Finland
massimo.menichinelli@aalto.fi

Abstract. The Maker Movement emerged from a renewed interest in the physical side of innovation following the dot-com bubble and the rise of the participatory Web 2.0 and the decreasing costs of many digital fabrication technologies. Classifying concepts, i.e. building taxonomies, is a fundamental practice when developing a topic of interest into a research field. Taking advantage of the growth of the Social Web and participation platforms, this paper suggests a multidisciplinary analysis of communications and online behaviors related to the Maker community in order to develop a taxonomy informed by current practices and ongoing discussions. We analyze a number of sources such as Twitter, Wikipedia and Google Trends, applying co-word analysis, trend visualizations and emotional analysis. Whereas co-words and trends extract structural characteristics of the movement, emotional analysis is non-topical, extracting emotional interpretations.

Keywords: Maker movement · Internet science · Taxonomy · Development · Co-word analysis · Clustering · Emotion profiling

1 Introduction

Taxonomies are central elements to support the conceptual, methodological and scientific exploration of emerging phenomena such as making and the Maker Movement. The Maker Movement emerged from a renewed interest in the physical side of innovation following the dot-com bubble, the rise of the participatory Web 2.0, the diffusion of Open Source and the decreasing costs of many digital fabrication technologies. Simultaneously, the renowned publication venues Make magazine was launched in 2005 [1]. Neil Gershenfeld [2] calls the Maker Movement the next digital revolution as it enables personal fabrication on people's desks. The Massachusetts Institute of Technology's 'Bits to Atoms' program, which dates back to 2001, is often quoted as the first step of the Maker Movement. Open source and Web 2.0 did not only democratize knowledge production but also the means of design and invention by 'industrializing the Do It

© Springer International Publishing AG 2016
F. Bagnoli et al. (Eds.): INSCI 2016, LNCS 9934, pp. 189–204, 2016.
DOI: 10.1007/978-3-319-45982-0_17

Yourself (DIY) spirit' [3]. Today rapid prototyping is more accessible than ever before due to affordable computer-aided design software, 3-D printing, laser cutting and a knowledge community that is pushing the limits of what can be produced by individuals. For example, sales of goods on ETSY, an e-commerce marketplace specializing in crafts and maker products, reached a turnover of about 2.4 billion USD in 2015 [4].

Yet, it should be fair to say that the Maker Movement is primarily practice oriented, characterized in large parts by tacit knowledge and heuristics obtained through a continuous, problem-driven exchange within maker communities. However, in order to obtain a more robust and consolidated framework for analyzing the Maker Movement we argue that it is important to capture and systematize existing key concepts, semantic differences and changing connotations depending on geographical regions to advance and focus future research efforts.

In this context we look at the possibilities of Internet science as a field of research poised to support taxonomic developments. The authors of this paper aim for an explicitly interdisciplinary approach, combing the expertise of digital social innovation, digital fabrication and Natural Language Processing (NLP). This combination of diverse research domains is meant to strengthen the final taxonomy's pragmatic value as well as methodological efficiency in producing the taxonomy, a non-trivial challenge considering the epistemological differences inherent to interdisciplinarity [5].

Hence, our contribution to Internet science is a) to open the discussion in order to create a common understanding of terms and related implications; b) to suggest a first taxonomic structure for the Maker Movement (people, places and activities); and c) to explore the relevance and explanatory usefulness of social media in creating context.

This paper is organized as follows. First, we outline the benefits of pursuing a taxonomy of the Maker Movement. We then describe our methodology supporting the overall development of the taxonomy as well as specific data collection procedures (Sect. 3). In the fourth section we introduce some first basic components of a taxonomy around the Maker Movement, i.e. concepts related to Maker communities, spaces and activities. These concepts are then explored with the help of social media analysis (tweet mining) and access statistics (Wikipedia consultations, Goggle trends, related searches). This section also includes non-topical text analysis, extracting emotional interpretations from tweets. Here the aim is to enrich the meaning making process, exploring the possibility of attaching indications of joy or frustrations to Maker concepts, which can then be explored in more depth. The paper closes with a discussion of findings and an outline of next steps.

2   Why Having a Taxonomy Discussion?

Classifying concepts, i.e. building a taxonomy, is a fundamental practice when developing a topic of interest into a research field. For our purpose we are going to distinguish typologies and taxonomies, the former being deductive assignments into a priori defined groups (ideal types) whereas the latter are inductively determined memberships of a posteriori identified categories [6]. Put differently, *typologies* are intuitive classifications, which might turn out to be exhaustive or too restrictive. *Taxonomies*, on the other hand, start empirically, focusing on categorizing cases based on similarities between observed variables [7].

Our taxonomy discussion is embedded within the European funded H2020 research project 'MAKE-IT', exploring maker communities and their links with Collective Awareness Platforms for Sustainability and Social Innovation (CAPS). CAPS serve to raise awareness of problems related to sustainability or social injustice, with the aim that communities can develop solutions collaboratively and share the required design and implementation efforts among many. Typically, forms of communication, coordination, guiding ideologies within maker communities etc. depend on their histories and organizational setting. For example, in open source communities we know that collaboration is guided by fairly powerful community norms [8], e.g. deciding when forking an open source project is permissible or how to peer review and bug fix open source code. Hence, a taxonomy is a useful instrument to discuss the broad variety of community related phenomena in a systematic way and keeping it accessible for all participants.

Understanding changing meanings. An on-going observation of how concepts are used over time often reflects the development of a field as knowledge becomes increasingly more specific. An example shown in Sect. 4.2 refers to the relatively recent increase of 'maker spaces' as a search term, which became popular in 2011. We would assume that spaces dedicated to making existed before, but were simply subsumed under the concept of 'hacker spaces'. In fact, one of the early Hackerspaces, 'c-base' opened in Berlin in 1995, designing robotic devices that crossed the boundaries between the physical and the digital world [9].

Making research replicable and insights comparable. A further benefit of a taxonomy is that it provides some measure of unity to the description of research findings, which enables others to reconstruct and replicate the conditions under which a given method or procedure has been successful. Of course, this requires the taxonomy to be close to reality so that practitioners as well as researchers accept the taxonomy as a valid reflection of their experiences [10].

Working towards predictive and more general knowledge. A clear terminology is usually a sign of an established research area, where there is a sufficiently large body of knowledge describing the boundaries of a term and interdependencies between terms [11]. Hence, a common language will be a necessary precondition to better describe developments around the maker community in a European context, where earlier studies have already shown distinct characteristics between maker spaces and fab labs in terms of their network structures and interaction intensities [12].

3   A Methodology for Taxonomy Refinement

A taxonomy cannot be created with one swift move. The long-term goal is to start with a draft, which is progressively unified and becomes an increasingly accepted terminology that precedes comparable and eventually generalizable knowledge [7]. Working towards this taxonomy will comprise multiple stages (cf. Fig. 1, based on [11]).

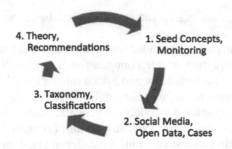

Fig. 1. Circle of taxonomy refinement

Since no new taxonomy can be created ex nihilo, we start with seed categories (*people*, *spaces* and *activities*) and seed concepts drawn from the existing literature (step1 1). In step 2, we monitor and explore available open and social media data in combination with observations and surveys from European maker communities. Aggregating multiple data explorations and experiences across multiple case studies will then allow us to draft a first version of the taxonomy, which is then published on the Web for further commenting (step 3). Eventually, some taxonomy entries can be linked with results from relevant research studies, for example concerning governance or value creation in maker spaces (step 4).

Seed categories and concepts. We found the combination of *people*, *spaces* and *activities* to be a recurring theme in many publications [13–15]. Although not all authors were using exactly the same labels and might refer to communities instead of people, or they focused on tools rather than the spaces, where people got access to the tools. The main purpose of working with these three areas was to have a first set of keywords, which could eventually lead us to related concepts. In this situation the exact naming is not overly relevant as long as the chosen seed concepts are broad enough and concepts within maker-related texts are sufficiently linked. A similar approach can be found when using folksonomies. A 'folksonomy' is a combination of the words 'folk' and 'taxonomy' and refers to the user-generated nature of a taxonomy based on social tagging, the public labeling or categorization of online recourses [16]. Folksonomies are likely to have flatter hierarchies than their scientific counterparts and have shown to converge towards smaller sets of frequently used tags, despite their decentralized and informal usage. The initial set of *seed concepts* used for exploring open and social media around the Maker Movement is as follows:

- *people*: maker, hacker;
- *places*: makerspaces, hackerspaces, fablabs;
- *activities*: DIY, 3d-printing, making, hacking, maker_education.

Not all concepts are equally useful as seed concepts due to their homonymic characteristics, e.g. you can hack into a computer, or hack a piece of wood. Even if the meaning stays the same, sometimes a word is used in a context that makes it less relevant for the intended analysis. For example, the DIY philosophy is said to define the Maker Movement [17], when the same term is also frequently used with wedding preparations.

Taxonomic structures. Once a taxonomy has an *empirical basis* –as in biological classifications–, hierarchies are build around central categories which branch out into sub-categories [7]. In a 'Maker Movement' context, that could concern additive making technologies which include different 3D printing technologies such as Stereolithography (SLA), Digital Light Processing (DLP), Fused deposition modelling (FDM), Selective Laser Sintering (SLS) etc. The same thinking could be applied to an activity such as 'maker education', here a first level differentiation might include the use of maker technology in formal, non-formal and informal education [18] and even further differentiation might then distinguish between electronic and fabrication kits, which enable different types of learning [19].

4 Experimenting with Categories and Seed Concepts

Before we could start experimenting with seed concepts, we explored a number of data sources. Main criteria were open access and a minimum of limitations for analyzing data going back in time, so that conceptual changes could be identified. Sources meeting these criteria included Twitter, Wikipedia, Google Trends, The Guardian (a UK newspaper with an open API) as well as a number of bibliographic databases including Scopus [20] and Web of Science [21]. The list is by no means complete and other sources such as Google Scholar can also be accessed through web scraping, see [22] for a comparison of different citation databases.

The experimentations described in this section reflects stage two 'social media and open data analysis' of our overall methodology (cf. Fig. 1) and aims at extracting related concepts as well as trending concepts. An additional experiment looks into the emotional profiling of groups of tweets, exploring the possibility of identifying concept related feelings such as 'joy' or 'frustration'.

4.1 Concept Identification: Tweet Mining and Google's Related Searches

Classification is key to conceptualizing a domain space, compare data, reason with data etc. As stated in Sect. 2, classification can be done through a typology or a taxonomy [7]. The former relies on classifying along theoretical dimensions (e.g. a makerspace might be for-profit or non-profit), and the latter relies on empirical observations leading to measurable similarities. Different cluster techniques can then group similar concepts [23].

Tweet mining. In this section we start with selecting tweets containing hashtags commonly describing people, places and activities in and around the Maker Movement. From that corpus we started extracting frequent co-words (i.e. words co-occurring with specified hashtags). Co-words can be used as indicators of a concept's cognitive structure, and changes in co-words may indicate a change of strategy in order to make a concept more appealing or successful [24, 25]. The amount of tweets that can be accessed on one day is limited to 13,000 and includes tweets no older than the last seven days. This relatively small window of analysis means that events in that week can have a strong impact on co-word appearances, as we will see in the case of #makerspace tweets. We used different R packages [26] for accessing tweets [27], data clean up and generation of co-word matrices [28] and visualization [29]. For data cleaning, we removed English stop words, but avoided stemming in order to maintain readability of concepts. We did some lightweight curating of the resulting tables of frequent co-words by removing the plural or a different spelling of the seed concept. We analyzed a total of 50,097 tweets: 12,180 for #makerspace, 1,614 for #FabLab, 4,370 for #makerEd (a prominent hashtag for making in education), 13,000 for #3dprinting, 11,269 for #hacker and 7,664 for #maker.

#makerspace. We collected the tweets early November, when two education conferences took place, which also focused on the question 'How to apply maker concepts to education?' Firstly, the 2015 California STEM Symposium, a gathering of 3,100 teachers and administrators looking into how robotics or 3D-printing could increase the attractiveness of science, technology, engineering and math [30]. Secondly, the 17th conference of the American Association of School Librarians, thematizing the use of libraries as maker-spaces [31]. Consequently Fig. 2 showcases makerspaces with a focus on education, i.e. more than 2,200 tweets referred to #makerspace and to library, STEM or school.

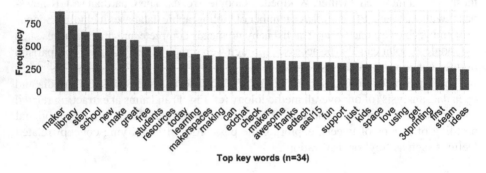

Fig. 2. Co-word analysis based on 12,180 tweets containing #makerspace (Nov 2015)

#makered. If we look at tweets explicitly referring to making in education, using the *#makered* hashtag, the term 'makerspace' appears first, confirming the high co-occurrence of both terms (Fig. 3).

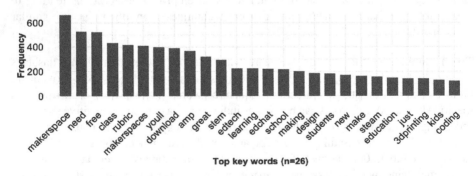

Fig. 3. Co-word analysis based on 4,370 tweets containing #MakerEd (Nov 2015)

However, the reminder of the co-word list looks quite differently indicating things like 'class', 'rubrics' and 'educational technology' (hashtag *#MakerEd*).

In this context we can infer elements of a 'making in education' discussion around the need for assessing learners maker qualities in the setting of a class and, with less frequency, concrete activities such as coding and 3D-printing (Fig. 4).

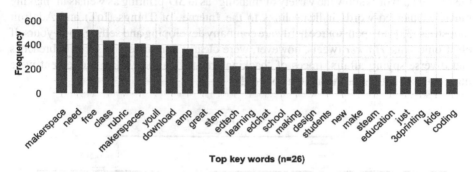

Fig. 4. Co-word analysis based on 4,370 tweets containing #MakerEd (Nov 2015)

#3dprinting. If we follow up on #3D-printing, the educational dimension almost disappears. Rather, what we see are structural elements of the 3D-printing process, including parts, design, 3D-printers and filaments (see Fig. 5). Additionally the co-word list indicates discussion around novel uses of 3D-prinitng for the production of human tissue and body parts. Again, the co-word list shows the impact of a highly visible event during the week of tweet collection, when toy maker Mattel announced ThingMaker, a low cost 3D-Printer for kids at around 300 USD [32].

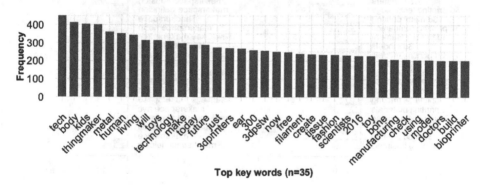

Fig. 5. Co-word analysis based on 13,000 tweets containing #3dprinting (Feb 2016)

A further dataset of 1,614 #fablab related tweets was manually filtered, since a single message (the opening of a FabLab in India) was retweeted 104 times (6.4 %) and dominated the co-word analysis. Although the retweets were certainly relevant, they were also very specific in a geographical sense. Comparing co-word rankings from 'makerspace' tweets (see Fig. 2) with 'FabLab' tweets, the former showed more educational key words, whereas the latter had more entrepreneurial tendencies. However, at this stage our focus is on getting a first impression of whether or not the presented analyses can generate some early hypotheses, which we can then revisit with larger data sets covering a span of several months.

Another visualization of co-word analysis is shown in Fig. 6, where we can see how the *Maker* co-words show the variety of 'making' as in 3D-printing as well as in 'making music'. Other co-words indicate links to the Internet of Things (IoT) and Adafruit Industries [33], an open-source hardware company developing and selling do-it-yourself electronics kits. *Hacker* tweets, however, were clearly dominated by security breaches and diverse spying affairs, traces of 'hacking' related to the Maker Movement were marginal and not under the top ten co-words.

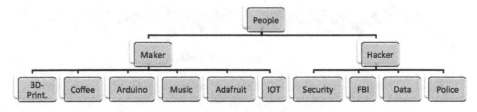

Fig. 6. Co-words for #maker and #hacker

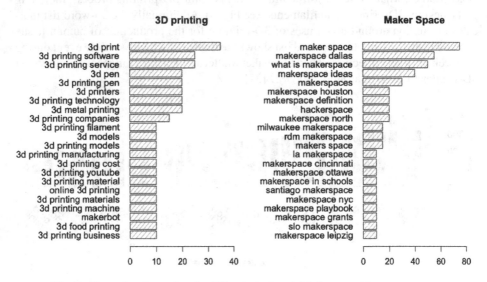

Fig. 7. Top related searches for '3D printing' and 'Makerspace' in percentages

A better understanding of how today's Internet community conceptualizes given phenomena such as 'maker space' or '3D-printing' can also be gained through an analysis of related Google searches also offered through the Google Trend service. For example, people who searched 3D-printing between January 2015 and February 2016, also searched related software or the possibility of 3D-printing metal (8th place in Fig. 6).

Interestingly, printing metal also appeared on 5th place within our co-word analysis in Fig. 5, indicating makers' interest in extending 3D-printing towards more durable materials. Another possible reason for 3D printing's increasing popularity could be higher awareness of novel application areas such as printing food and an increasing commercialization of 3D printing, indicated by search terms such as 'services', 'companies', 'costs' and 'businesses'. It's also informative to look back, as for example in 2013, one of the most popular related searches was '3d printing stocks', indicating people's interest in 3D printing as an investment option.

4.2   Concepts Over Time: Trends in Google Search and Wikipedia

Another possibility to extract the taxonomic structure of the Maker Movement is to look into data provided by Google Trends, namely popularity of Google searches as well as most frequent co-occurring queries per search session [34]. Data are available since 2004 and were collected using the R package gtrendsR [35]. The data obtained from Google Trends represent total searches for a term relative to the total number of searches done on Google over time, mapping the development of a search term's relative popularity (no absolute search volumes are shown) (Fig. 8).

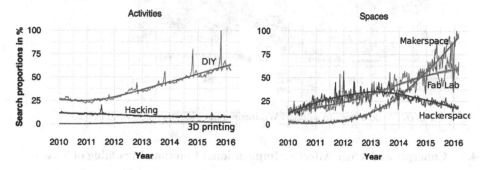

Fig. 8. Google Trends Data about 'maker activities' (left) and 'maker spaces' (right)

Figure 7 (left side) shows prototypical activities such as DIY, Hacking and 3D-printing. At this point we would argue that the diagram indicates primarily semantic differences, with DIY used for the most diverse purposes – 'making' being only one among many – and the terms 'hacking' and '3D-printing' becoming increasingly more specific and consequently referring to smaller target groups. The graphs on the right side, however, offer a clearer picture for comparison, as Makerspaces and FabLabs are on the rise and hackerspaces are declining in search popularity. With Wikipedia being one of the 'go to' sources for people who seek a first idea what a concept means, we had another proxy for general interest levels related to the 'Maker Movement'. Data have been collected through the R package 'WikipediaTrend' [36] and visualized in Fig. 9. Similar to Google Trends, consultations of the 'hackerspace' page are declining after 2013 and visits to 'maker_culture' page are increasing. Overlapping with the raise of 'maker_culture' page visits is the publication of Chris Anderson's [14] *Makers: The New Industrial Revolution.* Whereas the substantial rise of 3D-Printing overlaps with another key publication, when the cover story from the Economist in February, 2011 said "Print me a Stradivarius" [37]. Other tendencies such as the decline of 'hacker-spaces' (Fig. 7) or the decline in access statistics of the 'DIY' Wikipedia page (Fig. 9) cannot yet be explained and need further exploration, cross-referencing data from other sources (e.g. looking into relevant social media in and around the year 2013).

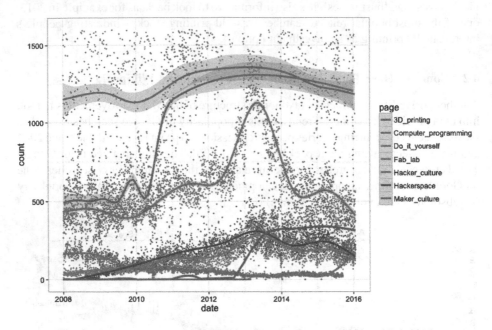

Fig. 9.   Access rates to selected Wikipedia pages between 2008 and Feb 2016

4.3   Concepts and Their Affective Implications: Emotional Profiling of Tweets

Beside identifying networks of related terms and trending terms, we were also interested in exploring the emotional dimensions of the tweets we had already analyzed in 4.1. Emotion analysis is already widely used in areas such as customer satisfaction [38] or the popularity of political parties. In general, the aim of characterizing the feelings present in a text can be achieved either through word-list associations (affective diction-aries and databases of common-sense knowledge) or machine learning [39]. For our purposes we used the SentiProfiler, an emotional analysis system described in [40, 41]. The SentiProfiler uses an ontology, i.e. a hierarchy of emotions derived from WordNet-Affect as the main source of emotional knowledge [42]. The WordNet-Affect ontology contains four main categories of emotions: negative, positive, ambiguous and neutral. Under each category exist several classes containing a list of emotion words. The WordNet-Affect, combines 1,316 words in 250 classes. For example, a positive feeling could fall into the class of 'liking', identified through words such as 'approval', 'sympathy' or 'friendliness' [41]. Additionally, text classification is supported by a number of disambiguation rules to exclude instances where words implying positive feelings are negated or an emotion bearing word fulfills a different role, such as the use of 'like' as a preposition, meaning 'similar to'. Analyzing the sets of tweets, we found between 0.77 and 3.3 percent of all words were emotion-bearing words (see Table 1).

Table 1. Emotion bearing words per Twitter data set

Twitter Data	Number of words	Emotion words	Emotion words in precent	Positive to negative ratio
3d-printing.txt	113.481	2.275	2,00 %	0,87
makerspace.txt	152.33	4.134	2,71 %	0,94
fablab.txt	20.193	156	0,77 %	0,77
hacker.txt	149.544	4.178	2,79 %	0,87
maker.txt	57.4	1.894	3,30 %	0,87

From the table we notice that in general the tweets have been very positive, with a positive to negative emotion ratio of above 0.85 (with the exception of FabLab tweets). Following a more detailed comparison of #makerspace and #fablab tweets. Figure 10 shows a section of emotional expressions contained in 'makerspace' tweets compared to 'fablab' tweets, positive emotions are under 'joy' (left side) and negative emotions are under 'despair' (right side). Red nodes indicate emotions that are less than the comparative profile, green nodes indicate the emotions are more than the comparative profile and blue nodes indicate emotions only found in the 'makerspace' profile.

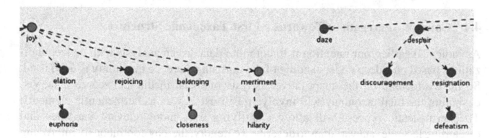

Fig. 10. Emotions in 'makerspace' tweets compared to 'fablab' tweets

To provide a more concrete picture of the analysis, we selected three tweets for positive and negative emotions, indicating a clear, an ambiguous and a false classification. Words in brackets indicate the class of emotions.

Positive emotions.

- Clear: 'Lots of *great* practical tips for creating a library makerspace' (Eagerness)
- Ambiguous: 'Huge thanks to Cargill Salt for *approving* our grant for our Makerspace. You're helping to foster creativity and innovation!' (Liking)
- False: 'Aleph objects opens new *fulfilment* center in Australia, offers free shipping – 3dprint.com ' (Fulfillment)

The ambiguous classification is not outright false, but the emotion word 'approving' refers to funding granted, rather than an approved design or a maker activity. The false classification is due to the multiple meanings of 'fulfillment' indicating either a feeling of satisfaction or the execution of a shipping order.

Negative emotions.

- Clear: 'thanks! I want to try... *afraid* I'll do most of the work in the minimal makerspace time we have!' (Distress)
- Ambiguous: *'Worried* about #makerspace logistics? #fallcue is here to help! It's not just for tech! ' (Distress)
- False: 'If kids can imagine it, they can build it! Perfect for blasting away *boredom* #makerspace' (Weariness)

In the above case, the classification is ambiguous because the distress is anticipated, not actual. Still, one could argue that makerspace logistics is characterized as a worrisome issue. Assigning the emotional class of 'weariness' to 'blasting away kids boredom' is a false classification as it is missing the negation of boredom.

Based on this first experience with analyzing the emotional value of tweets, we see a promising application area in filtering tweets and other social media expression about specific equipment, make spaces or events in order to get an impression of what might cause frustration or joy. Specific messages around logistics in maker spaces could then lead to a more targeted analysis, possibly including a wider range of social media beyond twitter, eventually leading to improved conditions for Makers.

4.4   Concept Aggregation: Towards a First Taxonomic Structure

As indicated earlier, our intention is to combine data driven analysis with conceptualizations based on a deep understanding of the domain where the taxonomy is to be used. So far we focused on the quantitative analysis of social media and search data, yet, designing the final taxonomy will involve positivist as well as hermeneutic elements. The hermeneutic process will allow identifying ever more relevant variables and discarding less relevant variables to describe a category, thereby continuously improving categories as well as the resulting taxonomy. Identifying suitable variables that can help to structure a domain more effectively is far from trivial and categorization becomes more complex if a concept is characterized by a high number of dimensions (e.g. variables describing different types of makers) [7]. One could imagine an iterative design process, where more and more categories are empirically scrutinized.

Five types of analyses have been presented in this paper: (a) Twitter co-word analysis, (b) Google co-search analysis, (c) Wikipedia access, (d) Google search terms and (e) emotional profiling. The first two (a and b) support the identification of related concepts - the basic building blocks of the future taxonomy -, analyses (c) to (e) are more suitable to support the narratives around the identified concepts (e.g. how their popularity changed over time or whether they occur in a primarily positive or negative context). What is still missing is a technique that can aggregate single concepts into bottom-up categories.

For this, hierarchical or k-means clustering techniques can be used. Clustering algorithms group concepts in accordance to their distance to each other, i.e. "given a representation of n objects, find K groups based on a measure of similarity such that the similarities between objects in the same group are high while the similarities between objects in different groups are low." [43]. However, given the limited size of the dataset, the following cluster can only illustrate the value of clustering as an aggregation mechanism. Eventually, clusters based on a larger dataset are likely to look differently. Figure 11 takes 12 co-occurring keywords in tweets including '#makerspace' (cf. Fig. 2) - omitting less domain specific words such as 'new', 'great' or 'today' - and clustered those keywords according to their retweeting and favoring values. The underlying rationale for analyzing not only the frequency of co-occurring keywords but also the amount of re-tweets, for example, is the idea that retweeting is a form of joining a public discourse, publicly agreeing with someone or simply disseminating the message to new audiences [44], which are activities particularly relevant to promote the discourse needed for a more widely accepted set of categories.

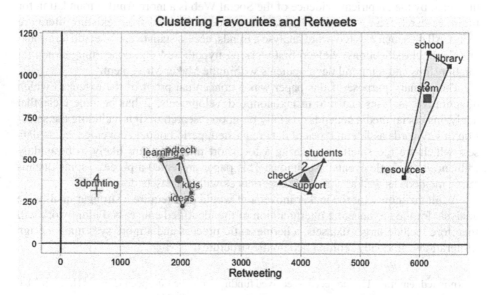

Fig. 11. Clustering '#makerspace' tweets according to their retweeting and favoring values

A first, though loose, interpretation of Fig. 11 suggests three clusters: (1) kids learning in maker spaces, (2) supporting students in maker spaces, (3) schools and libraries as maker spaces and (4) 3D-printing as a single item cluster. Hence, starting from words most frequently co-occurring with the concept of '#makerspace', we identified a shared understanding of maker spaces as places where young people learn. '3D-printing' as one of the characterizing activities in maker spaces is still present but appears to be less instrumental if judged by the number of times 3D-printing related tweets had been retweeted or favored. Future research foci are primarily expected at the junction of specific activities and spaces, such as 'libraries as maker spaces of the future' or 'integrating a maker culture with education'. In this sense, we will aim for taxonomies which are initially limited in scope but effective in terms of their applicability as they can be created with particular spaces and activities already in mind.

5   Discussion of Findings and Future Research

With the growth of the Social Web and participation platforms such as Wikipedia.com (creating knowledge collectively), Thingiverse.com (sharing digital designs) or Twitter (sharing generic messages), a world defined by the few is transformed into a world where almost everyone can participate [45]. Accompanying these platforms are emerging cultures of participation that offer powerful mechanisms to raise awareness of some of today's most pressing societal problems. Working towards a closer connection between empirical evidence of what potential makers are interested in and what determines current research agendas has been the broader context of this paper.

A first step towards such a nexus has been the evaluation of different data sources (Twitter, Wikipedia and Google Trends) in combination with descriptive statistics and corresponding visualizations. Based on the work presented, we suggest that a taxonomy informed by the empirical evidence of the Social Web is a more fruitful foundation for future research than a taxonomy based on concepts derived from existing literature alone. All data analyses (co-word analyses, trends, access statistics, co-search terms and emotional classifications) yielded first working hypotheses, e.g. concerning structural relationships and temporal developments within the Maker Movement.

The primary purpose of this paper was a conceptual proof of the extent to which quantitative analyses can inform taxonomic developments. It has become clear that analyzing social media depends crucially on good research design including the selection of keywords as filters, types of data requested, period of data coverage, etc. – data sets which are too small or covering a too short time span are likely to be unduly influenced by single events or opinions. This paper presented a process, some quantitative methods as well as some first experiences with a datasets describing

A full taxonomy for the maker movement would also require a stronger qualitative analysis for the hermeneutic interpretation of the identified concepts. Future work will therefore include larger datasets, a hermeneutic process and a more systematic design of iterations, gradually refining taxonomic structures.

Acknowledgement.   This project has received funding from the European Union's Horizon 2020 research and innovation programme under grant agreement 688241.

References

1. The Blueprint Talks with Dale Dougherty. https://theblueprint.com/stories/dale-dougherty
2. Gershenfeld, N.: How to make almost anything: the digital fabrication revolution. Foreign Aff. **91**, 43 (2012)
3. Gershenfeld, N.: Fab: the coming revolution on your desktop–from personal computers to personal fabrication. Basic Books (2008)
4. Etsy's total annual merchandise sales volume from 2005 to 2015. http://www.statista.com/statistics/219412/etsys-total-merchandise-sales-per-year
5. Dini, P., Iqani, M., Mansell, R.: The (im) possibility of interdisciplinarity: lessons from constructing a theoretical framework for digital ecosystems. Culture Theory Critique **52**, 3–27 (2011)

6. Fiedler, K.D., Grover, V., Teng, J.T.: An empirically derived taxonomy of information technology structure and its relationship to organizational structure. J. Manage. Inf. Syst. **13**, 9–34 (1996)
7. Bailey, K.D.: Typologies and taxonomies: an introduction to classification techniques. Sage (1994)
8. Stewart, K.J., Gosain, S.: The impact of ideology on effectiveness in open source software development teams. MIS Q. **30**, 291–314 (2006)
9. Baichtal, J.: Hack this: 24 incredible hackerspace projects from the DIY movement. New Riders (2011)
10. Rich, P.: The organizational taxonomy: definition and design. Acad. Manag. Rev. **17**, 758–781 (1992)
11. Horan, T., Lafky, D.: Toward an empirical user taxonomy for personal health records systems. In: AMCIS 2006 Proceedings, p. 341 (2006)
12. Menichinelli, M.: Mapping the structure of the global maker laboratories community through Twitter connections. In: Levallois, C., Marchand, M., Mata, T., Panisson, A. (eds.) Twitter for Research Handbook 2015–2016, pp. 47–62. EMLYON Press, Lyon (2016)
13. Ratto, M.: Critical making: Conceptual and material studies in technology and social life. Inf. Soc. **27**, 252–260 (2011)
14. Anderson, C.: Makers: the New Industrial Revolution. Crown Business, New York (2012)
15. Dougherty, D.: The maker movement. innovations. **7**, 11–14 (2012)
16. Trant, J.: Studying social tagging and folksonomy: a review and framework. J. Digital Inf. **10** (2009)
17. Carmody, T.: Big DIY: The Year the Maker Movement Broke. Wired, 6 August 2011. http://www.wired.com/20
18. Eisenberg, M.: Educational fabrication, in and out of the classroom. In: Society for Information Technology & Teacher Education International Conference (2011)
19. Bull, G., Haj-Hariri, H., Atkins, R., Moran, P.: An educational framework for digital manufacturing in schools. 3D Printing Additive Manuf. **2**, 42–49 (2015)
20. Scopus (bibliographic database). www.scopus.com
21. Web of Science. http://ipscience.thomsonreuters.com/product/web-of-science
22. Meho, L.I., Yang, K.: Impact of data sources on citation counts and rankings of LIS faculty: Web of Science versus Scopus and Google Scholar. J. Am. Soc. Inform. Sci. Technol. **58**, 2105–2125 (2007)
23. Small, H., Sweeney, E.: Clustering the science citation index® using co-citations: I. A comparison methods. Scientometrics **7**, 391–409 (1985)
24. Leydesdroff, L.: Words and co-words as indicators of intellectual organization. Res. Policy **18**, 209–223 (1989)
25. Callon, M., Courtial, J.-P., Turner, W.A., Bauin, S.: From translations to problematic networks: An introduction to co-word analysis. Soc. Sci. Inf. **22**, 191–235 (1983)
26. The Comprehensive R Archive Network. https://cran.r-project.org
27. Package "twitteR". https://cran.r-project.org/web/packages/twitteR/twitteR.pdf
28. Package "Text Mining in R". https://cran.r-project.org/web/packages/tm/vignettes/tm.pdf
29. Package "ggPlot2". https://cran.r-project.org/web/packages/ggplot2/ggplot2.pdf
30. 3,100 teachers, administrators gather to promote STEM education. http://edsource.org/2015/3100-teachers-administrators-gather-to-promote-stem-education/89840
31. American Association of School Librarians. http://national.aasl.org
32. Mattel is Using 3D Printing to Resurrect an Old Hit. http://fortune.com/2016/02/12/mattel-3d-printing-toys
33. Adafruits Industries. https://www.adafruit.com/

34. Google Trends. https://www.google.com/trends/
35. Package "gtrendsR."
36. Package "WikipediaTrend." https://cran.r-project.org/web/packages/wikipediatrend/wikipediatrend.pdf
37. Technology: Print me a Stradivarius. http://www.economist.com/node/18114327
38. Bougie, R., Pieters, R., Zeelenberg, M.: Angry customers don't come back, they get back: the experience and behavioral implications of anger and dissatisfaction in services. J. Acad. Mark. Sci. **31**, 377–393 (2003)
39. Ortony, A., Clore, G.L., Foss, M.A.: The referential structure of the affective lexicon. Cogn. Science. **11**, 341–364 (1987)
40. Kakkonen, T., Kakkonen, G.G.: SentiProfiler: creating comparable visual profiles of sentimental content in texts. In: Language Technologies for Digital Humanities and Cultural Heritage, vol. 62 (2011)
41. Suero Montero, C., Munezero, M., Kakkonen, T.: Investigating the role of emotion-based features in author gender classification of text. In: Gelbukh, A. (ed.) CICLing 2014, Part II. LNCS, vol. 8404, pp. 98–114. Springer, Heidelberg (2014)
42. Strapparava, C., Valitutti, A.: WordNet affect: an affective extension of WordNet. In: LREC. pp. 1083–1086 (2004)
43. Jain, A.K.: Data clustering: 50 years beyond K-means. Pattern Recogn. Lett. **31**, 651–666 (2010)
44. Boyd, D., Golder, S., Lotan, G.: Tweet, tweet, retweet: Conversational aspects of retweeting on twitter. In: System Sciences (HICSS), 2010 43rd Hawaii International Conference on. pp. 1–10. IEEE (2010)
45. Fischer, G.: Understanding, fostering, and supporting cultures of participation. Interactions **18**, 42–53 (2011)

Semantic Integration of Web Data for International Investment Decision Support

Boyan Simeonov[1], Vladimir Alexiev[1], Dimitris Liparas[2(✉)],
Marti Puigbo[3], Stefanos Vrochidis[2], Emmanuel Jamin[4],
and Ioannis Kompatsiaris[2]

[1] Ontotext Corp, Sofia, Bulgaria
{boyan.simeonov,vladimir.alexiev}@ontotext.com
[2] Information Technologies Institute,
Centre for Research and Technology Hellas, Thermi-Thessaloniki, Greece
{dliparas,stefanos,ikom}@iti.gr
[3] PIMEC, Barcelona, Spain
mpuigbo@pimec.org
[4] Everis, Barcelona, Spain
emmanuel.jean.jacques.jamin@everis.com

Abstract. Given the current economic situation and the financial crisis in many European countries, Small and Medium Enterprises (SMEs) have found internationalisation and exportation of their products as the main way out of this crisis. In this paper, we provide a decision support system that semantically aggregates information from many heterogeneous web resources and provides guidance to SMEs for their potential investments. The main contributions of this paper are the introduction of SME internationalisation indicators that can be considered for such decisions, as well as the novel decision support system for SME internationalisation based on inference over semantically integrated data from heterogeneous web resources. The system is evaluated by SME experts in realistic scenarios in the section of dairy products.

Keywords: Decision support · Indicators · Heterogeneous web resources · SME internationalisation · Semantic integration

1 Introduction

Given the current economic situation and the financial crisis in many European countries, SMEs have found internationalisation and exportation of their products as the main way out of this crisis. To this end, SMEs need to find relevant information that will facilitate this process such as: (i) spending habits of consumers in potential markets, (ii) economic fundamentals of the countries (micro and macro indicators), (iii) geographic and entry barriers (legislation, certifications, etc.), (iv) consumer behaviour, (v) domestic and foreign competition, (vi) distributors of its product to export in the selected markets, (vii) contact information of potential customers. In order to find this information, SMEs have to access foreign trade offices in each country (e.g. Chambers of Commerce), dedicated databases (e.g. Market access database, Eurostat, etc.), as well

© Springer International Publishing AG 2016
F. Bagnoli et al. (Eds.): INSCI 2016, LNCS 9934, pp. 205–217, 2016.
DOI: 10.1007/978-3-319-45982-0_18

as the web by using general purpose search engines (e.g. Google). In these resources, the companies expect to find the competence, potential clients and all the information that is required to take a decision for exporting products to the right country. However, this method is time consuming and complicated, because information is distributed and heterogeneous and there is no existing platform that provides access to all the necessary information. The additional problem faced by many SMEs is the language barrier, since this information is usually provided only in the language of the host country. To deal with this problem, we need technologies that provide unified access to multilingual and multicultural economic material across borders in order to guide the international investments of SMEs.

In this paper we present a decision support system that semantically aggregates information from many heterogeneous web resources and provides guidance to SMEs for their potential investments. The main contributions of this paper are the introduction of SME internationalisation indicators that can be considered for such decisions, as well as the novel decision support framework for SME internationalisation based on inference over semantically integrated data from heterogeneous web resources. The SME internationalisation indicators are chosen between different data sources. They provide information about four main domains - Economy, Social, Politics and Product. These domains represent the most important aspects of the current situation in a given country. Each individual indicator is picked with the help of people working in SME organisations and is ranked according to their view on its importance. To the best of our knowledge, this is the first attempt to develop a decision support system for SME internationalisation.

The rest of this paper is organised as follows: Sect. 2 provides some theoretical background, as well as an overview of the related work. Section 3 describes the SME internationalisation indicators that are utilised by the decision support framework presented in this paper. The different components of the proposed framework are described in detail in Sect. 4. In Sect. 5, the experimental results from the application of the framework to data collected from several resources are presented and discussed. Finally, some concluding remarks are provided in Sect. 6.

2   Related Work

Decision support systems (DSSs) can be broadly defined as computer-based applications that support people and organisations in their decision-making processes. Research on this very important scientific field has spanned 50 years and many different kinds of systems have been presented. According to [1], DSSs can be divided into the following main categories:

- Model-driven DSSs: These include computerised systems that employ accounting and financial models, representational models, and/or optimisation models to assist in decision-making [2]. One representative system in this category is ILOG JRules [3]. Using model-driven DSSs can lead to substantial benefits, such as the reduction in decision process cycle time.

- Data-driven DSSs: These systems aim at accessing and processing large amounts of data. Simple file systems accessed by query and retrieval tools provide the most elementary level of functionality in this category [4]. A nice example of data-driven DSSs is the Geographical Information System (GIS), which can be used to visually represent geographically dependent data using maps. Among other things, data-driven DSSs can provide improved data accessibility and fact-based decision making.

- Document-driven DSSs: Multimedia document collections serve as the backbone of the decision-making process in document-driven DSSs. Document analysis and information retrieval (IR) systems are simple examples from this category [5]. Improved information flow and flexible document retrieval are some of the advantages of using document-driven DSSs.

- Communication-driven DSSs: These systems aim at supporting groups of people working on a given task, by focusing on the interaction and collaboration aspects of decision-making. At its basic form, a system of this type can be a simple threaded e-mail and in its complex form, it can be an interactive video or a web communication application.

- Knowledge-driven DSSs: They actually recommend or suggest actions to the users, rather than just retrieve information relevant to a certain decision, i.e., these systems try to perform some part of the actual decision-making for the user through special-purpose problem-solving capabilities [5]. The use of knowledge-driven DSSs can result in more consistent decisions and can reduce the time needed to solve problems [6]. It should be noted that the framework proposed in this paper belongs to the knowledge-driven DSS category.

A detailed literature review with respect to the research conducted on the use of semantic web technologies in the DSS context can be found in [5]. A large number of semantic web-related studies have focused on the medical and healthcare domains. For instance, [7] explore the use of Web Ontology Language (OWL) reasoning services, in order to execute clinical practice guidelines (CPG) in clinical decision support systems (CDSSs). In [8], a generic architecture for the semantic enhancement of CDSSs, which also considers the reutilisation of knowledge embedded in a CDSS, is proposed. Another prominent research domain is e-commerce. In this context, [9] introduce a Semantic Web Constraint Language (SWCL) based on OWL and utilise it, in order to implement a shopping agent in the Semantic Web environment. Furthermore, [10] design and develop a shopbot that can help customers compare products located in e-stores, using different languages. In order to achieve this, they propose a semi-automatic method for constructing multilingual ontologies, as well as a semantic searching mechanism based on concept similarity. In another approach [11], the authors present a system that provides high quality environmental information for personalised decision support based on reasoning.

In addition to the aforementioned works, there have been some DSS-related studies that deal specifically with financial management and investment decision-making. More specifically, [12] employ the Object Oriented Bayesian Knowledge Base (OOBKB) design to develop a real-time DSS that supports managing of investment portfolios. In another work, [13] presents a hybrid intelligent system that consists of a

DSS based on portfolio management rules, as well as a fuzzy inference system. Finally, a detailed analysis of tools implemented in DSS to support individuals in their financial management and investment decisions is provided in [14].

This paper, inspired by the ontology-based decision support systems, such as [9] and [11], presents a knowledge-driven DSS for SME internationalisation based on semantic integration of heterogeneous internet data.

3   SME Internationalisation Indicators

In most cases, the main issue for SMEs that want to internationalise is to assess the different countries that could be potentially interesting for exporting their products. The selection of the correct country for the international investment depends on a number of indicators, which allow a comparison and therefore help the decision-making process. In order to build the decision support tool, we need to combine the indicators in a sophisticated manner. To do that, we have to initially conduct a screening and establish a categorisation of the indicators so that we can prioritise and weight them according to their relevance. The grouping considered captures a framework of macroenvironmental criteria, which is considered in the strategic management of SMEs when assessing opportunities or threats. The study for the definition of indicators has taken place in the context of the MULTISENSOR project[1] with the support of the SME internationalisation department of PIMEC[2].

To analyse, select and organise the indicators and its grouping, we considered the PEST analysis together with other indicators more focused on the product. SMEs export managers with a vast experience on internationalisation assessment participated in the elaboration of the methodology of the decision support. As we are dealing with targeting foreign markets, the decision support tool needs to consider external factors. In this sense, the PEST analysis refers to the combination of Political, Economic, Social and Technological factors which can affect the business. Within this framework, contrasted indicators from reliable sources – e.g. Eurostat[3], World Bank[4] – were selected so that the final result is robust.

For a more complete and personalised feature, SME export managers agreed that there was a need for another group of indicators, directly related to the small company that is looking for a market to export. For this purpose, we included UN COMTRADE data, which measures trade for every product and set of products. Thus, the decision support tool incorporates the selected product data, so that the outcome is not only the result of general factors but also, and decisively, of the concrete product commercialisation.

Altogether, we obtain a combination between product specific and personalised data with country factors. Indeed, every indicator and group of indicators have a

---

[1] http://www.multisensorproject.eu/.
[2] http://www.pimec.org/.
[3] http://eurostat.linked-statistics.org.
[4] http://worldbank.270a.info/.

different importance when making a decision. Hence, the decision support tool integrates a differentiated weight for every indicator and category according to SME export managers' criteria. More specifically, we grouped the indicators into 4 categories:

- **Product**: This is the key category in our system, as it is directly connected with the product the SME is producing or offering. Here, we include UN COMTRADE data, which gives precise information of the export/import flows between countries worldwide, segmented by product in the Harmonised System code. We take into account trade between the targeted country and the rest of the world, as well as trade with the country of origin of the SME running the decision support tool. Furthermore, we incorporate a variable that measures the trade flows per capita. Lastly, distance between countries is included; a value that is given different weight depending on the product, as the type of good conditions the importance of a fast delivery. In all, the Product category captures the balance of trade of the specific products and brings very relevant information to compare the economic and consumption specificities of the countries in relation to SMEs' commercialisation.

- **Economy**: This category combines macroeconomic data that gives an input of the countries' recent economic performance – e.g. GDP growth, GDP per capita – together with general balance of payments data. Also, the Easiness Doing Business ranking from the World Bank plays a major role, as it describes how difficult it is to do business in every country, including variables such as how complex it is to start a business, to what extent contracts are enforced and how easy it is to trade across borders. Hence, we obtain a sum of the health of the countries' economies, its overall trade flows and its business and legal culture.

- **Politics**: Membership of economic areas and trade agreements constitute the basis of this category. In this sense, the main variables are being or not a member of the European Union, the European Economic Area or the European Free Trade Association, which affect commercial trade decisively. Additionally, legal certainty aspects are also included with indexes that reflect political stability, levels of corruption and government effectiveness.

- **Social**: This category has the lowest weight and embraces societal aspects that describe the market dimension and consumption possibilities of the countries with selected indicators such as their total population, their level of education or their unemployment rate. Furthermore, we include an economic perception index that gives us the sentiment that consumers have towards their economy, which can affect their preferences when it comes to consumption.

In Table 1, we present the most important SME internationalisation indicators categorised and weighted.

4 Decision Support Framework

The proposed decision support framework consists of the following main components (Fig. 1): Indicator information mining from the web, semantic integration of this data in a semantic repository (database) and the decision support mechanism. Guidance is provided through a user friendly interface.

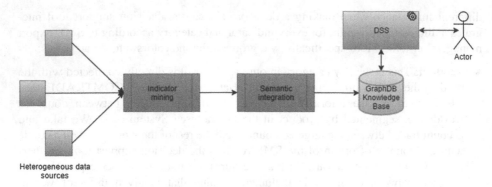

Fig. 1. Data flow (architecture) of the decision support system

4.1 Web Retrieval and Mining of Indicator Data

In order to extract information on the SME internationalisation indicators we use dedicated APIs from specific websites, such as Eurostat and WorldBank and UN COMTRADE (see Sect. 5.1 for details). Then we convert this information to RDF and load it in Ontotext GraphDB[5], a semantic repository that can store and query semantic data. We selected the W3C CUBE ontology [15], which is the most appropriate way of representing statistical data in RDF format. We also use elements from SDMX [16], which provides terms for some common dimensions such as population, GDP, etc.

Some sources are already available in the required format [17], e.g. data from Eurostat and World Bank. For other sources, e.g. UN COMTRADE[6], we had to use a specific API. We used the free API that has the following limits: 100 requests per hour per IP address, and maximum of 50 k records per request (each record represents one trade flow between two partners for one product group). This provides plenty of data (potentially up to 120 million records per day), but we had to make sure that none of our requests exceeded the limit of 50 k. We did this through judicious selection of dimensions and downloading strategies (e.g. how many Year series and Product codes to retrieve at a time). To collect this data, we developed a program that queries the COMTRADE API repeatedly by varying the parameters, downloads the data and converts it to the required format.

In addition, we used the Google distance API to extract information regarding city distances in various transportation modes (e.g. air vs surface). We used the Google distance API, which is free for 2.5 k requests. In order to stream-line the process, we prefetched the distances between the capitals of all source and destination countries, and converted them to RDF using a custom program.

With respect to other indicators such as Government type, Political instability, Corruption percent index, Human development index, etc., we downloaded them

---

[5] http://ontotext.com/products/graphdb/.

[6] http://comtrade.un.org/.

manually from various web sites and converted them using a mix of automatic and manual approaches. They are provided in different tabular formats, such as web pages, csv, tsv, etc.

4.2   Semantic Data Integration

Data integration from disparate data sources is often required for online analytical processing (OLAP) and DSS analysis. In recent years, semantic data integration [18] has emerged as the most promising integration approach, because of the simple and uniform data model that it uses (RDF). RDF is a graph data mode, in which data is broken into atomic facts called triples. URLs are used to identify every block of data and every property (relation or attribute). This allows data sharing on a global scale.

Reusing property names and values defined by others (in our case, SDMX and Eurostat) is one of the benefits of the semantic web. It both reduces the time required to create data models and the chance of data reuse by others. After converting the data to RDF and loading it to Ontotext GraphDB, we calculate some derived data using SPARQL UPDATE [19]. Then, for each pair of observations IMP/EXP with matching dimensions, we calculate two derived observations: TOT (total trade) and BAL (trade balance). We record them in URLs that mirror the original URLs, where the Indicator part is replaced appropriately.

Since we deal with many indicators in different areas, each indicator has its own value range and direction of growth (for some indicators increase is desirable, for others decrease is desirable). To perform meaningful calculations over this heterogeneous data, we need to normalise data to the range between 0 and 1. For this purpose, we are using the following formula for the normalization of rates, which adjusts values measured on different scales to a notionally common one:

$$z_i = x_i - min(x)/(max(x) - min(x)) \tag{1}$$

where $x = (x_1,...,x_n)$ and z_i is the ith normalised data.

4.3   Query-Based Decision Support

After indicator data is converted to RDF and loaded into the knowledge base, it can be used for decision support. The user selects his country of origin and the desired commodity code (product group). We use the following mechanisms:

- Weighted sum of indicators (see Table 1 for the weights, which are empirically selected).
- Simple inference rules, e.g. "If the commodity code indicates a perishable product then use air cargo distance; if it's a heavy product then use surface cargo distance; else compare both".
- Simple decision tree elements.

We implement these mechanisms using parameterised SPARQL templates to retrieve the appropriate data from the cube. This backend processing allows us to implement the following features:

- Rank target countries (and select the most appropriate one)
- Compare two countries across all indicators, e.g. we can compare Germany and Austria across GDP, GDP per capita, corruption level, human development index, population purchasing power, etc.
- Display various charts to illustrate the time dimension (we use the Google Charts API).

Using these functionalities, a manager or entrepreneur can obtain solid information to support the decision making process. It should be noted that at the moment the system has some limitations. It can't be customised, which means that there is no way the user can change the weights of the indicators or disable some of them.

5   Experiments

In this Section, we present the experimental results from the application of our decision support framework to data collected from several heterogeneous web resources.

5.1   Dataset Description

The following data were retrieved and integrated for this experiment: (i) Eurostat dataset, including statistical indicators for Europe; (ii) World Bank dataset, which includes World Bank Indicators; (iii) UN COMTRADE, which offers comprehensive data on cross-country trade volumes for various types of goods, collected since 1962.

Other indicators obtained from specific sources such as: (i) Political Stability and Absence of Violence/Terrorism measures perceptions of the likelihood of political instability and/or politically motivated violence, including terrorism. (Worldwide Governance Indicators - The World Bank Group); (ii) Corruption perception index (Transparency International). In the table below you can see the main selected indicators, divided by categories. Each indicator is assigned an individual weight (w1). The indicator categories are also assigned with weight w2.

5.2   Experimental Setup

In order to evaluate the proposed system, we have conducted experiments at PIMEC in Barcelona [20]. The evaluation involved a focus group of five (5) export managers, as well as one-on-one interviews with three (3) SME export responsible and one (1) export manager. This focus group was used to test the individual functionalities of the system. First we provided an initial explanation of the status of the system and described the functionalities that were available for testing. The participants were asked to evaluate a table of indicators where the user could compare two countries and visualise their differences. Specifically, the task was the following: each user was supposed to be the

Table 1. SME internationalisation indicators and weights

Group	Indicator	w1	w2
Economy	GDP growth	6	6
	GDP per capita PPP	5	
	Exports in % of GDP	3	
	Imports in % of GDP	3	
	Inward FDI stocks in % of GDP	3	
	Export Import ratio	3	
	Harmonised Index of Consumer Prices (Inflation)	3	
	Easiness of Doing Business	10	
	Average days to Export	6	
	Average days to Import	6	
Social	Total population	8	3
	% Tertiary education	3	
	Unemployment rate	5	
	Economic Sentiment Indicator	3	
	Human Development Index	3	
	% Internet at Home	3	
Politics	EU Member	10	5
	EEA Member	10	
	EFTA Member	10*	
	Political Stability Index	3	
	Corruption Perception Index	3	
	Government Effectiveness Index	3	
Product	Export (partner: World)	5	10
	Import (partner: World)	10	
	Export (partner: country of origin)	5	
	Import (partner: country of origin)	8	
	Product balance (Export – Import/Partner: World)	8	
	Product balance (Export – Import/Partner: country of origin)	6	
	Imports per capita (Import divided by Population/Partner: World)	8	
	Imports per capita (Import divided by Population/Partner: Country of origin)	6	
	Distance	10	

CEO of an SME selling dairy products (e.g. cheese and yogurt), who wanted to decide which country is the most convenient for exporting products. In Fig. 2, we illustrate the user interface of the decision support system, in which different indicators for Slovenia and Czech Republic are compared[7].

---

[7] A demo of the DSS is available at: http://grinder1.multisensorproject.eu/uc3/.

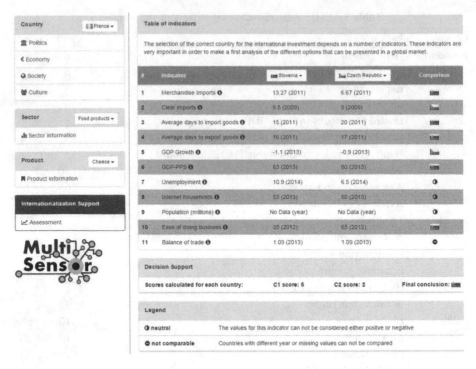

Fig. 2. Decision support interface

5.3 Results

In general, the results of the performed evaluation can be considered satisfactory. Regarding the task, where the users were asked to evaluate the table of indicators and the decision support functionality (Fig. 3), the table was valued positively by all users as a way to compare the situation in two countries more easily. In addition, all users valued its relevance positively. Larger number of indicators and more concrete ones were mentioned as a way to better capture the advantages or disadvantages of exporting to one country or another.

After evaluating concrete aspects of the different tasks, the evaluators were required to value the overall efficiency of the decision support platform. The results are visualised in Fig. 4 and as we can see, the opinions were highly positive. All the participants felt that the system was easy to use and, also, that it allowed them to save time compared to alternative ways of looking for similar information.

Regarding satisfaction (Fig. 5), all the participants (100 %) felt in control when using the decision support system and thought that it was intuitive and easy to use. Participants appreciated the user interface and navigability of the platform. A vast majority (80 %) considered its use as a satisfying experience and a way to be more productive (65 %). The majority of the participants (55 %) would also recommend the system to colleagues. In general, most of the participants acknowledged that the MULTISENSOR DSS has a good potential, but could also benefit from additional

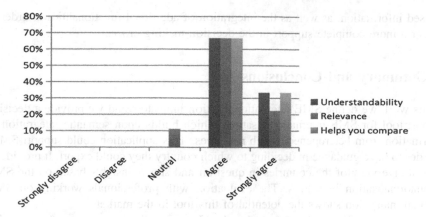

Fig. 3. Evaluation of indicators table and decision support functionality

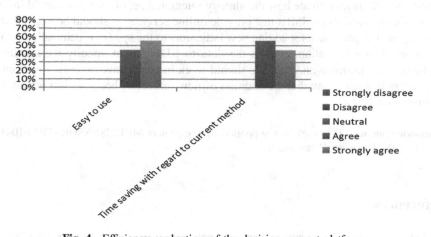

Fig. 4. Efficiency evaluation of the decision support platform

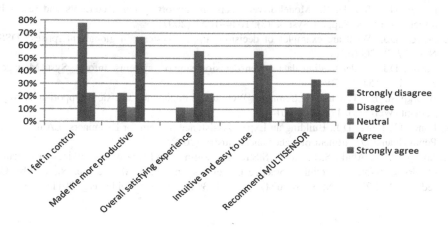

Fig. 5. Satisfaction evaluation of the decision support platform

focused information, as well as the integration of additional functionalities in order to deliver a more complete support in the decision-making process.

6 Summary and Conclusions

In this work we define SME internationalisation indicators and we provide a decision support tool for SME internationalisation, which builds upon semantic integration of information from heterogeneous web resources. This application could support SMEs in order to have guidance in deciding to which country they could export. It provides a comparative view of the countries in question and shows insights based on the SME internationalisation indicators. The evaluation with professionals working on SME internationalisation shows the potential of this tool in the market.

As future work, we plan to crawl and add more indicators to the system, employ additional techniques for providing guidance, such as decision trees and fuzzy reasoning, as well as investigate how the already integrated set of indicators could inform Internet-based services or SMEs that provide online services. It should be noted that at the moment, the procedure for adding new indicators to the system is demanding (crawl new sources of information, convert the data to RDF format, modify the system to handle the new information, etc.). We should work in perspective, in order to automate the process in such a way that the addition of new indicators becomes just a matter of configuration.

Acknowledgments. This work was supported by the project MULTISENSOR (FP7-610411), funded by the European Commission.

References

1. Power, D.J.: Decision support systems: concepts and resources for managers. Stud. Inform. Control **11**(4), 349–350 (2002)
2. Power, D.J., Sharda, R.: Model-driven decision support systems: concepts and research directions. Decis. Support Syst. **43**(3), 1044–1061 (2007)
3. Power, D.J.: What are examples of decision support systems in global enterprises? DSS News, **7**(7) (2006)
4. Power, D.J.: Understanding data-driven decision support systems. Inform. Syst. Manage. **25**(2), 149–154 (2008)
5. Blomqvist, E.: The use of Semantic Web technologies for decision support–a survey. Semant. Web **5**(3), 177–201 (2014)
6. Pontz, C., Power D.J.: Building an Expert Assistance System for Examiners (EASE) at the Pennsylvania Department of Labor and Industry (2002)
7. Jafarpour, B., Abidi, S.R., Abidi, S.S.R.: Exploiting OWL reasoning services to execute ontologically-modeled clinical practice guidelines. In: Peleg, M., Lavrač, N., Combi, C. (eds.) AIME 2011. LNCS, vol. 6747, pp. 307–311. Springer, Heidelberg (2011)

8. Sanchez, E., Toro, C., Carrasco, E., Bueno, G., Parra, C., Bonachela, P., Graña, M., Guijarro, F.: An architecture for the semantic enhancement of clinical decision support systems. In: König, A., Dengel, A., Hinkelmann, K., Kise, K., Howlett, R.J., Jain, L.C. (eds.) KES 2011, Part II. LNCS, vol. 6882, pp. 611–620. Springer, Heidelberg (2011)
9. Kim, H.J., Kim, W., Lee, M.: Semantic Web Constraint Language and its application to an intelligent shopping agent. Decis. Support Syst. **46**(4), 882–894 (2009)
10. Huang, S.L., Tsai, Y.H.: Designing a cross-language comparison-shopping agent. Decis. Support Syst. **50**(2), 428–438 (2011)
11. Wanner, L., Rospocher, M., Vrochidis, S., Johansson, L., Bouayad-Aghae, N., Casamayor, G., Karppinen, A., Kompatsiaris, I., Millee, S., Moumtzidou, A., Serafini, L.: Ontology-centered environmental information delivery for personalized decision support. Expert Syst. Appl. **42**(12), 5032–5046 (2015)
12. Tseng, C.C., Gmytrasiewicz, P.J.: Real time decision support system for portfolio management. In Proceedings of the 35th Annual Hawaii International Conference on System Sciences, HICSS, pp. 1348–1356. IEEE (2002)
13. Casanova, I.J.: Portfolio investment decision support system based on a fuzzy inference system. In: Madani, K., Dourado Correia, A., Rosa, A., Filipe, J. (eds.) Computational Intelligence. SCI, vol. 399, pp. 183–196. Springer, Heidelberg (2012)
14. Weber, B.W.: Financial DSS: Systems for supporting investment decisions. Handbook on Decision Support Systems 2, pp. 419–442. Springer, Heidelberg (2008)
15. Cyganiak, R., Reynolds, D.: The RDF Data Cube Vocabulary. W3C Recommendation, 16 January 2014 (2014). http://www.w3.org/TR/vocab-data-cube/
16. Statistical Data and Metadata eXchange. Official site for the SDMX community. https://sdmx.org/. Accessed Apr 2016
17. Capadisli, S., Auer, S., Riedl, R.: Towards linked statistical data analysis. In: Proceedings of the 1st International Workshop on Semantic Statistics, Sydney, Australia, 11 October 2013. CEUR vol. 1549 (2013). http://csarven.ca/linked-statistical-data-analysis
18. Alexiev, V., Breu, M., de Bruijn, J., Fensel, D., Lara, R., Lausen, H.: Information Integration with Ontologies: Experiences from an Industrial Showcase. Wiley, New York (2005). ISBN 978-0-470-01048-8
19. Gearon, P., Passant, A., Polleres, A.: SPARQL 1.1 Update, W3C Recommendation 21 March 2013 (2013). https://www.w3.org/TR/sparql11-update/
20. Heise, N., Wagner, T., Eckhoff, M., Vrochidis, S., Peleja, F.: MULTISENSOR Second Prototype Evaluation Report (2015) http://www.multisensorproject.eu/wp-content/uploads/2013/09/D8.4_2ndPrototypeEvaluationReport_v1.0.pdf

An Analysis of IETF Activities Using Mailing Lists and Social Media

Heiko Niedermayer$^{(\boxtimes)}$, Daniel Raumer, Nikolai Schwellnus, Edwin Cordeiro, and Georg Carle

Chair of Network Architectures and Services, Department of Informatics, Technical University of Munich, Munich, Germany
{niedermayer,raumer,cordeiro,schwelln,carle}@net.in.tum.de

Abstract. The Internet Engineering Task Force is an open organization that produces Internet Standards. In this paper we look at Twitter and IETF mailing lists to answer questions on IETF participation and social media usage and IETF reaction to societal events: Are Internet Standards discussed on Twitter? Who is involved in the process? Do external events like Snowden revelations in 2013 correlate with related IETF activities? To answer this, we look in particular at security-related activities at the IETF like in the TLS working group. With respect to the Snowden leaks, we quantify the impact in terms of increase in activity and contributors in related areas.

Keywords: Social media · Standardization · Twitter · IETF · Social informatics · Internet science

1 Introduction

From its beginnings as dedicated research network, only accessible by an exclusive subset of people, the Internet has developed into an important catalyst to societal development. Being open to everyone and everyone's technology mankind is still witnessing new applications and fields to apply them each day. Within these efforts different stakeholders put efforts into refining the Internet and its underlying technologies by defining new standards. Although the definitions of what can be named a standard differs vastly, all have in common to define technologies and techniques to be used in order to achieve a level of agreement that improves interoperability for some purpose.

Standards may be called de facto standard with a positive attitude or quasi standard with a rather negative attitude. Powerful stakeholders, often market leaders or exclusive groups controlling a market, may either explicitly or implicitly define standards to protect their position. An explicit standard definition is usually made by open or closed documents and implicit standards are made by designing technologies or programs that are used by a majority. By ironing or adopting those standards, the masses constitute the second step for defining a standard: adoption.

© Springer International Publishing AG 2016
F. Bagnoli et al. (Eds.): INSCI 2016, LNCS 9934, pp. 218–230, 2016.
DOI: 10.1007/978-3-319-45982-0_19

Within this setting, the Internet Engineering Task Force (IETF) and its affiliated organization the Internet Research Task Force (IRTF) are an institution that is focused on improving the Internet and related technologies [2]. The IETF does that by means of RFCs that define standards and best practices. Formed in 1986 the IETF started as quarterly meetings of researchers funded by the US government. The meetings were opened to the general public in October 1986 and remained this way ever since [5,7]. The IETF remained supported by government funds until 1997 [5], since July 1995 the IETF is supported by the Internet Society (ISOC) [6].

Without formal membership or membership requirements, their contributors work on a voluntary basis. Anyone interested in contributing to the development of Internet protocols may participate. Despite that open policy, missing technical expertise or financial background, companies' policies, and formation of exclusive subgroups within the IETF may limit the openness. These factors may crush the will and possibility to contribute and constitute a problem for the IETF [8].

Our contribution is as follows: In this paper we want to put a flash light onto certain activity aspects of the IETF. We use social media data from Twitter as well as mailing list data from the IETF mailing lists. These mailing lists are the work horse of the IETF. With that we analyse interest into certain topics over time as well social media usage and presence of IETF standards. In Sect. 2, we discuss basic processes of the IETF standardisation. Section 3 contains the analysis and we conclude with Sect. 4.

2   Background

To coordinate the work of the people active in the IETF, the activities are organized in working groups, each one collaborating on a specific topic. We also have areas, like Internet or routing areas, that are formed by groups with similar topics. Working groups are created with a specific goal detailed in its charter, each has their own discussion mailing list and one or more working group chairs. After having fulfilled its purpose, a working group closes or is rechartered to add new goals [4].

A reasonably complex structure is necessary to guarantee the quality of the standards created by this open organization. The first part is the Internet Society, that provides financial and legal support for the IETF and the Internet Research Task Force (IRTF). The Internet Engineering Steering Group (IESG) is composed of all area directors and the IETF chair, all chosen for a 2-year renewable term by the Nominating Committee, in a complex but publicly verifiable process [12]. The IESG is responsible to manage the process of creating Internet Standards and other IETF activities. It works by guiding the process and not by making decisions about the standards.

The Internet Architecture Board (IAB) is an oversight committee for the "architecture of the Internet and its protocols" [11]. The IAB reviews the charters of new working groups before they are created and is expected to keep an eye on the "big picture" [10].

Over their lifetime working groups will produce a number of documents, with the goal to publish them as *Internet Standards* or *Informational* documents. The Standards Process of the IETF works as follows:

A working group – after reaching rough consensus with its participants – forwards a draft document to the IESG. The IESG issues a last call for comments for all working groups in all areas and if the ideas or methods of the draft document reach rough consensus in the IETF as a whole and the IESG does not have concerns, the draft will be forwarded to the RFC-Editor for publishing. If the IESG or the IETF community does have concerns or the ideas and methods of the draft document did not reach rough consensus in the IETF, the working group will update the draft document to begin the process again, or the document is abandoned by the working group.

3   Analysis of IETF Activities

The IETF follows a well defined process based open discussions. With its open architecture the IETF constitutes an eligible example to study actors of standardisation. Discussions in the IETF mostly take place on mailing lists, but also at the IETF meetings (three times per year) and interim working group meetings (scheduled when a working group needs it). Therefore, different events can be counted to determine the activity of a working group whereof mail exchange is the most continuous event. Jari Arkko [3] maintains a website with numerous statistics based on the official IETF documents. As this database only contains the official sources and the website only presents selected statistics we have built up our own database that we want to extend by further sources of social activity.

3.1   IETF Mailing Lists

Figure 1 shows the top 5 mailing lists IETF working groups of 2015 and their activity according to the absolute number of sent mails. Here and in following graphs we extrapolated the activity for 2016 on behalf of the first 3 month activities.

Each mailing list serves for discussion of its working group. Each group has an agenda published on the web. Shortly summarized the working groups listed in our graphs focus on the Transport Layer Security standard (TLS), the operation of Domain Name Systems (DNSop), data modeling for information exchange with network devices (netmod), deployment and operation of IPv6 networks (v6ops), maintenance and development of the Hypertext Transfer Protocol (httpbis), the Kerberos authentication protocol (krb-wg), the HTTP extension WebDAV (webdav), and mail security (ietf-dkim).

A few words on methodology: As spam mails do not reflect any activity we have filtered these in our analysis. We based the decision which mails to filter on the X-SpamScore field. This field is generated by Apache SpamAssassin [13] and present in the public mail archives. In the following top 5 selections we excluded special purpose IETF mailing lists like *iesg-agenda-dist* and *ipr-announce* as these do not reveal any insights into the interactions of individuals.

The SpamScore is generated by Apache SpamAssassin using rules that if hit, add or subtract a certain value to the score. Rule examples include message body with mostly blank lines, containing keywords like 'valium', 'viagra' or 'million dollars' and user is in the whitelist (Full list: [14]). On the other hand, mails with subject containing 'draft-' are very unlikely to be spam as it is probably a reference to a certain draft document, same holds true for squarebrackets referring to a certain mailing list, as well as network related keywords. Additionally, we did not interpret mail that are an answer to a non spam mail (denoted by the 'RE: *' subject) as spam.

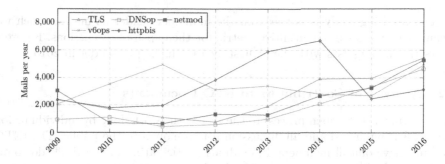

Fig. 1. Top 5 mailing lists of 2015 with most mails

Pure mailing list activity only constitutes one dimension of activity. Therefore, we also plotted the top 5 (of 2015) activity charts for counts of unique senders in Fig. 2. The degrading of the netmod (from position 3 to 17), v6ops (4 to 11), and httpbis (5 to 8) mailing list are an artefact of the different types of discussions on the mailing lists: focused discussions that are conduced by a small group but in a quite interactive and responsive manner and statement-based discussions with different authors.

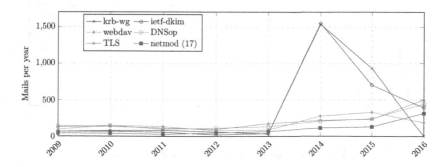

Fig. 2. Top 5 mailing lists of 2015 with most unique senders

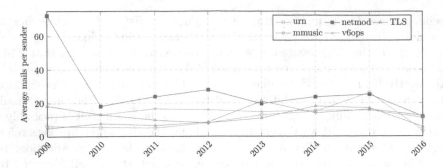

Fig. 3. Top 5 mailing lists with highest average mails per sender in 2015

The resulting analysis is based on the mails per sender as metric which we assume to be a naive but intuitive metric for the depth of discussions. Figure 3 shows the mailing lists with the highest average mails per sender in 2015.

3.2 Analysis of Contributors to RFC Documents

To conform with the open process in the IETF, the work done by individuals for the IETF is published and all discussion lists are open to the public. The IETF does not have official members. Individuals participating in the IETF do so on a voluntary basis. Therefore, a pay check affiliates contributors to a company or institution and entails a certain possibility of influence by the affiliated company or institution. Figure 4 shows the affiliation of document authors as stated in the RFCs. The interpretation of these authorship affiliations has to be made with care for two reasons: Specification of affiliation is voluntary according to the IETF and some persons may be affiliated twice, e.g. a professor affiliated to a company and a university. Similar statistics are generated by Jari Arkko [3].

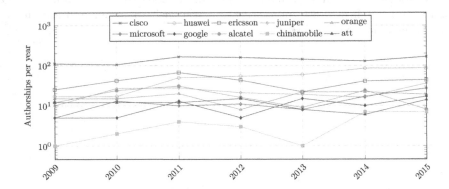

Fig. 4. Actors in the IETF standardisation

Cisco is constantly over many years the company that contributes most to the writing of Internet Standards. In recent years Huawei has taken over the 2nd place, similar to its own rise to global player in the market. Now, the authors of actual standards tend to have such company affiliations and usually there are 2 to 4 such authors. If you look at the IETF mailing lists that discuss these standards, you will find much more activity. Figure 12 shows the number of distinct persons contributing to the TLS mailing list at the IETF. The TLS mailing list is currently (2014–2016) defining the new standard TLS 1.3 for secure communication over the Internet. The discussions on the TLS mailing list involves on average 60 and up to 100 distinct people each month. Here we see the amount of other volunteers who may have completely different affiliations. So, while important companies may seem to be relevant for the actual writing of a standard, many more volunteers from industry, academia, or independent contribute to how the standard solution will look like.

3.3 IETF-Related Activity on Twitter

Despite the official channels for discussion it is likely that discussions will also happen in side channels that are not directly made available by the IETF. Social networks are possible side channels for such discussions. Questions to be answered in that context include: Are there any IETF-related posts or tweets? Do they get liked or shared? Therefore, we utilized that IETF standards documents are called RFCs and they have a name and a number. E.g. RFC 791 is the original standard for the Internet Protocol (IP) from 1981. Recent standards have numbers between 7000 to 8000.

For our analysis we monitored the visibility of Internet Standards (IETF RFCs) on Twitter. We searched Twitter for tweets that contain the terms IETF, RFC-EDITOR, or RFC and a number, and then evaluate if it is an IETF-related tweet and if we can infer an RFC number from it. For the evaluation we use a combination of whitelisting (direct link to IETF or RFC-EDITOR sites) and blacklisting (common terms found in non-IETF tweets, e.g. from Football and Rugby clubs, user names referring to RFCs). Since Twitter only returns the most recent tweets for a search term, we regularly monitored Twitter to continuously log the most recent tweets since September 2015.

Figure 5 shows the activity of tweets related to Internet Standards in the time from September 2015 to February 2016. The average number of tweets per day is approximately 33. The peak early November is during IETF 94 meeting in Yokohama on November 1–6, 2015. No other IETF meeting occurred during the reported period.

Considering the size of Twitter and the large number of Internet Standards 33 tweets per day, do not seem a lot. If favourites and retweets are considered the activity doubles. There are on average 12 retweets and 18 favourites per day. Nonetheless, Twitter is not the center of IETF activity.

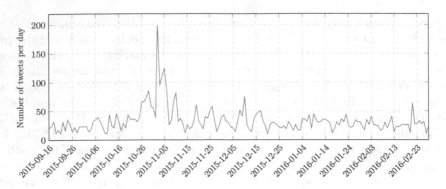

Fig. 5. Tweets per day; between September 2015 and March 2016

3.4 RFC Popularity on Twitter

Since we now know there is activity with respect to IETF standards on Twitter, we take a look at the standards that are discussed. Table 1 shows which Internet Standards got the most activity in February 2016. The first column shows the sum of tweets, retweets, and favourites.

Table 1. Twitter Statistics in February 2016

Activity	RFC	Name
109	6920	Naming things with hashes
103	822	STANDARD FOR THE FORMAT OF ARPA INTERNET TEXT MESSAGES
64	2324	Hyper text coffee pot control protocol (HTCPCP/1.0)
58	3546	Transport layer security (TLS) extensions
47	4204	Link management protocol (LMP)
44	7366	Encrypt-then-MAC for transport layer security (TLS)
38	1149	Standard for the transmission of IP datagrams on avian carriers
36	7748	Elliptic curves for security
34	2549	IP over avian carriers with quality of service
34	2119	Key words for use in RFCs to indicate requirement levels

A closer look at the tweets related to the RFC documents reveals the following relations: RFC 6920 is a proposed standard (not yet full Internet Standard) from April 2013. Most of the tweets simply name and link to the standard. Others additionally refer to the standard as being useful for their privacy protection. RFC 822 gets the tweets from people implementing related mail software. RFC 2324 is one of the famous April Fools RFCs, here a web protocol for the control of a coffee pot. The same is true for RFC 1149 and RFC 2549 on Internet over avian carriers like birds. RFC 3546 is present because of one tweet being favourited and retweeted a lot. It embraces the 13 year old standard and refers to discussions (e.g. in the context of new TLS versions) to abandon the standard, which the tweet authors is an opponent of. RFC 4202 is present due to a Twitter bot that keeps tweeting about the standard and other arbitrary things regularly. The bot is a follow-back bot. RFC 7366 is a standard from late 2014, a single tweet asks about TLS implementations supporting the standard and rest is retweets and favourites. RFC 7748 is a recent standard from January 2016. RFC 2119 is a relevant old standard about word usage at IETF.

Similar listings can be produced for other periods of time in our measurement. In summary we see the following categories of RFC-related activity on Twitter:

- RECENT: promoting a comparably new document or standard (e.g. also for further discussion to make it full standard)
- RELEVANT: discussing an older RFC due to some current relevance
- FUN: sharing April Fools RFCs for fun (e.g. Hyper Text Coffee Pot Control Protocol (HTCPCP/1.0))
- BOT: a bot posting random tweets (one of which refers to RFC 4204). For the further analysis, we ignore the category BOT as it is only one bot posting older RFCs over and over again.

Socially-relevant RFCs like RFC 6920 on privacy-friendly naming and RFC 7686 on Tor's .onion domain name (leading in October 2015 when the RFC was published) seem to be more visible in the top lists than arbitrary other recent RFCs.

In the following, we dive into the three main categories. For methodology: From 4-week-intervals ranging from October 2015 to end of February 2016, we took the top 10 most active RFCs of each month and further analysed their activity (tweets, retweets, favourite).

Figure 6 shows the events of all tweets from the RECENT category that made it into at least one 4-weekly top 10. 17 RFCs fall into this category. Dots indicate an event. The additional line indicates the months of publication of each RFC. As we do not have the data when tweets get retweeted and favourited, these are not contained in Fig. 6.

A closer look at the activity data reveals: While some of the RFCs still show little activity after their initial peaks, these subsequent Twitter activities are very few and orders of magnitude lower than the initial peak. So, while the overall interest shows that these new RFCs are interesting to many people, there is yet no sign of viral activity. Compared to the initial peak, the activity afterwards declines to close to 0 in comparison. The large peaks, however, make it the category with most activity (2439 activities in the time interval, 143 per RFC).

Figures 7 and 8 show activities for the RELEVANT categories which includes older RFCs that receive some interest. In this case, only 7 RFCs made it into the list. In contrast to the tweets in the RECENT category that mostly announce RFCs, here there is usually a reference to an older RFC due to some relevance to a topic of interest, e.g. which IP addresses are private (RFC 1918 in the

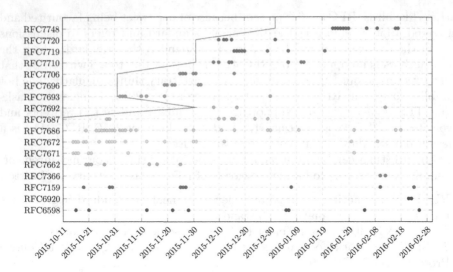

Fig. 6. Publication date of RFC and times when tweets about RFCs in category RECENT occurred

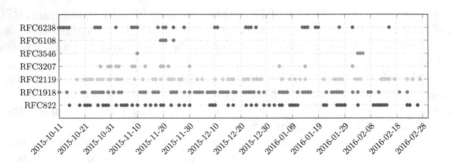

Fig. 7. Times when tweets about RFCs in category RELEVANT occured

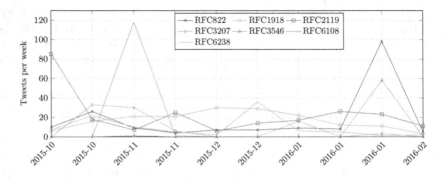

Fig. 8. Tweets per week in category RELEVANT

graph). The dots in Fig. 7 show a much more continuous pattern over time than the related plot from the RECENT category. However, Fig. 8 shows significant spikes of interest for some and more continuous interest for other RFCs. From the three categories, RELEVANT is the one that has the lowest activity (1119 activities in the time interval, 160 activities per RFC).

Figures 9 and 10 are the related graphs for the FUN category that consists of standards published as April Fools RFCs on some April 1st. There are also 7 RFCs that made it to this list during the time we observed Twitter. Results displayed in Figure 9 follow a similar pattern as those in Fig. 7. However, the graphs in Fig. 10 reveal a more continuous interest into the individual RFCs than in the RELEVANT category. Here, both figures indicate a viral behaviour (for 6 of the 7 RFCs in the graph). From the three categories it has the 2nd highest activity (1314 activities in the time interval, 187 activities per RFC).

Figure 11 shows the amount of tweets, retweets, and favourites for each of the three categories. The RECENT category received a lot of retweets and favourites, but fewer actual tweets. The RELEVANT category has a similar percentage of favourites, but more actual tweets. The FUN category is dominated by tweets.

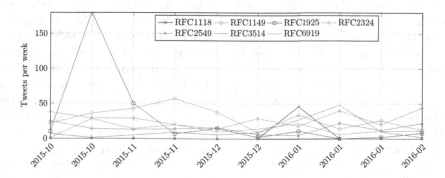

Fig. 9. Times when tweets about RFCs in category FUN occurred

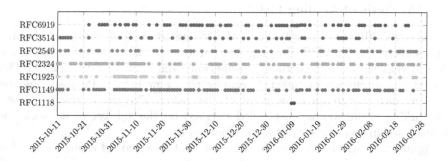

Fig. 10. Tweets per week in category FUN

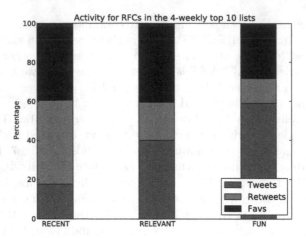

Fig. 11. Twitter activity type distribution concerning RFCs from the top 10 lists

3.5   Impact of Snowden Revelations on the IETF (Analysis of Mailing Lists)

On June 06, 2013 the Guardian and the Washington Post published the first articles about the NSA PRISM program. On June 09, 2013 the information was released that they obtained their information from Edward Snowden [9]. This brought the discussion of Internet security to a new height. Pervasive passive attackers have become a focus point of discussion. The IETF founded a new non-working group mailing list called PerPass on the subject of network and protocol design to mitigate pervasive monitoring [1].

Figure 12 shows the PerPass mailing list that was started a few weeks after the revelations and the security related working group mailing lists of DANE and TLS. We marked August 2013, the month of the Snowden leaks by dashed lines (right ones). As Fig. 12(a) shows, the number of mails immediately went into the orders of an important active mailing list like the TLS mailing list. In comparison the Snowdon leaks have no direct impact on the DANE mailing list. Furthermore, Fig. 12(b), below shows the number of different people that contributed each month. And indeed, a lot of different people contributed which shows the general interest into the subject. In the first half of 2014, however, the interest slowed down. There is still some activity, but it is minor compared to the initial peaks in late 2013. Nonetheless, the figure also shows increasing activities in the TLS working group that sets the standards for the most important security protocol of the web (TLS, which is, e.g. used as security in HTTPS).

When the previous version of TLS (version 1.2) was standardised in 2008, the mailing lists activities were only around 50 mails per month. Thus, the increase in interest cannot be explained with the standardisation of the next version (version 1.3). However, a similar peak with even up to 1097 mails per month (November 2009, left dashed line in Fig. 12) was reached for a short period of time end of 2009 when a severe security flaw was found in TLS which was fixed

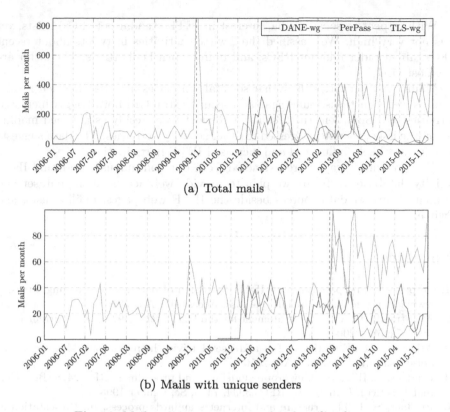

(a) Total mails

(b) Mails with unique senders

Fig. 12. Mail activity on security-related mailing lists

with RFC 5746 in February 2010. Interestingly, the number of distinct senders (different people) that contributed on the mailing list in the interval only peaked at 64 different senders, which is roughly the number it currently converged to as normal number of distinct contributors. The all-time monthly peak was 100 in April 2014.

The IETF had to react to the Snowden revelations. However, concluding from the statistics of our observation, we can see that IETF reaction to the revelations of Edward Snowden was indeed profound and spawned a lot of interest and participation in related IETF activities. Using the TLS mailing list again, Fig. 12 shows that its activity before and around 2008 was much smaller, even though the previous TLS standard was produced back then.

4   Conclusions and Outlook

The authors of IETF standards are usually associated with an Internet-related tech company, Cisco and Huawei being currently the largest contributors. Looking only at the actual authors is, however, misleading as a much larger amount of people takes part in the discussions of the standardisation process.

With respect to social media, interest in IETF standards (RFCs) exists, yet it is not very high. We classified the Twitter activities in tweets about recent RFCs, about older relevant RFCs, and funny April Fools RFCs. The latter are the most viral.

Mailing list activities show that societal influences like the Snowden revelations as well as attacks found against a security standard both trigger increased activities. For Snowden, the first increase went to a freshly established mailing list on the issue, then it progressed to other security activities. The data suggests that it most likely influenced and increased IETF participation.

The presented graphs are extracted from our continuously updating IETF activity database. In future we plan to provide web access to the dataset and to include further data sources beside the IETF web pages, mailing lists, and Twitter.

References

1. Perpass - the perpass list is for IETF discussion of pervasive monitoring. https://www.ietf.org/mailman/listinfo/perpass. Accessed 1 Apr 2016
2. Alvestrand, H.: A mission statement for the IETF. RFC 3935 (Best Current Practice), October 2004
3. Arkko, J.: IETF statistics: fun and useful trivia. http://www.arkko.com/tools/docstats.html. Accessed 1 Apr 2016
4. Bradner, S.: IETF working group guidelines and procedures. RFC 2418 (Best Current Practice), Updated by RFCs 3934, 7475, September 1998
5. Bradner, S.: IETF structure and internet standards process. In: Presentation at the 81st IETF (2011)
6. Cerf, V.: IETF and the Internet Society. Webpage, July 1995. http://www.internetsociety.org/internet/what-internet/history-internet/ietf-and-internet-society
7. Ockman, S., DiBona, C.: Open Sources: Voices From the Open Source Revolution. O'Reilly Media, Sebastopol (1999)
8. Crocker, D., Clark, N.: An IETF with much diversity and professional conduct. RFC 7704 (Informational), November 2015
9. Gidda, M.: Edward snowden and the NSA files timeline, August 2013. http://www.theguardian.com/world/2013/jun/23/edward-snowden-nsa-files-timeline Accessed 1 Apr 2016
10. Hoffman, P., Harris, S.: The tao of IETF - a Novice's guide to the internet engineering task force. Webpage as specified in RFC 6722 (2012)
11. Hovey, R., Bradner, S.: The organizations involved in the IETF standards process. RFC 2028 (Best Current Practice), Updated by RFCs 3668, 3979, October 1996
12. Kucherawy, M.: IAB, IESG, and IAOC selection, confirmation, and recall process: operation of the nominating and recall committees. RFC 7437 (Best Current Practice), January 2015
13. The Apache SpamAssassin Project: SpamAssassin. http://spamassassin.apache.org. Accessed 1 Apr 2016
14. The Apache SpamAssassin Project. Tests performed: v3.3.x. http://spamassassin.apache.org/tests_3_3_x.html

Assessing Media Pluralism in the Digital Age

Iva Nenadic[✉] and Alina Ostling

Centre for Media Pluralism and Media Freedom (CMPF),
European University Institute, Florence, Italy
{iva.nenadic,alina.ostling}@eui.eu

Abstract. The purpose of this paper is to discuss challenges of measuring media pluralism and freedom in the digital age. We do this while presenting the updates of Media Pluralism Monitor implemented by the Centre for Media Pluralism and Media Freedom in three consecutive cycles (2014, 2015 and 2016). The paper explores methodological limitations and other pressing issues in regard to optimal assessment of risks to media pluralism in the digital environment.

Keywords: Media pluralism · Digital age · Methodology · Internet

1 Introduction

Media pluralism cannot be measured without taking into account the developments in the digital sphere. However, with all the complexity and ongoing change of the digital environment, it is very challenging to select the main indicators and to apply the right method in assessing the level of media pluralism and freedom. Usually the approach and tools used for traditional media settings do not apply in new participatory environment. Hence, the conduct exercised in the digital sphere should be taken into account while analysing and measuring various digital phenomena.

To address the need to measure and assess the extent of media pluralism and freedom in the digital environment, the Centre for Media Pluralism and Media Freedom (CMPF) – to which the two authors are affiliated - has developed a number of indicators that are being tested through the Media Pluralism Monitor,[1] a tool for assessing risks to media pluralism in the EU and beyond. The Monitor adopts an interdisciplinary approach by examining media pluralism from legal, socio-political and economic perspectives.

This paper has two objectives. Firstly, it discusses the progress made so far regarding the development and testing of the Monitor indicators in the field of digital media pluralism in 28 EU member states and in two candidate countries (Montenegro and Turkey). Even though our analysis necessarily maintains a macro perspective, assigning values to countries, the Monitor indicators aim at focusing on the complex and often less visible aspects of digital transition, such as the evaluation of digital safety of journalists and national applications of the net neutrality principle. During the process of MPM extension in the digital sphere, some pressing issues and key methodological challenges were noted. The aim of the paper is to describe the monitoring process and

---

[1] http://monitor.cmpf.eui.eu/.

© Springer International Publishing AG 2016
F. Bagnoli et al. (Eds.): INSCI 2016, LNCS 9934, pp. 231–238, 2016.
DOI: 10.1007/978-3-319-45982-0_20

to discuss the particular opportunities and challenges of assessing the digital aspects of media pluralism.

1.1 Methodology

In the last three years the Centre for Media Pluralism and Media Freedom (CMPF) has been re-designing the Media Pluralism Monitor, originally developed in 2009[2], in order to improve its applicability and to update the indicators considering the growing digital influence. The media pluralism in online environment is not assessed as a separate phenomenon by the CMPF but addressed by capturing digital risks throughout the different indicators of the Monitor. The Monitor addresses digital dimensions of pluralism in all of its four domains, which represent the main dimensions of media pluralism: Basic protection, Market plurality, Political independence, and Social inclusiveness.

The data collection is carried out by a network of experts in the Member States, Montenegro and Turkey, and is coordinated by the CMPF. The country experts carry out desk-based research by consulting and referencing relevant documentation (e.g. academic literature, civil society reports and legislative acts) and, in some cases, by interviewing relevant experts. For certain indicators, the country teams have to make an evaluative assessment based on their expert knowledge (e.g. to assess how effective a specific policy is). A number of particularly sensitive and complex indicators also go through an external peer review. The country experts enter the answers and sources in an online database. The CMPF research team verifies and carries out an analysis of the inserted data.

A standardised formula is used to calculate risk on each indicator and area. The quantitative and qualitative answers are calibrated on a scale from 0 to 1, with scores closer to 0 pointing to a low risk assessment, and those closer to 1 to a high risk assessment. The answering options also include 'not applicable' and 'no data'. The answers coded as 'not applicable' are excluded from analysis and do not contribute to the risk scores. The 'no data' coding was introduced in 2015 primarily since previous MPM implementation showed that economic data was missing across many EU member states. Following the choice of the 'no data' answer, country teams are asked to evaluate whether the lack of data represents a transparency problem within their national context since there is a variety of reasons why certain data is not available across EU member states, and not all of these reasons are causes for concern.

---

[2] Media Pluralism Monitor (MPM) was originally developed by the Katholieke Universiteit Leuven – ICRI, Central European University – CMCS and Jönköping International Business School – MMTC, together with a consultancy firm, Ernst & Young Belgium and subcontractors in all Member States. Subsequently, the European Commission has awarded several grants to the Centre for Media Pluralism and Media Freedom to conduct pilot implementations of the MPM.

2   The Approach of the Media Pluralism Monitor to Assessing the Digital Dimensions of Media Pluralism

The participatory nature of the Internet has a potential to facilitate bottom-up forms of content production and hence to contribute to pluralism of information and viewpoints. However, an excessive concentration of control over the network infrastructure or content production, dissemination and consummation poses a risk to Internet's pluralistic potential.

During the three testing rounds of the Media Pluralism Monitor since 2014,[3] the digital dimensions of media pluralism have grown in importance. In the 2016 edition of the Monitor, variables with a digital dimension span over all four of the Monitor areas, ranging from issues such as basic protection of freedom of expression online and of the journalistic profession to ownership, net neutrality and access to as well as skills necessary to use digital media. The following sections discuss the content of the specific Monitor variables and the approach used to measure them.

2.1   Freedom of Expression Online

Freedom of expression (FoE) is considered to a fundamental aspect of media pluralism. The Media Pluralism Monitor assesses regulatory safeguards for FoE and focuses especially on whether restrictions to FoE in the law are defined in accordance with international and regional human rights standards, and if they are proportionate. The current debate in academia, international organisations and among practitioners, as well as the provisions on rights and freedoms set out in the European Convention on Human Rights, emphasize that FoE should apply equally online and offline. In line with these provisions, the Monitor assesses whether FoE online is limited on the same grounds as FoE offline (see variables 10 and 11 in Appendix).

2.2   Safety of Journalists

An issue closely related to FoE is the safety of journalists, and especially online threats to journalists are a matter of growing concern over the past years. In MPM2015, most countries showed high or medium risk both in terms of physical and digital safety of journalists. The MPM2016 assesses the presence of threats to the digital safety of journalists defined as illegitimate surveillance of their searches and online activities, their email or social media profiles, hacking and other attacks by state or non-state actors.

---

[3] In 2014, the Monitor assessed the following nine countries: Belgium, Bulgaria, Denmark, Estonia, France, Greece, Hungary, Italy and the UK. In 2015, 19 EU countries that were not covered during the first pilot phase in 2014 were assessed: Austria, Croatia, Cyprus, Czech republic, Finland, Germany, Ireland, Latvia, Lithuania, Luxembourg, Malta, Netherlands, Poland, Portugal, Romania, Slovakia, Slovenia, Spain and Sweden. http://monitor.cmpf.eui.eu/mpm2015/results/#download.

2.3   Net Neutrality and Intermediaries

The Monitor also examines the legal basis for net neutrality and does a reality check of what is actually happening in practice. Firstly, we ask whether there are regulatory safeguards regarding net neutrality in the countries examined. In particular, if there are safeguards for impartial transmission of information without regard to content, destination or source. Regulatory safeguards are defined in a broad sense as laws, regulations or case law, and decisions of authorities, e.g. as policy measures to avoid blocking of certain Internet content and applications; policies to avoid quality discrimination between content and service providers; and regulation on the information of the quality of the services offered by the Internet service providers (ISPs); and transparency obligations concerning discriminatory practices in ISP services.

The 2015 results showed that most countries (13 of 19) do not have any regulatory safeguards regarding net neutrality. We are asking the same question in MPM2016 in order to assess whether there has been any regulatory improvement over time in EU:19 and whether the other EU:28 countries and the two candidate countries show similar results on net neutrality regulation.

To do a reality check on net neutrality, MPM2016 assesses if the state and the ISPs are filtering, monitoring, blocking or removing online content in an arbitrary way. In addition, the extent of concentration of ISPs in the country is evaluated by considering market shares of the top 4 ISPs. The risk of ISP concentration is tangible; the 2015 implementation revealed over 80 % concentration in most of the 19 examined countries.

2.4   Internet Access and Digital Skills

Moving from the basic safeguards for FoE online by the state and the role of digital intermediaries, the Monitor also assesses the Internet access and digital skills of the general population. Firstly, the MPM2016 examines Internet coverage and access, as well as trends in Internet usage. The indicators examine how much of the population is covered by broadband, in particular in rural areas, and how much of the population subscribe to broadband. The quality of the connection is also taken into account by assessing the average Internet speed in the countries. In terms of Internet usage, the MPM2015 findings[4] shows that the number of individuals regularly using the Internet and using mobile devices to access Internet has increased over the past years.

MPM2015 revealed limited digital skills among the populations examined. Only 26 % of the EU:19 population has basic digital skills (Eurostat 2015). In MPM2016, the digital skills variables have been fine-tuned based on expert consultations. They now examine two dimensions of digital competencies: (1) usage skills, defined as the share of population who have basic software skills, information skills, and problem solving skills; and (2) communication skills includes the following elements: sending/receiving emails, participating in social networks, telephoning/video calls over the Internet, uploading self-created content to a website.

---

[4] Most of the Internet access, coverage and usage variables rely on statistics from Eurostat.

2.5 Political Independence

Political independence area of the MPM2016 assesses political control over the media management and funding, and political interference with editorial autonomy. The development and growth of digital media environment is often not encompassed with adequate regulation and/or self-regulation. Self-regulation offers more flexibility than state regulation in ever changing environment but the traditional journalistic codes of conduct sometimes do not apply to online sphere. Therefore the variable introduced in 2016 edition of the MPM explore whether the existing self-regulatory measures consider online activities of media and individual journalists or media organizations are developing new digital-specific self-regulation to increase accountability and prevent political interference.

3   Outlook: Monitoring Media Pluralism in 2017

The CMPF has tested the Media Pluralism Monitor over two consecutive years (2014 and 2015) and the next results are due at the end of 2016. Over these years, one of the most challenging aspects of monitoring has been the measurement of the digital aspects of media pluralism. The CMPF has discussed digital indicators with a range a stakeholders, including the British media regulator (Office for Communication (Ofcom)), the London School of Economics, the Political Science Media Policy Project at the Reuters Institute for the Study of Journalism, commercial enterprises that specialise in digital audience research, and various experts in the field. Several changes to the Monitor were made as a result of these discussions. However, measuring pluralism in the digital world remains a very challenging task and the optimal methods of measurement are still out of reach. One of the most pressing issues is the lack of reliable and comparative data that would enable assessment of particularly important issues for online pluralism in recent years, including:

- The role of digital intermediaries: There is a lack of comparative data on the citizens' use of intermediaries in news consumption. At the same time, individual studies show growing importance of Google News and Facebook as sources of news (Craufurd Smith et al. 2012). Moreover, policy decisions and content strategies of these intermediaries impact different actors and power relations even beyond the digital sphere.
- Ownership concentration: In particular, it is difficult to get data about market shares of Internet service providers (ISPs) and market shares of Internet content providers (ICPs). For the purpose of MPM, ISPs are defined as companies offering access to the Internet; while ICPs include traditional news media with a presence online, native digital news media, and digital intermediaries (e.g. social media and news aggregators).
- The quality of Internet connection: The sources of data on Internet speed have turned out to be volatile. In 2015, the Media Pluralism Monitor assessed the speed of broadband (upload and download) by using the data of the company NetIndex/Ookla. The data provision has now been discontinued. MPM2016 uses the freely available speed

data from the content distribution management company Akamai[5], which has servers around the world and is reported to handle a large percentage of global Internet traffic.[6] However, given that Akamai is a private company, also this service can be discontinued or be charged for in the future.

- Media and digital literacy is very complex and interdisciplinary issue, which also complicates its measurement. The majority of research at the European level has focused on what is easier to measure, including technical skills, online access skills and media consumption, at the expense of the more complex and critical issues, such as the capacity of people to evaluate and produce media messages (Celot 2015). Hence, comparative data on the ability to interpret and to critically assess online content, as well as on skills needed to contribute the content production and dissemination is lacking.

A pressing issue is also the lack of universal understanding of the key concepts relevant to the media pluralism and freedom in the digital age. In the context of net neutrality, ISPs and ICPs play a significant role. ISPs have the technical capability to monitor user activities and are able to apply blocking of particular content. Under certain conditions, ICPs manage to agree high-speed delivery of their content with ISPs (Belli and De Filippi 2016) and in some cases ISPs can also act as ICPs. Due to the changing nature of these phenomena there is no operational definition and no clear distinction between ISPs and ICPs, which presents a great challenge for measuring pluralism online.

Many other terms, such as digital media, also lack comprehensive understanding. Considering the rapid evolution of media systems it is hard to expect more clarity in the future. Hence, future research should focus on the development of interdisciplinary methods that reflect the very nature of the digital environment. In general, more engagement from diverse stakeholders is needed, and further efforts from regulatory agencies and other authorities in collecting and systematizing relevant data.

Appendix

The MPM2016 Variables Focusing on the Digital Aspects of Media Pluralism

Basic protection

10. Are restrictions upon freedom of expression online clearly defined in law in accordance with international and regional human rights standards?

11. Are the restrictions to freedom of expression online 'proportionate' to the legitimate aim pursued'?

---

[5] Akamai's connection speed measures how quickly (in kilobits per second) data can be transferred from the Internet to a local computer. The data averages all of the connection speeds calculated over a period of time from the unique IP addresses determined to be in a specific country. Faris and Heacock Jones (2013) suggest that the sampling structure of Akamai's data, based on a large proportion of Internet connections, has a more reliable measurement of speeds. Akamai source: https://www.akamai.com/us/en/our-thinking/state-of-the-Internet-report/.

[6] https://www.akamai.com/us/en/our-thinking/state-of-the-Internet-report/.

12. Does the state generally refrain from filtering and/or monitoring and/or blocking and/or removing online content in an arbitrary way?

13. Do the ISPs generally refrain from filtering and/or monitoring and/or blocking and/or removing online content in an arbitrary way?

29. Are there threats to the digital safety of journalists?

47. What percentage of the population is covered by broadband?

48. What percentage of the rural population is covered by broadband?

49. What is the percentage of broadband subscription in your country?

50. What is the average Internet connection speed in your country?

51. What is the percentage of market shares of the TOP 4 ISPs in your country?

52. Are there regulatory safeguards regarding net neutrality in your country?

Market Plurality

84. What is the market share of the Top4 Internet content providers?

85. What is the audience concentration for Internet content providers in your country?

105. Has expenditure for online advertising increased over the past two years?

107. Has the number of individuals regularly using the Internet increased over the past two years?

108. Has the number of individuals using mobile devices to access Internet on the move increased over the past two years?

Social Inclusiveness

143. What is the percentage of population that has at least basic digital usage skills?

144. What is the percentage of population that has at least basic digital communication skills?

Political Independence

166. (165. Are there self-regulatory measures that stipulate editorial independence in the news media?) Do these self-regulatory measures consider online news media?

References

Braman, S.: Where has media policy gone? Defining the field in the twenty-first century. Commun. Law Policy **9**(2), 153–182 (2004)

Belli, L., De Filippi, P.: Net Neutrality Compendium. Human Rights Free Competition and the Future of the Internet. Springer, Switzerland (2016)

Breindl, J.: Internet content regulation in liberal democracies. A literature review. Working Papers Digital Humanities 2. Göttingen Centre for Digital Humanities (2013). http://www.gcdh.de/files/1113/6549/2342/YBreindl_Literature_Review_Mar2013_final.pdf

Celot, P.: Assessing media literacy levels in Europe and the EC pilot initiative. In: EAVI (2015). http://www.eavi.eu/joomla/images/stories/About_EAVI/assessing.pdf

Craufurd Smith, R., Tambini, D., Morisi, D.: Regulating media plurality and media power in the 21st century. In: Broughton Micova, S., Tambini, D. (eds.) LSE Media Policy Project Series, Media Policy Brief 7. The London School of Economics and Political Science, London (2012)

Eurostat: Individual's level of digital skills [isoc_sk_dskl_i] (2015). http://ec.europa.eu/eurostat/data/database?node_code=isoc_sk_dskl_i

Faris R., Heacock Jones R.: Measuring Internet Activity: A (Selective) Review of Methods and Metrics. Harvard University - Berkman Center for Internet & Society (2013). http://papers.ssrn.com/sol3/papers.cfm?abstract_id=2353457

Henrichsen, J.R., Betz, M., Lisosky, L.M.: Building Digital Safety for Journalists: A Survey of Selected Issues. Unesco Series on Internet Freedom. UNESCO, Paris (2015)

Karppinen, K.: Rethinking media pluralism and communicative abundance. Observatorio J. **11**, 151–169 (2009)

Krueger, C.C, Swatman, P.M.C.: Who are the internet content providers? Identifying a realistic taxonomy of content providers in the online news sector. In: The Proceedings of IFIP I3E, Sao Paolo, Brazil (2003). http://www.cimne.com/simweb/formacion/ifipi3e.pdf

Murdoch, S.J., Anderson, R.: Tools and technologies of internet filtering. In: Ronald, J.D., Palfrey, J.G., Rohozinski, R., Zittrain, J. (eds.) Access Denied: The Practice and Policy of Global Internet Filtering. MIT Press, Cambridge (2008)

Responsible Research and Innovation in ICT – A Framework

Žiga Turk[✉], Carlo Sessa, Stephanie Morales, and Anthony Dupont

RRI-ICT Forum Project, Brussels, Belgium
ziga.turk@fgg.uni-lj.si
http://rri-ict.eu/

Abstract. The issue of ethics and responsibility is gaining attention among the creators of scientific policy, funding agencies and society at large. Responsible research is defined as research that aligns both its process and its outcomes with the values, needs and expectations of society. In the EU-funded project "Responsible Research and Innovation (RRI) in Information and Communication Technology" (RRI-ICT Forum), a four dimensional framework for defining and monitoring responsibility has been defined. The four dimensions are (1) actors who are responsible or to whom research is responsible, (2) kinds of responsibility – in what way are they responsible, (3) how much they are responsible and (4) in what area of ICT responsible research and innovation can take place. In this paper we present the simplified version of the framework and initial observations about the ways and degrees in which the researchers in ICT are adhering to it, discuss the future uses of the framework and the dilemmas about responsibility that it is opening.

Keywords: Research policy · Innovation · Responsibility · Social responsibility · Scientific method

1 Introduction

With the increasing power of scientific development and impacts of the new discoveries on the planet in general and on society and humans in particular, the issue of research ethics and responsibility is gaining attention among the creators of scientific policy, researchers and society at large. The topic has been particularly relevant in some fields of life sciences, which are tackling the very fabric of life. But the concept is spreading to other fields as well. Among the reasons is the need of funding agencies and policy-makers to present the case for societal value of research that is being publicly funded.

Responsibility is a broader concept than that of ethics that has been present in life sciences for decades if not centuries. The idea of responsibility has been emerging in European and national research programs for a long time but more intensely in this century. In 2001, the "Science and Society Action Plan" was created. In 2010, "Science in Society (SiS)" emerged. In 2010, the Responsible Research and Innovation (RRI) concept was defined as a response to aspirations and ambitions of European citizens as a part of the effort to better justify the public investment in research and innovation. In

© Springer International Publishing AG 2016
F. Bagnoli et al. (Eds.): INSCI 2016, LNCS 9934, pp. 239–243, 2016.
DOI: 10.1007/978-3-319-45982-0_21

2014, the idea of RRI in ICT made it into the programmatic document of Horizon 2020. The concept made big advances from the baseline idea, which would claim that the only responsibility of research and innovators is to do good quality research.

2   Responsible Research and Innovation

The European Commission defines responsible research as "an inclusive approach to Research and Innovation (R&I), to ensure that societal actors work together during the whole research and innovation process. It aims to better align both the process and outcomes of R&I, with the values, needs and expectations of European society. In doing so, it fosters the creativity and innovativeness of European societies to tackle the grand societal challenges that lie before them, while at the same time pro-actively addressing potential side-effects" [1].

It goes on in saying that "In general terms, RRI implies anticipating and assessing potential implications and societal expectations with regard to research and innovation. In practice, RRI consists of designing and implementing R&I policy that will: engage society more broadly in its research and innovation activities, increase access to scientific results, ensure gender equality in both the research process and research content, take into account the ethical dimension, and promote formal and informal science education".

On the other hand, another European institution, the Economic and Social Committee, stated a concern [2] that RRI might in fact harm the freedom of the mind achieved by the Enlightenment and wrote "What is needed is a fundamental change in social attitudes, so that innovations are not seen primarily as a risk or a threat, but rather as an opportunity for further progress, more jobs and European economic strength and competitiveness, and for shaping the European social model".

The authors are involved in a project called "Responsible Research and Innovation (RRI) in Information and Communication Technology (ICT)", funded by the European Commission as a part of the Horizon 2020 program. The project has been set up acknowledging the major impact that ICT research has on society and aims at monitoring, analyzing, supporting and promoting RRI approach in ICT research in Europe, particularly in Horizon 2020 projects [3]. The goals of the project are to (1) promote a contribution of social sciences and humanities to ICT R&I under Horizon 2020, (2) curate the RRI-ICT domain in H2020 empowering projects and other stakeholders, (3) facilitate the interaction for the emerging RRI-ICT community and (4) create a networking platform – real and in cyberspace – where stakeholders would meet and exchange views.

3   RRI Framework

We are defining the RRI concept through mapping. We present a 4-dimensional map of RRI in ICT. The dimensions define (1) who and to whom are actors responsible, (2) what the responsibility is about, (3) how much responsibility there is and (4) to what topic of ICT the responsibility applies.

The **actors** (who) include researchers, funding agencies, policymakers, educators, students, society at large, all either as individuals or in groups or institutions.

The **kinds** (what about) include epistemic responsibility, procedural responsibility, social responsibility, ethical responsibility and finally the legal & financial responsibility [4]:

- **Epistemic** responsibility is to deliver good science, a responsibility that the community of scientist should take care of for their own deontology and career, by making a proper use of the scientific method and source of knowledge in the research; also includes freedom of thought and pursuit of ideas unlimited by religious or political limitations.
- **Social** responsibility is responsibility to the needs of society and their challenges and about the outside impact of research and innovation. It is primarily a responsibility towards citizens and society that is sometimes channeled into research program priorities, topics and research project goals.
- **Ethical** responsibility is towards a set of established values and norms that in principle represent a "higher being" (i.e. they are beyond the interests and stakes represented by any single actor), but in practice may be identified with the norms and values prevailing in the societal context where research and innovation is done (e.g. with fundamental rights and safety protection levels set by the EU, the UN Chart of Human Rights, etc.). It is responsibility towards the planet, living beings, life etc.
- **Procedural** responsibility is responsibility to ensure an open, inclusive, transparent and fair engagement of all stakeholders (including the citizens) affected by research and innovation activities, the latter in particular to be involved because they contribute as tax payers (when the research is publicly funded) and as prospective users/customers of the research results. It includes openness of research findings and openness of research processes for example to women and minorities.
- **Legal and financial** responsibility is about the contractual obligations the research institutions may establish with funding agencies, about regulations and laws needed to introduce new products and services on the market, and new evidence-based policies enforced through legal acts taken by public authorities.

Levels of responsibility range from being unaware of RRI (as a distinct concept) to establishing systematic procedures to maintain and increase the levels of responsibility within an actor [ibid.]:

- **Not aware** about the idea of responsibility, but perhaps doing something about it intuitively. An example of that would be natural tendency to disseminate research, report it on conferences and events, thus making it more open.
- **Aware** of the idea of responsibility and using that occasionally to improve on it. An example would be to analyze the big societal challenges and include the tackling of them in research that is unfolding.
- Aware and **systematically practicing** responsible ways of doing things. An example of that would be organizational efforts to maintain compliance with rules and regulations.

ICT **topics** are an important dimension because particularly the social and ethical aims may be quite different for different kinds of technologies being developed. Big data research, for example, would have different responsibility implications than, for example, high performance computing.

The four dimensions would come together in charts like the one in Fig. 1 below that would show the **levels** of responsibility across the five **kinds** of responsibility for a given **topic** or **actor**, comparing it to others.

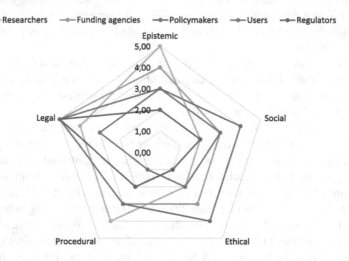

Fig. 1. RRI footprint in ICT structured according to the RRI framework and based on informal impression of the consortium members.

4   Conclusions and Discussion

The concept of responsible research and innovation has been presented. It includes a framework for mapping responsibility along four different dimensions. The concept has been verified so as to think about the responsibility in ICT research along those dimensions. It has been found out that the scheme is generic enough to do so.

Through surveys, concertation events and the first annual forum, we have found out that initial awareness of the RRI concept is **low but exists**. The only aspect that has been seen as internalized (part of the research and innovation process itself) has been the responsibility to do good science. Projects are aware of the formal Commission demands for ethical compliance and are doing it as a part of the project management. They are also aware of the major societal challenges and dilemmas raised by ICT and associate the most prominent ones (e.g. privacy protection) with responsibility.

Nevertheless some **good practices are emerging** such as facilitating public discussion with different kinds of stakeholders in AA1000 standard and MATTER initiative, engaging and listening to citizens to take into account their values, needs and expectations in Engage 2020, VOICES and SOCIENTIZE projects, bringing together multidisciplinary teams to boost creativity and maximize impact in e-Olive project, going for open access, transparency and disclosure of research results in VOA3R project and considering social, ethical and environmental impacts in the DREAM project.

RRI is a research topic in itself. While some work has been done in conceptualizing RRI, it is far from finished. In the future we will need to improve, harmonize and **validate the conceptualization of RRI - the 4D model of RRI**. We would like to establish a method and use it to **measure and benchmark** responsibility in ICT-related H2020 projects. Finally, it will be important to investigate the importance of the various aspects of responsibility and their impact on the public funding of research. When is it enough that good science is made regardless of other aims of responsibility and when not? What are the threshold values and how do the aims of responsibility add up or do they, as vectors, multiply – in which case it is the area covered by the various aspects of responsibility that matters?

Acknowledgements. The work presented has been in part funded by the RRI-ICT Forum EU-funded project.

References

1. European Commission, Science with and for Society. https://ec.europa.eu/programmes/horizon2020/en/h2020-section/science-and-society. Accessed 1 Mar 2016
2. Wolf, G.: Opinion of the European Economic and Social Committee on the communication from the Commission to the European Parliament, the Council, the European Economic and Social Committee and the Committee of the Regions — Research and innovation as sources of renewed growth, (COM(2014) 339 final — SWD(2014) 181 final) (2015/C 230/09), Official Journal of the European Union C 230/59. http://eur-lex.europa.eu/legal-content/EN/TXT/?uri=uriserv:OJ.C_.2015.230.01.0059.01.ENG. Accessed 1 Mar 2016
3. RRI-ICT Forum project webpage. http://rri-ict.eu/. Accessed 1 Mar 2016
4. Sessa, C., Turk, Z., Morales, S.: Deliverable D2.1 - Interaction with projects and analysis. http://rri-ict.eu/wp-content/uploads/2016/01/D2.1-Interaction_with_projects_and_analysis-report_1.pdf. Accessed 1 Mar 2016

End-to-End Encrypted Messaging Protocols: An Overview

Ksenia Ermoshina[1(✉)], Francesca Musiani[2], and Harry Halpin[3]

[1] i3-CSI, MINES ParisTech, Paris, France
ksenia.ermoshina@mines-paristech.fr
[2] ISCC, CNRS/Paris-Sorbonne/UPMC, Paris, France
francesca.musiani@cnrs.fr
[3] INRIA, Paris, France
harry.halpin@inria.fr

Abstract. This paper aims at giving an overview of the different core protocols used for decentralized chat and email-oriented services. This work is part of a survey of 30 projects focused on decentralized and/or end-to-end encrypted internet messaging, currently conducted in the early stages of the H2020 CAPS project NEXTLEAP.

Keywords: Decentralization · End-to-end encryption · Messaging · Protocols · NEXTLEAP

1 Introduction

This exploratory paper first gives an overview of the different core protocols subtending the development of end-to-end encrypted internet messaging (chat and email-oriented) services. In its second part, the paper outlines initial findings of a survey of thirty decentralized and/or end-to-end encrypted projects. The paper also presents the methodological opportunities and challenges of studying such systems with social sciences tools.

Currently, end-to-end encrypted messaging has risen to prominence, with the adoption of end-to-end encrypted messaging by large proprietary applications such as WhatsApp or Facebook Messenger, and the interest in securing communication privacy provoked by the Snowden revelations. In "end-to-end" encrypted messaging, the server that hosts messages for a user or any third-party adversary that intercepts data as the message is *en route* cannot read the message content due to the use of encryption. The "end" in "end-to-end" encryption refers to the "endpoint," which in the case of messaging is the client device of the user rather than the server.

However, open standards for encrypted e-mail and chat are still not seeing widespread use, and a new generation of end-to-end encrypted messaging protocols offering better security properties are rapidly gaining traction, although most are not yet standardized or decentralized. The academic cryptographic community has renewed impetus post-Snowden to rigorously engage with the "secure messaging problem in the untrusted-server model," a problem that until recently "feels almost intentionally pushed aside" although the problem is perhaps "the most fundamental privacy problem in cryptography: how can parties communicate in such a way that nobody knows who said

© Springer International Publishing AG 2016
F. Bagnoli et al. (Eds.): INSCI 2016, LNCS 9934, pp. 244–254, 2016.
DOI: 10.1007/978-3-319-45982-0_22

what" [1]. While the security research community has already began to overview the technical details of these protocols [2], what is missing from the technical work currently in progress is the needs and expectations of users when using encrypted end-to-end messaging applications.

An ongoing study on the use of encryption and decentralized communication tools is being conducted via the H2020 CAPS (Collective Awareness Project) project NEXt-generation Techno-social Legal Encryption Access and Privacy (NEXTLEAP).[1] NEXT-LEAP seeks to address, in an interdisciplinary manner, the recent erosion of public trust in the Internet as a secure means of communication that has been prompted by the Snowden revelations. The core objective of NEXTLEAP is to improve, create, validate, and deploy communication protocols that can serve as pillars for a secure, trust-worthy and privacy-respecting Internet able to ensure citizens' fundamental rights. For this purpose, NEXTLEAP seeks to develop an interdisciplinary internet science of decentralization as the basis on which these protocols can not only be built, but become fully (and meaningfully) embedded in society. In this regard, the social aspect of end-to-end encryption must be included in the overall analysis of trust and decentralization at the heart of Internet Science.

The two main kinds of projects that we seek to examine are related to email and chat clients (also called "instant messaging" clients), both of which are considered to be particular cases of messaging. Historically, e-mail is considered asynchronous messaging, where a user does not have to be online to receive the message, while chat is considered asynchronous messaging, where a user has to be online to receive the message. However, in general these distinctions are blurring as any popular chat protocols now support asynchronous messaging. The remainder of this paper presents a genealogy of these fundamental protocols used for email and chat-oriented services, and then moves on to present preliminary findings and open questions.

2 E-mail Protocols

Email is based on standardized and open protocols descended insofar as the fundamental protocols allow interoperability between different email servers, so that a Microsoft server can send email to a Google server. Classically, as revealed by the PRISM program of the NSA, e-mail is sent unencrypted and so the server has full access to the content of e-mail. Thus, there has been a long-standing program to send email "end-to-end" encryption.

SMTP (Simple Mail Transfer Protocol) is the protocol originally used for transferring email and as such is one of the oldest standards for asynchronous messaging, first defined in 1982 by the IETF[2] and by default not having provision for content confidentiality using end-to-end encryption. PGP (Pretty Good Privacy) was created to add end-to-end encryption capabilities to e-mail in 1991 by Phil Zimmerman. In part due to pressure from the U.S. government, and in part due to patent claims by RSA Corporation,

---

[1] https://nextleap.eu.
[2] https://tools.ietf.org/html/rfc821.

Phil Zimmerman pushed PGP to become an IETF standard. The OpenPGP set of standards was finally defined in 1997, to allow the open implementation of PGP. OpenPGP is implemented in open-source software such as Thunderbird with the Enigmail plug-in as well as in mobile apps, such as the IPGMail for iOS and the Openkeychain key management system for Android and F-Droid. GPG (GnuGPG) is a free software implementation of the OpenPGP standards developed by Free Software Foundation in 1999 and compliant with the OpenPGP standard specifications, serving as the free software foundation for most modern PGP-enabled applications. Recently, the IETF has recently opened up the OpenPGP Working Group in order to allow the fundamental algorithms to be upgraded and to use more modern cryptographic primitives, such as larger keys.

S/MIME[3] is another IETF standard addressing the need for encrypting e-mail. In contrast to PGP that is based on a decentralized "Web of Trust" between users who accept and sign each others keys (and so "offloads" the complexity of key management to the end-user), S/MIME uses a centralized public key infrastructure to manage keys. Thus, while it has been adopted by some large centralized institutions, it has had much less success among the general public and so is not part of the study.

The problem with implementations of OpenPGP such as GPG is that they are difficult for most users to understand and use, especially in terms of usage and key management [3]. These problems extend to security: if an adversary compromises a user's private key, this allows all encrypted messages to be read. In general, while OpenPGP has had a resurgence of interest since 2013, it has not had as much deployment by ordinary users due to the aforementioned issues around user-friendliness and the fact that OpenPGP expects the users to understand the fundamentals of cryptography, such as public and private keys.

3   Chat Protocols

Unlike e-mail that is started as high-latency and asynchronous messaging, chat protocols began as low-latency synchronous messaging, although recently the line has become more blurred as many chat protocols allow asynchronous message delivery. There has long been an intuition that more and more messaging is moving from e-mail to messaging, although it seems that e-mail is still widely used.

XMPP (eXtensible Message and Presence Protocol) became an IETF (Internet Engineering Task Force) standard in 2004 for chat, and is probably the most widely used standardized chat protocol. XMPP is a federated standard that "provides a technology for the asynchronous, end-to-end exchange of structured data by means of direct, persistent XML streams among a distributed network of globally addressable, presence-aware clients and servers."[4] There are many implementations of the XMPP specifications, with the XMPP Foundation giving a list of 70 clients and 25 servers using the XMPP protocol. [5] Jabber.org is the original instant messaging service based on XMPP, and it is now one

---

[3] https://tools.ietf.org/html/rfc3851.

[4] https://tools.ietf.org/html/rfc6120#page-13.

[5] Ibid.

of the biggest nodes on the XMPP network. XMPP traffic or content are not encrypted by default, although network-level encryption security using SASL and TLS has been built into the core. In addition, as claimed by the XMPP foundation, a team of developers is working on an upgrade of the standard to support end-to-end encryption.[6]

The OTR (Off-the-Record) protocol released in 2004 is an extension to XMPP to provide end-to-end encryption. It also provides deniable authentication for users, unlike PGP messages, which can be later "used as a verifiable record of the communication event and the identities of the participants" [4]. OTR is a security upgrade over PGP at least insofar as it does not have long-term public keys that can be compromised. The original paper that defines OTR is called "Off-the-Record Communication, or, Why Not To Use PGP" [4]. The first OTR implementation was a popular Linux IM client, GAIM. At the present moment it is said to be used by 14 instant messaging clients,[7] including earlier versions of Cryptocat (in-browser Javascript client), Jitsi, and ChatSecure (XMPP client for Android and iOS). However, OTR is designed for synchronous messaging between two people, and so does not work for group messaging or asynchronous messaging. There seems to be a move away from OTR; the IM+ app for Android, even though having good user ranking and between 5 and 10 million downloads, is reported by users on Google Play Market as "abandoned" (last update in 2014). A further inquiry will be conducted in order to understand whether the reasons of abandonment are due to usability issues, to cryptographic failures or to other factors such as financial problems, maintenance costs, team conflicts or fusion with bigger projects.

Recently, a number of variations and alternatives to XMPP have been developed:

Matrix.org, released in December 2014, is designed as an "open specification for decentralized communication" using JSON rather than XML. Like XMPP, it is an application-layer communications protocol for federated real-time communication. It is unencrypted by default. However, using the Olm library (Axolotl ratcheting from the Signal Protocol, described below) encryption can be optionally achieved. Among the innovative features of Matrix.org compared to other standards is its interoperability, as underlined in several articles: the main goal of the project being to "create an architecture that tackles the interoperability problems that were not addressed by previous approaches" [5]. Others underline its attractiveness for users that results from this interoperability: "where IRC has a high barrier to entry, requiring you to know exactly what server you're connecting to and configure accordingly, Matrix would let you associate with as many public identities as you're willing to share (phone number, email address, Facebook, Google, and so on), as long as they support the Matrix standard. Otherwise requires no setup – it's just like if you were using any consumer messaging service" [6]. However, Matrix.org seems to have few users since none of the mainstream IM clients relies on it yet. The website lists 17 clients and 6 servers using Matrix.org.[8]

The Signal Protocol, the non-federated protocol developed in 2013 by Open Whisper Systems, provides end-to-end encryption for groups. Moxie Marlinspike, the co-author

---

[6] http://xmpp.org/about/technology-overview.html.

[7] https://otr.cypherpunks.ca/software.php.

[8] And the "Matrix Console" messaging app for Android reportedly has "between 1000 and 5000 downloads".

of Signal, was inspired with some OTR features, such as the idea of ephemeral key exchange [7], but also added additional security features such as future secrecy, support for asynchronous messaging and group messaging, going above and beyond OTR by allowing also clients to be offline. The Signal Protocol uses the "Axolotl" key ratchet for initial key exchange and the ongoing renewal and maintenance of short-lived session keys, so there is not only no long-term key that can be compromised. This provides forward secrecy so that the compromise of a short-term key does not compromise past keys (so that an adversary can decrypt past messages) as well as "backwards secrecy" (also called "future secrecy") so that the compromise of a key does not endanger future messages. It could be a standard, but is not yet recognized as such. The Signal protocol is said to be widely used in mobile messaging applications such as Signal (formerly TextSecure and RedPhone), WhatsApp[9], Secure Chat (by GData). Silent Circle uses a version of the Signal Protocol since 2015 in its Silent Phone. Recently Facebook announced the implementation of Signal Protocol for their Messenger[10]. The first step towards "standardization" of the Signal Protocol so far has been the creation of OMEMO.

OMEMO is a new encrypted extension of XMPP protocol developed in 2015 that effectively copies the Signal Protocol and adopts it to XMPP. It has been presented to the XMPP Standards Foundation but not yet approved in any official manner.[11]10 OMEMO builds upon the work of the Signal Protocol as OTR is said to have "inter-client mobility problems" and can only work when all conversation participants are online, while OpenPGP "does not provide any kind of forward secrecy and is vulnerable to replay attacks" [8]. The software implementations of OMEMO are growing such as conversations, an open-source application for Android that counts over 5000 downloads via Google Play Market, and an unknown number of installs via F-Droid.

4   Network-Level Anonymity

While this work is mostly focused on the application level, it seems important to mention the network-level initiatives, such as P2P routing services or anonymous remailers that can add supplementary privacy properties to end-to-end encrypted messaging. For example, end-to-end encryption does not usually allow a user to be anonymous to the server or third-party without additional network-level encryption. There seems to be no functional standards on this level; however, some solutions, such as Tor or I2P, tend to serve as references for different projects.

The Tor hidden service protocol offers a platform to develop decentralized and encrypted instant messenger servers. It is used by default by projects such as the Tor Messenger, Pond and Ricochet. Another example is the decentralized and end-to-end encrypted mobile messenger Briar that relies on the Tor network when available, but could also work over Bluetooth in case of emergency off-the-grid situations.

---

[9] WhatsApp turns to end-to-end encryption by default in April 2016.
[10] https://whispersystems.org/blog/facebook-messenger/.
[11] https://xmpp.org/extensions/inbox/omemo.html.

Tor provides only anonymity for network addresses, but not metadata such as the sender, recipient, and time of message such as are kept in the email header in the time of email or can be deduced by the server. There has also historically been work on anonymous high-latency remailers to fix these transport meta-data leaks in federated messaging, falling under three types: Cypherpunk Anonymous Remailer, Mixmaster, Mixminion. The latter is not currently active, according to the statement on the official website.[12] The statistics on the website show there are currently 18 Mixminion nodes running - compared to almost 1.2K of Tor routers.

There has been a number of experimental tools developed on the network level that, while not guaranteeing anonymity, provide some level of encryption. Zero Tier One is an end-to-end encrypted, peer-to-peer virtual network that provides static network addresses which remain stable even if the user changes physical WiFi/networks. CJDNS implements a virtual IPv6 network in which all packets are encrypted to the final recipient, using public key cryptography for network address allocation and a distributed hash table for routing.[13]

5   Towards a Set of Criteria for Categorization of Messaging Projects

While some projects are products of wide and well-known communities (such as Open Whisper Systems and Tor), new services either re-use the protocols or infrastructure independently by smaller groups and non-institutionalized developing teams. When standards are not available or not satisfying, there is a tendency to (re)use not yet officially standardized protocols and tools as standards such as Signal's Axolotl ratchet. That is why, taken in consideration this moving nebula of standards and non-standardized projects, we have proceeded with a mapping based on a defined range of criteria. We do not include the case-by-case mapping details in this paper for lack of space[14] but we briefly introduce the criteria that guided the analysis, and discuss some of its preliminary findings.

All of the 30 projects that are included in the mapping[15] are either centralized, with encrypted messages stored on (but not readable by) a central trusted authority, or decentralized, and so not having no central trusted authority for even storing messages. Decentralized systems are either federated (allowing multiple servers, including users setting up their own servers), or peer-to-peer (allowing direct communication between client devices). For the purpose of subsequent investigations with social science methods of ethnography, in-depth interviews and documental research, there are a number of features we seek to identify. We pay particular attention to open source projects,

---

[12] http://mixminion.net/.

[13] https://github.com/cjdelisle/cjdns/blob/master/doc/Whitepaper.md.

[14] The full 30 case studies can be downloaded from https://nextleap.eu.

[15] Briar, Caliopen, ChatSecure, CoverMe, CryptoCat, Equalit.ie, GData, i2P, Jitsi, Mailpile, Mailvelope, ParanoiaWorks, Patchwork, Pidgin, Pixelated, Pond, Protonmail, qTOX, Ricochet, Scramble, Signal, SilentCircle, SureSpot, Teem/SwellRT, Telegram, Threema, TorMessenger, Vuvuzela, Wickr, Wire.

however, business closed-source solutions are also of interest. We take into consideration the kinds of data collected by the applications, as well as the purpose of this collection. For instance, some applications (e.g. Wickr) collect user statistics: anonymous
information about basic usage statistics, such as the number of messages sent by all users
daily, what types of messages users tend to send (e.g., voice messages more often than
text), and so forth. The number of users, their geo-location and the targeted user-groups
must as well be defined (whether the app is optimized for anarchists, journalists, human
right defenders, power-users or developers, enterprises, government...).

An important caveat concerning terminology must be acknowledged here. As
regards (de-)centralization and federation, for the time being, we are referring to technology and algorithms. We should thus distinguish it from the "social federation", i.e.
the question about who controls, at a socio-political level, the instances of servers. For
instance, from this standpoint, Bitcoin is mostly technically decentralized but socially
centralized: there is a single core group creating and delivering the software, while users
effectively run the same software that calculates transactions in a decentralized way. In
order to analyze the (de)centralization of governance/power structures in messaging, we
have to conduct an in-depth investigation. The further ethnographic and sociological
analysis will aim at a deeper understanding of different models of socio-economical
federation these protocols and tools produce. It is an aspect that will be thoroughly
examined in the three in-depth case studies, and we open it up for further investigation
in the conclusions here.

6   Preliminary Findings and Methodological Concerns

This diversity poses a methodological challenge of representation and accuracy, which
will be further delved into as the research progresses; however, for the time being, this
research opens the way to a number of preliminary socio-technical observations of the
end-to-end encrypted messaging field.

Despite the prevalence of free and open source software projects, proprietary software is not absent in this landscape, revealing both a potentially fruitful 'business-to-
business' market for end-to-end encryption and a lack of open-source and standards
adoption by mainstream applications. Open source itself is multi-layered and sometimes
hybrid, with the code on the client side being open source and the server side being
proprietary. Perhaps unsurprisingly, the proprietary features are more important in
applications destined to a business-to-business use, while free and open source software
is predominant for tools destined to activists and tech-savvy users. This transparency of
code and encryption protocols is aimed not only at improving the project, but also at
creating an emulation around the project producing communities of peer reviewers,
experts, beta-testers and advanced users who participate in a collective reflection on the
future of privacy-enhancing technologies.

As we had the occasion to observe in previous mapping research on P2P services [9],
part of the reason why there is such a great diversity and complexity in this field is the
relatively short life span of several projects. While our mapping covers only projects
that are currently active (with one exception, Pond, 'in stasis' albeit not deactivated),

our preliminary research revealed countless others that, after two or three years of pre-beta phase, and sometimes less, stopped development with no evident explanation. While in more than a few cases, the motives behind this are primarily related to a technical experimentation that did not deliver as hoped or expected, a number of additional factors may also be responsible, including the failure to develop an economic model, the internal governance of FOSS development groups, and the inability to rally a critical mass of users around the app (possibly due to a lack of ease-of-use, as discussed below). These socio-technical factors will be useful to observe in the cases eventually selected for the in-depth ethnographic analysis, as a precious source of 'lessons learned' in terms of user recruitment and governance models.

A social perspective is necessary for the design and refinement of technical protocols, with a focus on whether or not users understand and value the various security properties of the protocols. For example, do users understand what a "key" is and forward secrecy? Often protocol designers make assumptions about whether or not ordinary users can understand the security and privacy properties of their protocols. For example, almost all protocols from PGP to Signal use methods such as "out-of-band fingerprint verification" to determine whether or not the recipient of their message really is who they think they are. It is unclear if users actually use these techniques to verify the identity of their contacts. Another example that has been debated in the technical community is deniable authentication. While a protocol may be technically deniable, would this cryptographic deniability hold up socially, much less in court? Answering these kinds of questions influences the kinds of protocols that can be designed by the research community. Lastly, why do only some protocols enable decentralization via open standards? It is unclear if users prefer (or can even tell the difference between) peer-to-peer solutions and centralized services. Between these two extremes, there is the question of how users make trust decisions in open and federated environments such as PGP and XMPP where users could run their own software or delegate this to a trusted group. Answering these questions is vitally important to ground the design of new decentralized protocols and refine existing ones to become decentralized.

The interdisciplinary character of NEXTLEAP project provides us access to several important communities working on improving messaging protocols and encryption, such as the LEAP/Pixelated team, Cryptocat, Open Whisper Systems, Briar, CJDNS, Tor and others. We plan a set of interviews with the teams of three selected projects, as well as observations during important cryptography, decentralization and privacy-related events. We are focusing on both developers and users. Thanks to previous research conducted in the field of activist-targeted technologies, we have connections within several activist user communities in different countries (France, Germany, UK, Austria, Greece, Russia, Mexico, Tunisia, and Lebanon). We will focus on the patterns of adoption/rejection of different messengers/mailing clients, on users' "careers" (e.g. studying usages of encryption and privacy enhancing technologies in dynamic relations to the activist careers and life trajectories), with a specific interest in the so-called "digital migration problem" (shifting from a non-encrypted tool to using end-to-end encryption).

The target audience of the applications is far from being limited to tech-savvy and activist groups; several projects are aimed at widespread use, and user-friendliness appears to be the main issue that stands between this wish and its realization in practice.

Interestingly, in some instances where user feedback is visible on the App Store or Google Play, it shows the 'digital migration'-related issues faced by end-to-end encryption; for example, this model is perceived as problematic because both sender and receiver have to install the app for encryption to take place, which complicates usage.

In the case of civic mobile and web applications studied previously [10], the number of users is explicitly made visible on the websites of the projects. It becomes an important tool for building user communities and empowering the impact of such activist projects. Whereas our analysis of the 30 projects shows that very few projects openly give the number of their active users (possibly due to privacy issues). A further exploration of the three selected cases will investigate these specific politics. In this context, bringing methods of social science to the topic of secure messaging protocols may be useful to elucidate the underlying processes of building user communities.

The analyzed projects propose several solutions to the problem of data storage. Indeed, despite the guarantees of "no personal data collection", some projects still store important amounts of data on the servers (such as usage statistics, device information, keys, usernames or friend relations). Developers tend to explain it by technical requirements (e.g. proposing better user experience based on the collected usage statistics). However, this preliminary inquiry shows that developing communities are aware of the problem and are seeking for alternatives with minimal data storage, and opt for stronger decentralization. The analysis shows that it is the question of metadata that appears to be an area of active research, stimulating experiments with standards and architectures (e.g. Vuvuzela's usage of "noise" to obfuscate metadata discussed in Ref. [11]).

A look at visual aspects, such as the design of interfaces and the design of diagrams and graphics to explain the functioning of the applications, is also revealing of the different publics targeted by the applications and how the developers perceive them. General public-oriented systems use very 'politically neutral' imagery, resorting to the very classical 'Alice and Bob' while stressing that their tools are for 'everyone' (e.g. "sharing photos from holidays"), while tools meant for companies emphasize in both visuals and words the security aspect. Other narratives boast fictional anarchist leaders or real-life activists (e.g. 'Nestor Makhno' or 'Vera Zassulitch'), which also strongly inform the target audience.

A related issue is the powerful 'double' narrative on end-to-end encryption. If on one hand, the discourse on empowerment and better protection of fundamental civil liberties is very strong, several projects show in parallel a desire/need to defend themselves from the "encryption is used by jihadists"-type allegations [12]. This narrative is fueled by previous and current ones about decentralized technologies and peer-to-peer, with their history of allegedly 'empowering-yet-illegal' tools. These issues are taking place in the broader context of discussions about governance by infrastructure and civil liberties [13], some of them particularly related to encryption (or the breaking of it), such as the Apple vs. FBI case and WhatsApp proposing, since April 2016, encryption by default. Thus, the present research hints at something that we will thoroughly address in the in-depth case studies – something a large majority of the projects needs to take into account, and indeed is already taking into account: architecture is politics, but it is not a substitute for politics [14].

7   Conclusions

The overview of the protocols presented in this short paper is focused on stabilizing a list of potential case studies among decentralized internet messaging projects. A further selection of these will be investigated in depth in the future with qualitative methods, including ethnography and in-depth interviews. This is deemed necessary as the proliferation of projects addressing encryption, decentralization, or both, in the field of messaging has not led so far to massive adoption outside of a few large centralized companies such as WhatsApp, for a number of factors that go beyond technology to include difficulty of use, economic sustainability, and unclear socio-legal status of encrypted communication. Thus, the development of a related Internet science requires insight from both social science and ICTs to understand the successes and failures in the design of end-to decentralized protocols.

Considering the lively and constantly-evolving ecosystem of standardized and non-standardized projects in the field of decentralized and encrypted messaging, it is important that a multi-year interdisciplinary effort such as NEXTLEAP starts with a comprehensive mapping of relevant protocols first, relevant projects applying them next, based on a defined range of criteria. This short paper presents a first exploration in this regard, especially peculiar from an interdisciplinary standpoint, inasmuch as it is elaborated by social scientists and is meant to serve their needs in the first place, as a pre-requisite to an in-depth, case study-based inquiry. However, this social science research is ultimately meant to feed back into the development of technical protocols – protocols that are not only technically sound, but made for users and able to find their way into networked societies that are increasingly concerned about the security and confidentiality of their online communications.

References

1. Rogaway, P.: The moral character of cryptographic work. In: IACR Distinguished Lecture at Asiacrypt 2015 (2015). http://web.cs.ucdavis.edu/~rogaway/papers/moral.pdf
2. Unger, N., Dechand, S., Bonneau, J., Fahl, S., Perl, H., Goldberg, I., Smith, M.: SoK: secure messaging. In: 2015 IEEE Symposium on Security and Privacy, pp. 232–249. IEEE (2015)
3. Whitten, A., Tygar. J.D.: Why Johnny can't encrypt: a usability evaluation of PGP 5.0. In: Proceedings of the 8th Conference on USENIX Security Symposium (SSYM 1999), vol. 8, p. 14. USENIX Association, Berkeley (1999)
4. Borisov, N., Goldberg, I., Brewer, E.: Off-the-record communication, or, why not to use PGP. In: Proceedings of the 2004 ACM Workshop on Privacy in the Electronic Society (2004). https://otr.cypherpunks.ca/otr-wpes.pdf, doi:10.1145/1029179.1029200
5. Prokop, A.: Solving the WebRTC interoperability problem. NoJitter (2015). http://www.nojitter.com/post/240169575/solving-the-webrtc-interoperability-problem
6. Weinberger, M.: Matrix wants to smash the walled gardens of messaging. ITworld, 16 September 2014. http://www.itworld.com/article/2694500/unified-communications/matrix-wants-to-smash-the-walled-gardens-of-messaging.html
7. Marlinspike, M.: Advanced cryptographic ratcheting. OpenWhisperSystems, 26 November 2013. https://whispersystems.org/blog/advanced-ratcheting/

8. Straub, A.: "OMEMO Encryption", a protoXEP standards track proposed to XMPP, 25 October 2015. https://xmpp.org/extensions/inbox/omemo.html#intro-motivation
9. Méadel, C., Musiani, F. (coord.): Abécédaire des architectures distribuées, Presses des Mines. In: Musiani, F., Cogburn, D.L., DeNardis, L., Levinson, N.S. (dir.): The Turn to Infrastructure in Internet Governance. Palgrave Macmillan (2015)
10. Ermoshina, K.: Democracy as pothole repair: civic applications and cyber-empowerment in Russia. Cyberpsychol. J. Psychosoc. Res. Cyberspace 8(3) (2014). Article no: 1, doi:10.5817/CP2014-3-4
11. Van den Hooff, L., et al.: Vuvuzela: scalable private messaging resistant to traffic analysis. In: Proceedings of SOSP 2015 (2015). http://dx.doi.org/10.1145/2815400.2815417
12. Sanger, D., Perlroth, N.: Encrypted messaging apps face new scrutiny over possible role in Paris attacks. New York Times (2015). http://www.nytimes.com/2015/11/17/world/europe/encrypted-messaging-apps-face-new-scrutiny-over-possible-role-in-paris-attacks.html
13. Musiani, F., Cogburn, D.L., DeNardis, L., Levinson, N.S. (dir.): The Turn to Infrastructure in Internet Governance. Palgrave Macmillan, Basingstoke (2016)
14. Agre, P.E.: P2P and the promise of internet equality. Commun. ACM 46(2), 39–42 (2003)

Smart Cities and Sociotechnical Systems

Smart Cities and Socio-technical Systems

Making Computer and Normative Codes Converge: A Sociotechnical Approach to Smart Cities

Elena Pavan[1(✉)] and Mario Diani[2]

[1] Institute of Humanities and Social Sciences,
Scuola Normale Superiore, Florence, Italy
elena.pavan@sns.it
[2] Department of Sociology and Social Research,
University of Trento, Trento, Italy
mario.diani@unitn.it

Abstract. In this paper, we propose that a smart approach to city development must seek a continuous and recursive interplay between two levels of codes. On the one hand, computer codes, which rule the development and the functioning of the ubiquitous ICTs infrastructure. On the other hand, normative codes, which govern the practices through which social actors perceive ICTs and decide to exploit them in order to improve their lives. We thus take an exploratory standpoint and investigate to what extent key players in the EU smart cities policy domain are framed according to such a sociotechnical perspective. To this purpose, we first map an online issue network on the topic of smart cities in Europe and then explore the frames that circulate within its core. Our results suggest that, although smart cities are framed sociotechnically, EU key players tend to better converge around technological aspects rather than social ones.

Keywords: Smart cities · Sociotechnical systems · Frames · Online issue networks

1 Introduction

The label "smart city" has rapidly become a passe-partout concept to indicate collaborative projects aimed at virtuously integrating enhanced human and technological possibilities for developing the urban environment and improving the quality of life within it [1]. As noted by [2], different labels have been used to refer to cities "embracing ICTs as a development strategy". However, beyond semantic heterogeneity, smart cities can be broadly defined as "places where IT is combined with infrastructure, architecture, everyday objects and even our bodies to address social, economic and environmental problems" [3]. Thus, smart cities usually present five main characteristics: (i) the widespread embedding of ICTs into the urban fabric; (ii) business-led urban development and a neoliberal approach to governance; (iii) a focus on the social and human dimensions of the city from a creative perspective; (iv) the adoption of a smarter community agenda with programmes aimed at social learning, education and

© Springer International Publishing AG 2016
F. Bagnoli et al. (Eds.): INSCI 2016, LNCS 9934, pp. 257–277, 2016.
DOI: 10.1007/978-3-319-45982-0_23

social capital; (v) and a focus on social and environmental sustainability (Hollands 2008 quoted in [2]).

To be fair, even in the pre-digital age cities always tried to be as "smart" as possible [3]. In fact, governmental and institutional actors, often in partnership with business entities, have always pushed forward strategies to address and, possibly, solve great social and environmental challenges. As the abovementioned definitions suggest, what truly seems to characterize current struggles to urban smartness is the role played by ubiquitous information and communication technologies. In his review of the concept, [2] borrows an expression originally used by [4] and notes that a major understanding of smart cities has been that of

> "urban places composed of 'everyweare'; that is, pervasive and ubiquitous computing and digitally instrumented devices built into the very fabric of urban environments (e.g., fixed and wireless telecom networks, digitally controlled utility services and transport infrastructure, sensor and camera networks, building management systems, and so on) that are used to monitor, manage and regulate city flows and processes, often in real-time, and mobile computing (e.g., smart phones) used by many urban citizens to engage with and navigate the city which themselves produce data about their users (such as location and activity).".

As contemporary ICTs are characterized by a pervasive networked structure and are developed to foster interaction between individuals [5], their ubiquitous presence allows for a more cohesive urban environment and thus guarantees the continuous production of data streams, which preludes to a "real-time understanding" of our cities [2].

This nonetheless, ICTs *on their own* are not conductive of any effect [8]. As [6] note, the "smartness" of urban environments does not only require cutting-edge technology but it is linked to management and policy concerns. In this sense, it is not a "smart technology" but, rather, its "smart use" that makes cities thrive. Existing literature on smart cities translate this idea of smart use in a claim to take into explicit account other elements beside technologies – such as policy and management [6], human social capital and economic policies [1], governance mechanisms that underpin the development of "knowledge economies" [2]. Moreover, social sciences accounts on the dynamic interplay between science, technology and society underline that, in order to master their potential, ICTs need to be "set in motion" by social actors [6]. However, because social actors approach technology depending with a variety of different aims, expectations and skills, they can perceive the same technology as able to afford different functions and, therefore, exploit it in many different ways [4]. Therefore, at the crossroads between human and technological agencies, smart city projects can take many different forms depending on several factors: the type of technologies available and their level of embeddedness in the urban environment, the "future scenarios" that decision-makers envisage for their cities and crystallize within their policies, the role they assign to ICTs within these scenarios, how citizen approach ICTs in their daily lives and turn imagined scenarios into reality.

Because ICTs are a complex object that is defined at the crossroads of innovative artifacts, social activities and governance systems [5], the smartness of our cities depends from both technological and socio-political factors. In turn, such a twofold dependence requires much more than simply *acknowledging* that there are two sides to any smart city strategy or policy. Much more than this, it means rooting any successful smart city project into the *constitutive entanglement* between social and technological elements. As stated by [9]:

> A position of constitutive entanglement does not privilege either humans or technology (in one-way interactions), nor does it link them through a form of mutual reciprocation (in two-way interactions). Instead, the social and the material are considered to be inextricably related — there is no social that is not also material, and no material that is not also social.

In other words, to be genuinely "smart", a project of urban development must not be oriented by any *a priori* assumption on the effects that technology will generate, as its materiality is always mediated by social perceptions of available options, i.e., of affordances. In the same way, an excessive emphasis on social variability should also be avoided. Indeed, a specific materiality does invite users to adopt certain behaviors and, although infinite variations are possible, routines of uses are always likely to emerge.

Building on this background, we propose that a smart approach to city management and development should seek a continuous and recursive interplay between two levels of codes. On the one hand, *computer codes*, which rule the development and the functioning of the ubiquitous ICTs infrastructure that permeates our lives and has the potential to revolutionize them. On the other hand, *normative codes*, which govern the practices through which social actors (i.e., institutions in charge of steering policies but also citizens implementing them) perceive ICTs and decide to exploit them in order to change and, possibly, improve their lives. Thus, we claim that the more these two levels of coding will converge, the more smart cities will be shaped as consistent and coordinated *sociotechnical systems*, wherein ICTs are actually "embedded" within society and there is a constitutive relationship between technological and social networks [7]. Conversely, the more computing and normative coding activities will diverge, the more ICTs will remain simply "pervasive" without necessarily setting an active contribution in shaping (and thus enhancing) the social interactions and practices that innervate our cities.

In order to translate our proposition into empirical analysis, in this paper we propose to investigate the convergence of computer and normative codes according to an analytical framework that targets not so much *how smart cities projects have been implemented* in different contexts but, rather, how actors that are leading their drafting and implementation *understand smart cities*. In inviting to explicitly consider cognitive factors in the explanation of how smart cities projects are put in place and deployed, our framework deeply grounds within social sciences and, more precisely, within social movement and collective action studies (for an example, see [10]). Indeed, social movement scholarship has long insisted on the fact that cognitive elements are crucial for acting collectively in view of achieving social change [11]. In this context, the concept of "frame" – that is, the "schemata of interpretation that enables individuals to locate, perceive, identify and label" reality and its realizations [12] – has been largely employed as a preferred entry point to explain the construction of alliances and the probability of movements' success. More precisely, the more social actors share the same frames the more they agree on the same interpretations of the world as well as on how it should be transformed (e.g., [13, 14]). In this sense, the more frames are shared the higher the probability that actors will join forces and act in a coordinated manner to achieve their objectives [15].

When it comes to the domain of smart cities, we argue that the same mechanisms that have been studied in relation to other forms of collective political action are in place. In this regard, we believe that the way in which the issue of "smart cities" is framed does have an impact on the type of plans and the strategies of action that decision-makers and key players will propose and adopt.

We thus argue that the World Wide Web provides a privileged entry point to investigate actors' frames. Social actors can employ a variety of means to communicate to the rest of the world their opinions and interpretations of reality (for example, direct communication, appearing on mainstream media, making a press release, draft a report or a flier). However, digital communication tools like websites provide, in comparison to other means, a very prominent arena make visions, missions and commitments publicly accessible to citizens as well as to a variety of other stakeholders [16].

In what follows, we apply our framework to investigate from an exploratory stand the level of sociotechnicality that characterizes the discussion of smart cities at the European Union level. We being by tracing the online "conversation" that is established by means of hyperlinks amongst the websites of organizations, institutions and private sector entities that are participating in the governance of the smart city domain at the EU level. Implicit in our focus on online hyperlink structures is the assumption, typical of actor-network theory (ANT), that non-human technological agents, such as coding languages and the hyperlink system that sustains the Web, shape and, at the same time, are shaped by social courses of action [17]. Subsequently, we analyze how the issue of smart cities is framed within this online conversation. In a first step, we investigate whether key players (i.e., incumbents of central positions within the online hyperlink network) adopt a socio-technical frame or if, conversely, they tend to emphasize technological or social aspects. Secondly, we explore the extent to which these key players share the same frames and thus possess the potential to act collectively for translating their common vision into practice. Ultimately, while we do not perform any traditional hypothesis testing nor do we aim at generalizing our results to the whole EU strategy for smart cities, we seek to provide an innovative analytic framework to push forward research activities in the smart city domain.

The rest of the paper is organized as follows. In the next section, we briefly introduce the European case study we examine and provide an overview of how our proposed framework is operationalized. In section three, we illustrate our results – firstly about who key players in the EU online conversation on smart cities are, secondly on the type of frame they endorse, and finally on their levels of coordination based on frame-sharing. We conclude in section four by discussing our results and identifying future perspectives of research.

2   Investigating the Sociotechnicality of Smart Cities in Europe

Over the last few years, smart cities have actually become one of the most important components of the European Digital Agenda (EUDA) – the EU initiative for the full exploitation and integration of ICTs for enhancing Europe's economy and empowering its citizens. Quite interestingly, the official EUDA website defines smart cities as "a place where the traditional networks and services are made more efficient with the use of digital and telecommunication technologies, for the benefit of its inhabitants and businesses".[1] Moreover, the website specifies that the concept of smart cities "goes beyond the use of ICTs for better resource use and less emissions" and thus encompasses innovation within the procedures for governing and administrating urban environments.

On the overall, the above definition appears to be consistent with the great emphasis put on ICTs as the core element of contemporary smart cities. At the same time, it acknowledges that the employment of these tools must not be self-referential but, rather, geared towards generating social and economic benefits. However, to a closer look, the EUDA definition seems to suggest that efficiency and social benefits flow directly from a not-better-specified "use" of technologies whereas EU citizens (but also business entities) are depicted somehow as passive recipients. Is this definition a hint of the fact that the EU is adopting a mainly techno-deterministic perspective on smart cities? Or, rather, is it an attempt to provide a general conceptual framework within which computer and normative codes can actually converge? Ultimately, how sociotechnical is the EU approach to smart cities?

In order to begin answering this question, we lean on the analytic approach outlined in the previous section and explore how key players in the smart city domain at the EU level frame this issue by making use of their websites.

2.1 Data and Methods

In order to perform our exploration, we lean on a combined use of two research techniques: digital methods and network analysis. The former is a set of research tools that were developed precisely to analyze digital objects (e.g., hyperlinks, web pages, search engines, etc.) in order to maximize their informative potential about social dynamics [18]. The latter, instead, is a set of research techniques to study the patterns of relations amongst a set of actors, also called nodes [19].

In this specific case, digital methods allow us to map an "online issue network", i.e., a network of websites with a common thematic focus and tied together via hyperlinks, which can be considered as a proxy for the online conversation on smart cities that we are interested in. To obtain this network, we start from a list of 10 URLs associated with the key initiatives and networks identified by the European Innovation Partnership on Smart Cities and Communities (EIP-SCC), the core initiative launched by the European Commission to foster the European smart city project.[2] This list is then processed by a tool called Issue Crawler (IC), one of the first software designed to systematically crawl the Web and trace connections between websites.[3] For each "starting point" in the list, IC fetches all outgoing links. Whenever at least two starting points share outgoing links to a website or a Web resource that is not in the initial list,

---

[1] https://ec.europa.eu/digital-agenda/en/smart-cities.

[2] https://eu-smartcities.eu/about/useful_links. Starting points are available in Appendix A.1.

[3] http://www.issuecrawler.net.

the software adds a new node to the network. At the end of this process, called co-link analysis, the software traces hyperlinks amongst all nodes in the final list and returns a map of the online issue network (Fig. 1).[4]

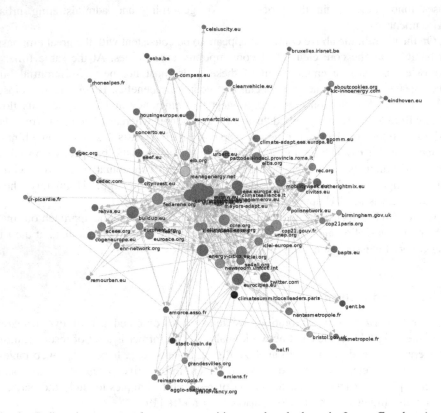

Fig. 1. Online issue network on smart cities produced through Issue Crawler (n. of iterations = 2; privileged starting points = on; crawl depth = 2)

Although the IC suffers (as much as any other crawler available) from important limits – above anything else, the fact that it cannot read and, hence, scrape, JavaScript (see [20]) – ultimately this online issue network provides us with a good approximation of "who is talking to whom" in the online space about smart cities.

Within this network, we identify those we have called above the "key players", that is, actors that occupy most "powerful" positions as they show higher centrality values. Indeed, following existing approaches (see [16]), we claim that most central actors, by virtue of their peculiar position, are able to "set the tone" of the overall conversation deploying within an online issue networks. Thus, we distinguish between three categories of key players:

---

[4] The final list of websites included in the online issue network is available in Appendix A.2.

- Programmers, i.e., nodes with a high indegree value and that, therefore, enjoy a wide recognition from other nodes in the online issue network about smart cities:
- Mobilizers, i.e., nodes with a high outdegree and that, therefore, are actively engaged in building the online network of discussion on smart cities;
- Switchers, i.e., nodes with a high betweenness and that, therefore, mediate indirect connections amongst other nodes in the online network of discussion.

Subsequently, we investigate how key players in three categories frame smart cities arguing that:

- frames supported by programmers are those which are taken as "points of reference" by other nodes in the online network;
- frames hold by mobilizers are those which motivate the construction of the online network of discussion;
- frames endorsed by switchers are those upon which coordination within the network can be achieved.

In order to track actors' frames, we employ a tool called Googlescraper, which queries websites for sets of keywords and returns the number of pages that contain every keyword.[5] Thus, we query central websites for keywords able to capture different facets of the two broad visions of smart cities we discussed above: on the one hand, a techno-centric vision based on the predominant role of ICTs; on the other, a socio-centric one that emphasizes social and human aspects.

We derived our set of keywords building on a thorough literature review realized by [21], who argue that all possible labels attached to the smart city concept provide a sort of variation on a theme and can be ultimately reduced to three main categories: technology, people and communities. Indeed, as the authors. The first category links back to what we have called above a "techno-centric" approach to smart city, one that privileges technological factors: it is the case of labels such as digital city, which emphasizes the ubiquity of the digital communication infrastructure; intelligent city, which instead highlights how technology can and should be used within the urban environment not only in view of an incremental change but, rather, of a radical one; ubiquitous city, that underlines the characteristic of universal access to the digital communication networks; wired city, which emphasizes explicitly the element of infrastructure; hybrid city, that points to the contamination between online and offline spaces of action; information city, which ties back to the abovementioned idea of the continuous production of data on how the city works and is lived [21] (Table 1).

On the other side, [21] group labels which emphasize the "human dimension" of smart cities, which, in our approach, can be associated to a socio-centric approach to their development. In this case, non-technological aspects play a major role – like creativity (creative city), human social capital (humane city), social learning (learning city), knowledge production and diffusion (knowledge city), and the virtuous relationship between institutions and citizens aimed at increasing the overall quality of the urban life (smart community) [21].

---

[5] https://tools.digitalmethods.net/beta/scrapeGoogle/.

Table 1. Techno-centric and socio-centric frames on smart cities

Dimension	Frame	Dimension	Frame
Techno-centric	Digital city	Socio-centric	Creative city
	Intelligent city		Learning city
	Ubiquitous city		Humane city
	Wired city		Knowledge city
	Hybrid city		Smart community
	Information city		

Building on the output produced by the Googlescraper, we first assess the extent to which actors adopt a sociotechnical perspective on smart cities by looking at how much they endorse techno-centric and socio-centric frames within their websites. We also explore the level of coordination these actors can reach amongst themselves when endorsing specific frames on smart cities. In this respect, we examine through network analysis techniques the extent to which most central actors tend to converge around common frames and compare their levels of coordination when they focus on technological or social aspects of smart cities.

3   Results

Online issue network composition and structure. The online issue network mapped through Issue Crawler (IC) is structured within one sole component, tying together 75 nodes through 550 hyperlinks (Table 2). As shown in Fig. 1, which depicts nodes in the issue network with different colors depending on the top-level domain they carry, the majority of nodes are colored in light green and thus belong to the big family of actors operating at the European level (i.e., URLs end with a.eu domain). Within this large group, we find actors such as the European Commission (ec.europa.eu); the European Covenant of Mayors for Climate and Energy (covenantofmayors.eu), launched in 2008 and today one of the largest initiatives for the governance of climate

Table 2. Online issue network on smart cities overall metrics

Measure	Value	Measure	Value
Size	75	Reciprocity	17 %
Ties	550	Core members	20
Density	0.10	Indegree Centralization	40,63 %
n. components	1	Outdegree Centralization	46,11 %

change and renewable energies in the world; the European Environment Agency (eea. europa.eu); Eurocites (eurocieites.eu), the network of major European cities.

The second large group in the network is formed by.org websites, which gathers several actors and initiatives that are mainly of institutional nature. For example, prominent nodes in this category are iclei.org and iclei-europe.org, the general and the European websites of the world-leading network of cities and towns committed to implement plans for a sustainable development. Other examples are klimabuendnis.org and climatealliance.org, the general and the English versions of the website of the alliance between European cities and indigenous rainforest people.

Beside these two main groups, the network gathers a plurality of websites representing local actors who, also in this case, are mostly of institutional nature. This is particularly true for French websites, which are associated mainly to municipalities engaged in the implementation of smart development plans such as Nantes, Lille, Amiens, St. Etienne and Reims; or to the main institutional events linked to the discussion of climate change (the Cop21 conference and the Climate Summit for Local Leaders). Other representatives of local institutions come from Germany (stadt-koeln. de, the website of the city of Cologne), Finland (hel.fi, the website of the city of Helsinki), UK (with the city portals birmingham.gov.uk and bristol.gov.it) and Belgium (bruxelles.irisnet.be, the portal of the city of Brussels). Italy does also enter the network, however only by providing national versions of international initiatives, in particular of the Covenant of Mayors (pattodeisindaci.provincia.roma.it) and the Climate Alliance (climatealliance.it).

The relatively low presence of.com actors, limited to kic-innoenergy.com and cedec.org – two networked companies active on the EU territory –, should not be misunderstood for an absence of the private sector from the issue network.[6] Indeed, "hidden" under other TLDs, there are several nodes pointing to initiatives concerned with financing and private corporations in the smart city "business". One example is provided by fi-compass.eu, a platform for advisory services on financial instruments linked to the European Structural and Investment funds (ESIF) as well as to the Programme for Employment and Social Innovation (EaSI).

On the overall, nodes in the online issue network are rather densely connected as 10 % of possible ties are activated (Table 2). However, amongst these, though, only 17 % are reciprocal and are mainly located within the core of the network, which gathers 20 actors that either belong to the EU context or to the big networked initiatives as the Climate Alliance.[7] Conversely, all local constituencies are located within network periphery (Fig. 2). Taken together, these elements suggest that, although a structure for discussing smart cities is in place and includes, either directly and indirectly, European and local constituencies, a true and diffused dialogue is not taking place yet.

---

[6] The third .com node is Twitter.com. The platform appears as a node on its own although webpages crawled by the Issue Crawler link to a specific account within it because of the overall incapability of the software to process deep links within social media platforms.

[7] Nodes in the network core are ec.europa.eu, klimabuendnis.org, eumayors.eu, covenantofmayors.eu, eea.europa.eu, iclei-europe.org, iclei.org, eurocities.eu, buildup.eu, eltis.org, eib.org, mobilityweek. eu, mayors-adapt.eu, managenergy.net, fedarene.org, energy-cities.eu, climatealliance.org, ccre.org, soglasheniemerov.eu, euroace.org.

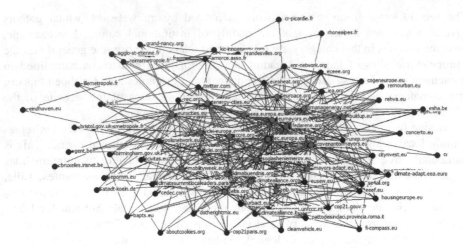

Fig. 2. Core-periphery representation of the online issue network (final fitness = 0.524). Red nodes = core; black nodes = periphery. Red lines = reciprocal ties. (Color figure online)

Key Players. Results in Table 2 also show that the online conversation is dominated by few actors who seem to catalyze the majority of links from others (see *Indegree Centralization*) but also to be particularly active in building connections (see *Outdegree Centralization*).

Table 3 identifies these most central nodes and classifies them as programmers, mobilizers and switchers. As it shows, there is a substantial overlap between the three categories, in particular between programmers and switchers – as these latter are a subgroup of the actors playing a programming function. Not very surprisingly, the main programmer in the network is the European Commission (EC), which receives hyperlinks from half of other nodes. Other programmers, which are all expressions of more specific initiatives, receive links from a more limited number of network members (between 15 % and 21 %).

Following the EC in the indegree ranking we find eumayors.eu, the sister website of the Covenant of Mayors, and buildup.eu, the European portal for Energy Efficiency in Buildings. These websites well represent the two general subgroups that can be found in the programmers category.

On the one hand, amongst programmers we find initiatives aimed at joining different localities and municipalities within networks often in view of fostering coordination in relation to environmental governance. In this first cluster sit actors like Energy Cities (energy-cities.eu); the European Federation of Agencies and Regions for energy and the Environment (fedarene.org); Eurocities, the network of major European cities (eurocities.eu); the Council of European Municipalities and Regions (ccre.org); ManageEnergy (managenergy.net), an initiative of support to public sector actors working in the field of renewable energy; and the abovementioned iclei.org. On the other hand, programmers are thematically focused initiatives that are centered on a specific topic. Within this group, we find the European Environment Agency but also the European

Table 3. Programmers, Mobilizers and Switchers in the online smart cities issue network

Programmers	In-degree	Mobilizers	Out-degree	Switchers	Betweenness
ec.europa.eu	37	covenantofmayors.eu	41	ec.europa.eu	843.24
eumayors.eu	16	eumayors.eu	41	eumayors.eu	509.643
buildup.eu	16	energy-cities.eu	33	energy-cities.eu	388.526
energy-cities.eu	15	managenergy.net	31	eurocities.eu	373.753
covenantofmayors.eu	14	soglasheniemerov.eu	27	covenantofmayors.eu	357.151
fedarene.org	14	klimabuendnis.org	23	managenergy.net	320.493
eurocities.eu	14	polisnetwork.eu	23	iclei.org	216.609
mobilityweek.eu	13	iclei-europe.org	23		
ccre.org	13	fedarene.org	22		
eea.europa.eu	13	climatealliance.org	22		
eusew.eu	13	clei.orgi	18		
managenergy.net	12	eurocities.eu	17		
iclei.org	12	eltis.org	17		
eltis.org	12				
birmingham.gov.uk	12				

Note: Mean Indegree = 7.33; S.D. = 5.2; Mean Outdegree = 7.33; S.D. = 9.87; Mean Betweenness = 62.42; S.D. = 137.87.

Sustainability Energy Week (eusew.eu), a month-long set of initiatives aimed at sustainable mobility; and Eltis (eltis.org), the principal EU observatory on urban mobility.

On the one hand, amongst programmers we find initiatives aimed at joining different localities and municipalities within networks often in view of fostering coordination in relation to environmental governance. In this first cluster sit actors like Energy Cities (energy-cities.eu); the European Federation of Agencies and Regions for energy and the Environment (fedarene.org); Eurocities, the network of major European cities (eurocities.eu); the Council of European Municipalities and Regions (ccre.org); ManageEnergy (managenergy.net), an initiative of support to public sector actors working in the field of renewable energy; and the abovementioned iclei.org. On the other hand, programmers are thematically focused initiatives that are centered on a specific topic. Within this group, we find the European Environment Agency but also the European Sustainability Energy Week (eusew.eu), a month-long set of initiatives aimed at sustainable mobility; and Eltis (eltis.org), the principal EU observatory on urban mobility.

The category of *mobilizers* is instead made almost entirely from websites representing networks of cities often with a focus on energy and climate-change related interests. The majority of links are sent by the European Covenant of Mayors for Climate and Energy as well as by its sister site eumayors.eu. Quite interestingly, one of the main mobilizers is also the Russian version of the European mayors' portal (soglasheniemeriv.eu). Other relevant mobilizers are city-network actors functioning also as programmers, such as Energy Cities, ManageEnergy, ICLEI and ICLEI-Europe, the Federation of Agencies and Regions for energy and the Environment and Eurocities. In comparison to the programmers list there are also few but interesting exceptions: first, the general and the English websites of the website of the alliance between European cities and indigenous rainforest people (klimabuendnis.org and climatealliance.org); second, POLIS (polisnetwork.eu), the network of European cities committed to the amelioration of local transportation plans. Both are again city-network actors but with specific thematic foci that enrich the mobilizers agenda – respectively, environmental sustainability in indigenous areas and local transportation and mobility.

Finally, a restricted number of websites, besides programming the contents of the online issue network, are also in a favorable position to coordinate its different parts as switchers. Thus, these websites are also representative of the different interests that animate the online discussion: the regional ones, with the European Commission; the local ones, with the Eurocities network of cities; the institutional concern for energy and climate-change issues, with the European Covenant of Mayors for Climate and Energy and Manage Energy; mobility and local transportation with ICLEI.

Smart cities frames. Figure 3 represents the level of endorsement the different smart city frames found in the webpages of the key players' websites. As it shows, all frames are adopted in the online conversation on smart cities thus suggesting that the European discussion is on the overall supported by a sociotechnical approach. However, the figure also reveals that not all frames are endorsed with the same emphasis. Key players tend indeed to emphasize predominantly techno-centered frames, which are present in a larger amount of pages in all websites functioning as *programmers*, *mobilizers* or *switchers* (respectively, 57 %, 58 % and 60 % of webpages).

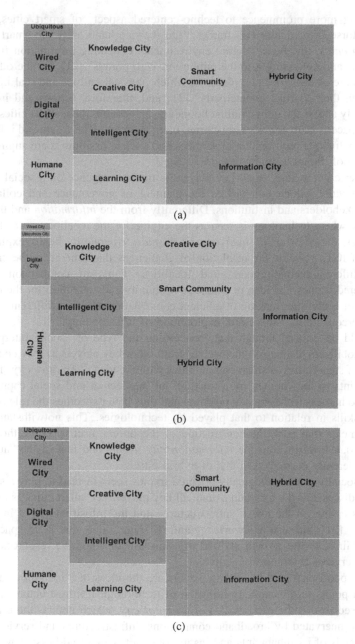

Fig. 3. Treemap of techno-centered and socio-centric frames endorsed by programmers (a), mobilizers (b) and switchers (c). Techno-centered frames are depicted in purple, socio-centered frames in gold. (Color figure online)

In giving more prominence to techno-centered aspects of smart cities, all key players endorse in particular two frames.[8] On the one hand, they see smart cities as *information cities*, that is, as "digital environments collecting information from local communities and delivering it to the public via Web portals" [21]. On the other hand, smart cities are also *hybrid*, i.e., places in which physical entities and real inhabitants enmesh with their "virtual counterparts" [21] and, therefore, actions and interaction deploy fluidly across the online/offline boundary. In general, then, smart cities are seen mainly as spaces where activities are carried on in a context where data and information allow the continuous proliferation of services and social possibilities are augmented by the presence of digital technologies.

A further commonality to all key players is the convergence on a social vision of *smart community*, which emphasizes the element of governance and collaboration between stakeholders and institutions. Differently from the *information* and the *hybrid* city frames, where technology is seen as the engine for the amelioration of the urban environment, the view of a smart city as a *smart community* bends explicitly the potential of ICTs towards the resolution of challenges that emerge at the crossroads between different neighborhoods and localities. Thus, in comparison to other socio-centered frames, the idea of a smart community calls attention for the *collective* dimension of social life, making of efficient coordination amongst different views and needs a prerequisite to a successful exploitation of ICTs potential.

It should be noticed though that, concerning the type of "social interpretation" adopted, mobilizers distinguish themselves from other key players as they complement the vision of smart cities as smart communities with the idea of *creative city*. This latter concept points to the "human infrastructure" of intellectual and social capital that is necessary to harness technological potential and thus better specifies the role of citizens and their skills in relation to that played by technologies. This notwithstanding, the main arguments that push the construction of the online discussion are those of the "power of data", intrinsic to the *information city frame*, and that of the "augmented reality", inherently connected to that of hybrid cities.

Other socially or technologically oriented frames seem instead to play a secondary role in the discussion. To a certain extent, all key players see smart cities as *intelligent*, that is, possessing all the latest "infrastructures and the infostructures of information technology" [21]; but also as *knowledge* and *learning* cities, that is, as spaces where innovation links to the growth and the transmission of cognitive resources to make skills and services continuously evolve.

Finally, poorly adopted by all key players are the ideas of a *ubiquitous* and *wired city*, which points to the capillary diffusion of technological infrastructure [21]. Scarcely endorsed is also the general idea of a *digital city,* which depicts the urban environment as innervated by broadband connections, infrastructures and services able to meet the needs of its inhabitants and institutions, yet without addressing where these needs come from. On the side of socio-centered frames, the less adopted one is that of a

---

[8] We base this section on the definitions given by [21] in Sects. 2.3.1 to 2.3.3.

humane city, a concept stressing the "multiple opportunities to exploit its human potential and lead a creative life" [21].

Sharing smart cities frames. Besides putting more emphasis on techno-centered interpretations of smart cities, key players also tend to coordinate more consistently around technological frames rather than around social ones.

Table 4 illustrates some key features of the affiliation networks that key players in their different roles form when sharing technological or social views of smart cities (see Fig. 4 for an example). As it shows, both programmers and mobilizers tend to converge more cohesively around technological interpretations. Not only networks based on sharing technological frames show a lower number of isolates, i.e., key players not systematically recognizing any specific frames (see values in column ISO).[9] Density values (column Δ), which indicate the proportion of existing ties on the total possible number of ties in the network, also indicate that both programmers and mobilizers form a much cohesive group when it comes to sharing technological interpretations rather than social ones. Finally, average degree values (column Av. Degree), which indicate the average number of connections established by nodes within a network, suggest that both programmers and mobilizers are more "active" in sharing techno-centered frames than social ones.

Table 4. Features of affiliation networks formed by key players based on sharing techno-centered and socio-centered frames

Key players	Frames	N	T	Δ	ISO	Av. degree
Programmers	Techno-centered	15	42	0.40	2	40.00
	Socio-centered	15	28	0.27	6	26.67
Mobilizers	Techno-centered	13	35	0.45	2	44.87
	Socio-centered	13	16	0.21	5	20.51
Switchers	Techno-centered	7	7	0.33	2	33.33
	Socio-centered	7	7	0.33	2	33.33

Interestingly, the coordination of the network provided by switchers seems not to be "structurally sensitive" to the different type of frames. This element may relate to the overall heterogeneity of this specific group in terms of interests represented. Indeed, as we noted above, in spite of its limited size, the group of switchers represents all interests brought in by programmers and mobilizers: the regional ones, the local ones together with the energetic and the mobility and transportation ones.

---

[9] In each affiliation matrix, there is a tie between any couple of key players if they shared a frame in a number of webpages higher than the average number of pages where any socio-centered or techno-centered frames could be found.

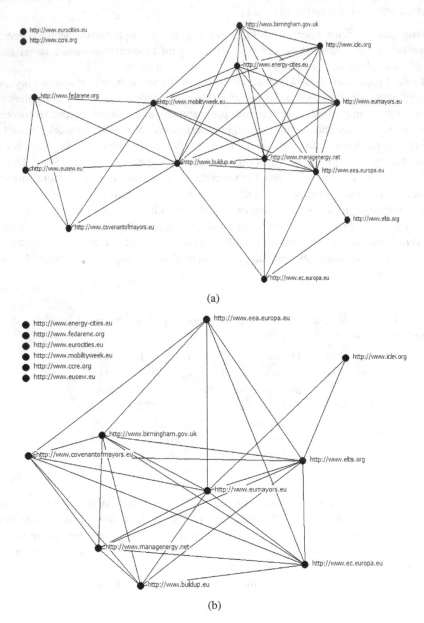

Fig. 4. Affiliation network between programmers sharing techno-centered frames (a) and socio-centered frames (b)

4   Discussion and Future Perspectives

In this paper, we claimed that contemporary efforts to make cities thrive through the strategic exploitation of ICTs can be successful to the extent to which they are carried on in a sociotechnical fashion. In fact, grounding innovative approaches into pervasive ICTs infrastructures necessarily entails depending also on social perceptions of technological affordances and from the scopes and the aims that social actors aim to achieve. In this sense, we claimed that it is necessary to build smart city projects that endorse a sociotechnical approach, i.e., that recognize and give value to technical as much as to social aspects. We thus explored how much the European approach to smart cities is carried on sociotechnically. Our results suggest that, although there is a formal recognition of both technological and social aspects, key players in this policy domain are keener to emphasize the former. More importantly, they are also better coordinated around techno-centered frames rather than around socio-centered ones. In this sense, it is more likely that the EU smart city strategy will be carried on along a technological perspective leaving somehow the social behind.

A number of factors may explain this dominance of technical factors over social frames. Right because ICTs are grounded on specific materialities, it is certainly easier to envisage the role they can play in relation to the process of social innovation. Conversely, social dynamics are much more complex and rather unpredictable – as the current economic crisis has very well showed us. It is therefore rather understandable that key players with such complex agendas do privilege in their policy action those aspects and theses upon which it is easier to find agreement.

However, our results also do suggest that a more challenging mechanism may be in place. Albeit it is non-representative of the current policy discussion in the smart city domain, the online issue network we analyzed points to the total exclusion of citizens' and civil society initiatives from the discussion. On the overall, ICTs specialists, whether they are from the private sector or of governmental nature, seem to have far greater access to resources and opportunities to influence policy processes in the smart city domain by comparison with other types of organizations – e.g., civil society initiatives working to educate citizens to a conscious use of technologies or to defend their freedom of expression and privacy. Thus, not much space seems to be left for the actual inclusion in the discussion of other disciplines, such as social and political sciences, besides engineering.

The overall emphasis that EU key players put on the "smart community" frame testifies a general acknowledgement of the fact that political innovation lays at the core of successful smart city strategies. And yet, the initial exploration we performed seems to suggest that this overall awareness is not accompanied by the adoption of a multi-stakeholder governance approach, where institutional, private sector and civil society actors are actual partners in the governance of our cities. A systematic and sustained effort to implement the "smart community" frame seems then to be the road to follow, so that a fertile ground to sociotechnical smart cities can be set through an actual inclusive and democratic collective effort.

A Appendix

A.1 List of Starting Points submitted to Issue Crawler

1. http://ec.europa.eu/eip/smartcities/index_en.htm
2. http://smartcities-infosystem.eu/
3. http://www.civitas.eu/index.php?id=69
4. http://www.eumayors.eu/index_en.html
5. http://www.eurocities.eu/
6. http://eit.europa.eu/eit-community/climate-kic
7. http://www.kic-innoenergy.com/
8. http://www.errin.eu/
9. http://www.polisnetwork.eu/about/about-polis
10. http://www.energy-cities.eu/

A.2 List of Websites Included in the Final Online Issue Network

1. aboutcookies.org
2. agglo-st-etienne.fr
3. amiens.fr
4. amorce.asso.fr
5. bapts.eu
6. birmingham.gov.uk
7. bristol.gov.uk
8. bruxelles.irisnet.be
9. buildup.eu
10. ccre.org
11. cedec.com
12. celsiuscity.eu
13. citynvest.eu
14. civitas.eu
15. cleanvehicle.eu
16. climate-adapt.eea.europa.eu
17. climatealliance.it
18. climatealliance.org
19. climatesummitlocalleaders.paris
20. cogeneurope.eu
21. concerto.eu
22. cop21.gouv.fr
23. cop21paris.org
24. covenantofmayors.eu
25. cr-picardie.fr

26. dotherightmix.eu
27. ec.europa.eu
28. eceee.org
29. eea.europa.eu
30. eeef.eu
31. egec.org
32. eib.org
33. eindhoven.eu
34. eltis.org
35. energy-cities.eu
36. enr-network.org
37. epomm.eu
38. esha.be
39. eumayors.eu
40. euroace.org
41. eurocities.eu
42. euroheat.org
43. eusew.eu
44. eu-smartcities.eu
45. fedarene.org
46. fi-compass.eu
47. gent.be
48. grandesvilles.org
49. grand-nancy.org
50. hel.fi
51. housingeurope.eu
52. iclei.org
53. iclei-europe.org
54. iea.org
55. kic-innoenergy.com
56. klimabuendnis.org
57. lillemetropole.fr
58. managenergy.net
59. mayors-adapt.eu
60. mobilityweek.eu
61. nantesmetropole.fr
62. newsroom.unfccc.int
63. pattodeisindaci.provincia.roma.it
64. polisnetwork.eu
65. rec.org
66. rehva.eu
67. reimsmetropole.fr
68. remourban.eu
69. rhonealpes.fr
70. se4all.org
71. soglasheniemerov.eu

72. stadt-koeln.de
73. twitter.com
74. unep.org
75. urbact.eu

References

1. Caragliu, A., Del Bo, C., Nijkamp, P.: Smart Cities in Europe. Series Research Memoranda 0048. VU University Amsterdam, Faculty of Economics, Business Administration and Econometrics (2009)
2. Kitchin, R.: The real-time city? Big data and smart urbanism. GeoJournal **79**, 1–14 (2014)
3. Townsend, A.M.: Smart Cities: Big Data, Civic Hackers, and the Quest for a New Utopia. W.W. Norton & Company Inc., New York (2013)
4. Greenfield, A.: Against the Smart City (The City is Here for You to Use). Do Projects, New York (2013)
5. Lievrouw, L.A,. Livingstone, S.: Introduction. In: Lievrouw, L.A,. Livingstone, S. (eds.) Handbook of New Media: Social Shaping and Social Consequences – fully revised student edn., pp. 1–14. SAGE Publications, London (2006)
6. Nam, T., Pardo, T.A.: Smart city as urban innovation: focusing on management, policy, and context. In: ICEGOV 2011 Proceedings of the 5th International Conference on Theory and Practice of Electronic Governance, pp. 185–194. ACM, New York (2011)
7. Pavan, E.: Embedding digital communications within collective action networks. A multidimensional network approach. Mobilization: Int. J. **19**(4), 441–455 (2014)
8. Leonardi, P.M.: Materiality, sociomateriality, and socio-technical systems: what do these terms mean? How are they different? Do we need them? In: Leonardi, P.M., Nardi, B.A., Kallinikos, J. (eds.) Materiality and Organizing: Social Interaction in a Technological World. Oxford University Press, Oxford (2012)
9. Orlikowski, W.: Sociomaterial practices: exploring technology at work. Organ. Stud. **28**(9), 1435–1448 (2007)
10. Della Porta, D., Diani, M.: Social Movement. An Introduction. Blackwell Publishing (2006)
11. Johnston, H.: Protest Cultures: performance, artifacts, and ideations. In: Johnston, H. (ed.) Culture, Social Movement and Protest, pp. 3–32. Ashgate, Farnham (2009)
12. Goffman, E.: Frame Analysis. Penguin Books, London (1974)
13. Snow, D., Rochford, E.B., Warden, S., Benford, R.: Frame alignment processes, micromobilization and movement participation. Am. Sociol. Rev. **51**(4), 464–481 (1986)
14. Benford, R.: Frame disputes within the nuclear disarmament movement. Soc. Forces **71**(3), 677–701 (1993)
15. Melucci, A.: Challenging Codes: Collective Action in the Information Age. Cambridge University Press, Cambridge (1996)
16. Padovani, C., Pavan, E.: Global governance and ICTs: exploring online governance networks around gender and media. Global Networks (forthcoming, 2016)
17. Latour, B.: Reassembling the Social. An introduction to Actor-Network Theory. Oxford University Press, New York (2005)
18. Rogers, R.: Digital Methods. MIT Press, Cambridge (2013)
19. Borgatti, S.P., Everett, M.G., Johnson, J.C.: Analyzing Social Networks. Sage, London (2013)

20. Bossetta, M., Dutceac Segesten, A.: Tracing eurosceptic party networks via hyperlink network analysis and #FAIL!ng: can web crawlers keep up with web design? In: #FAIL! Workshop at the Websci 2015 Conference, pp. 1–3 (2015). https://failworkshops.files. wordpress.com/2015/06/fail2015a_bossettadutceacsegesten.pdf
21. Nam, T., Pardo, T.A.: Conceptualizing smart city with dimensions of technology, people, and institutions. In: The Proceedings of the 12th Annual International Conference on Digital Government Research, pp. 282–291. ACM, New York (2011)

Smart Cities Tales and Trails

Athena Vakali[1]([⊠]), Angeliki Milonaki[2], and Ioannis Gkrosdanis[1]

[1] Department of Informatics, Aristotle University of Thessaloniki, Thessaloniki, Greece
`avakali@csd.auth.gr`
[2] Film Studies Graduate, Aristotle University of Thessaloniki, Thessaloniki, Greece

Abstract. Cities have been transformed to experimental platforms at which data produced capture everyday activities, pulse, and interactions. Placing humans at the centre of smart cities has motivated several efforts under the vision of having citizens at the forefront of the Internet of Everything. Cities though have been largely impacted by their own historic, cultural and past stories which drive todays city life and experiences. The proposed approach enables applications and platforms development which will merge the past with the present in an innovative manner by placing emphasis on the cultural content as a drive for today's dynamic city and social interacting. The focus is placed on people who navigate in the city and who are enabled to act as tales receptors and trails broadcasters. People receive cultural content (emphasis here is on film and cities) in the form of city relevant annotated storylines which trigger people's reactions expressed at the city's virtual spaces which may be enhanced by several dynamic (such as city trail reviewing, city offers outreaching, etc.). The proposed process targets an open platform which can be extended to integrate multi-domain (sensors, social networks, etc.) recommendations towards humanizing city experimentation and navigation experiences.

Keywords: Apps for society · Social innovation · City platform · Participation · Liveable city · Empowerment · Education apps · Public spaces · Geo information · Entrepreneurship · Data analysis · Cultural change

1 Introduction and Background

Most recent and groundbreaking scientific and technological innovations bear significant potential for evolving and redefining people's lives and interactions with cities. Smart cities have the potential to use technology-driven service provision to evolve rapid solutions for new challenges stemming from citizens. Embedding cutting-edge devices, networks and services into cities' centuries-old streets impacts citizens' daily lives in terms of their movement, habits and behavioral patterns. However, the design of such "smart city" solutions needs to be driven from the "bottom-up or citizen-led approach" as highlighted in the major theme of the recent "Re.Work Future Cities Summit" [9] and a related Guardian article [4]. It should be noted that Re.Work conference concluded in: "whatever the smart city might be, it will be acceptable as long as it emerges from the ground up".

© Springer International Publishing AG 2016
F. Bagnoli et al. (Eds.): INSCI 2016, LNCS 9934, pp. 278–290, 2016.
DOI: 10.1007/978-3-319-45982-0_24

The proposed work is inspired by the need to identify new qualitative criteria (such as attention, identity, and culture) which will support validation of culture's impact on today's cities dynamics, by utilizing information technologies and medium (such as mobile devices), in the daily experience of public spaces. Public spaces are all places publicly owned or of public use accessible and enjoyable by all for free and without a profit motive. Each public space has its own spatial, historic, environmental, social and economic features. Public spaces are the environments of shared living experience. For example, squares -also called piazzas- receive a focus in the everyday public life, because even since antiquity, they represent in European history the places where public life was staged and European culture and identity was and still continuous to be based. At such places, but also at other not so known ones many forms of cultural content (such as films) have captured daily life, architectural progress, societal interactions etc. At exactly those same places today, sensors are installed, social networks enable check in declarations and people interact socially at real and the corresponding virtual city spaces. A question arises therefore: how are these places experienced today in a heavy digitized world? To which extent these places are still exercising on us impact such as feelings of repose or identification or experiences of aesthetic excitement?

1.1 Cities of Today with People on the Move

People with mobile devices move on public spaces with a predefined view of the place, given recommendations and guidelines by others who also influence their opinions and sentiments. Moreover, public or private authorities and other stakeholders analyse and monitor social media communities as they emerge in Local Based Social Networks (LBSNs). Digitization and heavy hyper-connected reality is public spaces is primarily relevant with the next major research fields:

- **LBSNs Structure Analysis.** So far, there has been limited research on LBSN structure analysis. An early work [6] employed unsupervised clustering for finding groups of: (i) mobility patterns and (ii) users based on their activities, while in [11] graph analysis on some well-known LBSNs studied the correlation of geographic distance with the users' social network, considering only one static location per user. Same authors in [12] analysed the complete social network of Gowalla and identified the existence of the small world phenomenon and of a high clustering coefficient. However, analysis indicated that as time goes by, the frequency of making check-ins and visiting new places decreases much steeper than the frequency of making friends. This possibly indicates that it may be difficult for users to find new interesting places in their area. A statistical analysis [3] on check-ins from various LBSNs broadcasted to Twitter focused on how human mobility patterns vary in time and geographic area by correlating mobility patterns via social & content-based features.
- **Location-Aware Recommendations** in a broader perspective. GeoLife is a recent LBSN service that analyses users' uploaded GPS trajectories off-line to provide: (a) travel recommendations, by identifying interesting locations and travel sequences using a method similar to HITS with experienced users serving as hub and interesting locations as authority nodes [15], and (b) personalized location & friend

recommendations, combining content-based with user-based collaborative filtering and determining users' similarity from location history [16]. As GeoLife manages raw GPS trajectory data, it faces problems such as the identification of spatial areas constituting distinct landmarks.

- **Sentiment Analysis.** Humans throughout their everyday activities are experiencing a wide range of emotions. Inevitably such emotions are imprinted in the content generated by them in their online social activities, which is often geolocated or it refers to a specific city/region. Consequently, the capturing of the sentiments expressed in social media content could provide valuable insights about the city such content refers to. To this end, for instance, [2] proposed an argument-based approach, where via considering sentimental knowledge expressed in social media (Twitter, Facebook), arguments' extraction and policy making processes can be better supported. In order to successfully support such processes, they proceeded with text and opinion mining techniques for initially detecting the content generated in social media about a specific topic, and then analyzing the extracted content with respect to its opinion connotation. Finally, [5] exploited social media (Twitter, Facebook, Flickr, YouTube) content and applied sentiment and emotion detection approaches, for empowering the authorities' effectiveness when dealing with crisis situations within a city.

Several factors influence the residents of a city to visit a place, and can be leveraged for the dynamic segmentation of the city into functional regions. Intuitively, a segmentation of the city into geographic regions based on people's activities is expected to reflect more accurately the existing dynamics and behavioral/activity patterns, compared to a static city segmentation (e.g. based on population demographics, or fixed limits established by the municipality. In this sense, [1] proposed Livehoods, a clustering methodology for segmenting a city into dynamic areas, based on the everyday check-in activity of citizens on Foursquare and by exploiting both spatial (i.e. geographic proximity of places) and social (i.e. proximity of places based on the distribution of users that check-in to them) attributes of various places. Also, [10] proposed an approach for segmenting a city dynamically, based on temporal (i.e. temporal distributions of check-ins in a region) and spatial characteristics. Moreover, the data acquisition process (i.e. searching on diverse social media sources based on tags, keywords, and time / location based attributes) initiates when the city's authorities have an indication about a crisis event. Then, sentiment and affective analysis processes take place for classifying the obtained content based on the expressed sentiment and emotion, respectively.

1.2   Cities Identity and Cultural Profiles (Historic Background)

Cultural and historic content has been closely related to cities. It is important to notice that all European cinemas of the 20th century were primarily experienced as an urban phenomenon, which was both developed and consumed in an urban environment, as the majority of Greek movie halls were located in the cities). Furthermore, after the Second World War an important shift occurred in European cinema in certain locations such as in Greece which began to establish a functioning film industry (frequently characterized as the "golden era" for Greek national cinematography). Next, we focus on the Greek film industry impact on cities as an indicative paradigm of large impact on today's city transformations and shifts.

One of the most important issues raised by Greek popular cinema was the **representation of the city**, which reflected the emerging urban character of Greek society, the effects of economic reconstruction of the country, together with the rapid transformations of the cityscape. Following the paradigm of the city's depiction as a "**cinematic city**" with its own distinct qualities, Greek popular cinema of the 1950 s and the 1960 s served as a thorough study of the cinematic Greek city, since the adulthood of Greek cinema coincided with Athens' rebuilding and urban renewal. Greek feature films were dealing with the historic present of their time and were mainly shot in big cities (Athens and Thessaloniki), where thousands of internal migrants who have abandoned the countryside were in search of a profession and a better life.

Greek popular cinema of that time offered to its viewers an exceptional way to **tour around the history, culture and memory of the Greek city**, while witnessing existing and familiar urban settings, which were subject to the unprecedented transformations of the cityscape and the process of modernization in post-war Greece (new housing models, new consumer habits, new patterns of social behavior). Spaces in Greek films popular films of the 1950 s and the 1960 s "symbolized the post-war tourist growth of the city or, consequently became indicators of the tradition-modernity dichotomy, appointing optimism for the city's modernized image" [8]. New Greek cinema of 1970 s and 1980 s took a different approach from its predecessor in that it was not focusing on the city's representational value, but it would draw/direct its attention to questions of memory and it would thus make use of **dedramatized spaces**, which could operate as distinct memory places ("lieux de memoire"). During the 1990 s there was an important shift of gaze towards urban space on behalf of contemporary Greek filmmakers, who would return to the city and who would be interested to explore themes and places of the everyday life. Since then, Greek films continue to play an important role in depicting various aspects of the city image, in the sense that they emphasize on new cultural values of the ever-growing cityscape.

The idea of utilizing content, such as films archives, is presented in this paper. Given a public city space of historic and cultural value with parallel hyper-connected intensities, the proposed approach identifies, analyses and addresses reciprocal influences of city stories. The cultural content exploitation in the city context was motivated by the use case of Greek cinema and the city's strong cinematic profile. The motivation of this work originates from the fact that Greek (as other European and International) film archives embed valuable city stories of long lasting impact on cities formation and transformation. The novelty of this work is justified by its re-using of films content to generate innovative cultural digital assets via Web or mobile applications. Next section highlights the main characteristics of cities content, and proposes a flexible methodology which is exploited to deliver the CineMetro mobile application (discussed in Sect. 3), with details for its impact and feasibility at the city of Thessaloniki, Greece. Section 4 discussed the impact potential of the proposed work and finally Conclusions are summarized in Sect. 5.

2 City's Content as an Innovative Asset

Delivering, re-using and enriching content collections and archives (such as film archives) through the use of new technologies in cities brings city's stories, locations, and facts at the forefront of today's city navigation. The intention to make such archives easily accessible to new audiences and, at the same time, to highlight their influential role on predominant aspects of heritage communication, such as history, culture, memory and identity is crucial for advancing citizens' awareness and common sensing. To achieve so, emphasis in this work is placed on the following:

- *propose* to the users/viewers an innovative form of interpreting and exchanging film/audiovisual data, which will in turn bring new perspectives in the field of audiovisual collections that could be viewed in new, attractive ways;
- *provide* a new outlet of promotion and propagation of film archive material, and also highlighting the value of a multi-layered audiovisual heritage in the consolidation of a European cultural identity[1];
- *emphasize* on film's unique significance as "reflection and articulation of European cultural identities" and, in parallel, it would best showcase cinema's competence to "exploit the fundamental relationship between seeing and understanding, and recognizes the centrality of visual images to the formation of identity, whether personal, regional, national or European"[2].

This work's ambition is to proceed to the next advances in terms of:

- advancing earlier approaches which have considered aspects of **time and location separately**, so propose an innovative new approach which will jointly consider the criteria of location, time, emotion and frequency of public spaces visiting;
- provide a methodology for detecting citizen emotions and phenomena as they emerge in LBSNs via exploiting public spaces culture and history similarities, comments, etc., in an enjoyable and organized manner. User communities as **they emerge in the real life experimenting** may leverage on recommendation processes at which they can also interact;
- provide an **easily-deployable apps toolkit** that enables both trajectory analysis as well as people emotional standing for the case of humans physically moving and interacting within public spaces, as well as appropriate tools for searching in longer-term analysis. This will operate with full anonymization of estimated trajectory, demographic, and emotional personalized data.

---

[1] "This «cultural fortress Europe» and the nationalist reactions could be opposed by highlighting the history of Europe as a laboratory of especially equipped for cultural interchange, with the incidence of multiple extra-European elements, all historically consolidated and forming part of the "cultural identity of Europeans". Enrique Banus. "Cultural policy in the EU and european identity". In: M. Farrel, S. Fella and M. Newman (eds). European Integration in the 21st century. Unity in Diversity, London-Thousand Oaks: Sage, 2002, p. 171.

[2] W. Everett. "Introduction: European film and the quest of identity". In: W. Everett (ed). European Identity in Cinema. Bristol: Intellect Books, 1996, p. 8.

2.1   A Methodology for City Content and Its Dynamics Management

The proposed work is based on the need to capture real, virtual experiences and inter-actions in public spaces by primarily involving a methodology which will be used to leverage on archival cultural content (such as films) and at the same time it will enable data collecting and gaining input from citizens who are on the move and who can rate, interact, gain information, etc. Such an approach is proposed also in [13, 14] with emphasis on how different data sets can be merged towards delivering innovative appli-cations and services.

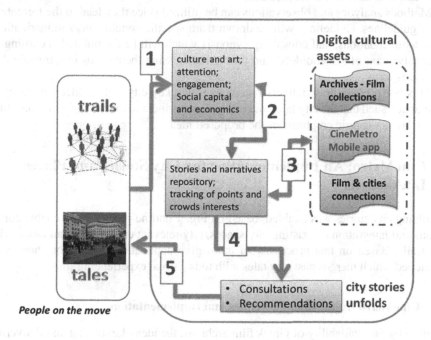

Fig. 1. A framework for cities tales unfolding

Figure 1 summarizes the methodology which can operate on some specifically defined building blocks which involve five specific flows in line with the objective to exploit city (past) stories with today's actual city navigation and experiencing. Next more details are given for each of these numbered flows.

1. Defining the public spaces real and virtual settings is required to enable city-driven trails and tales integration. Initially, specific data flows are needed as input to support relevant city's priorities (e.g. culture and art, attention, etc.) to address the city's data integration under a particular dimension focus. This is the initiating and triggering flow to set the public space's particular priorities which can also be related with specific city's areas at which cultural content can be located.
2. Then, a process to inter-connect the building blocks of disciplines and methods in a bi-directional fashion is followed since disciplines are mutually respected for iden-tifying the appropriate methods and for the completion of a city experiencing

activity. This flow is in line with both city's fundamental and technical objectives since the decisions made should be supported by appropriately designed technical solutions.

3. Methods are realized by specific multi-faceted instruments. The methods chosen for an activity along with their implementation procedures are employed in public space real and virtual settings, with people who are hyper-connected or physically interacting. This flow is also bi-directional since the methods can be refined and revised according to the actual activity and can run again in an iteration of an activity, for example at another public location.

4. Methods analytics and observations can be utilized since they lead to the formation of guidelines. Guidelines will be drawn from specific conclusions summarization, recommendations, and consulting synopsis which will be formulated according to involvement of stakeholders' and through addressing their needs in a transferable results form.

5. This last process flow delivers outcomes to the key city stakeholders, in an open manner, such that the city tales and trails can contribute to improving and promoting the human-centric emphasis of the proposed idea.

3   CineMetro: An Implementation for City Stories and Places Interactions

Unfolding city stories was enabled, based on Fig. 1 outline processes, to enable correlations and integration of existing city spaces, city relevant cultural content along with city trails. Based on this processes an innovative application ("app" for short) was developed which merges past city tales with today's real experiencing trails.

3.1   CineMetro Application Principles and Implementation

Inspired by the availability of Greek film archives, the idea of exposing and delivering film content appropriate text, images, and metadata in the city context has addressed needs to introduce to contemporary audiences the many different ways in which the image of the city and everyday life is depicted in film. Implementing a mobile application to do so was an immediate choice due to the penetration of mobile apps and devices to all city audiences. The goal is to further explore themes that are linked to the "cinema-city" relationship, which would not only give new research potentials to existing film archives, but it would also create new ways of reading our cultural heritage in terms of history, memory and identity.

The application developed was visualized as a so called CineMetro to deliver a familiar real metro-like experience with stops, hubs and people on the move trails. The virtual metro-like app informs and familiarizes the public about Thessaloniki's rich cinematic history through a modern navigation experience in the city. Various city's landmarks, which are linked to films shot in Thessaloniki, are used as "stations", where citizens can step by in order to see what's available (e.g. photos, texts, videos, podcasts, etc.). At those stops virtual spaces people on the move can deliver their experience summary (posts), their rating, their own suggestions, etc. At the same time social media interactions are enabled with connections to most popular social media check ins in the LBSNs manner of interacting.

Exploiting Film and City Innovative Application. Adding to this scope, the AUTH research group has already introduced an innovative application regarding Thessaloniki's film history, the "CineMetro"[3] app, with some of its screenshots depicted in Fig. 2.

The "CineMetro" app visualizes the **rich cultural heritage** of Thessaloniki's film history, which is represented in three major "metro"-like lines and their respective routes that pass through the city's historical center:

- Line 1 (films urban scenery). It deals with films that were shot in Thessaloniki and it thus promotes Thessaloniki through fiction film, proposing a filmography for the city.
- Line 2 (cinemas locations). It includes references to old and new cinemas (movie halls) in Thessaloniki, which provide a concise history of the film viewing experience in Thessaloniki.
- Line 3 (film festival timeline). It is exclusively dedicated to the Thessaloniki International Film Festival and its history since its foundation in 1960 as a local film festival until today. The Thessaloniki International Film Festival is "the top film festival of South Eastern Europe, the presentation platform for the year's Greek productions, and the primary and oldest festival in the Balkans for the creations of emerging film makers from all over the world"[4].

The "Cine Metro" app attempts to offer to its users a unique digital tour in Thessaloniki's film history, which **unfolds stories about the city and the cinema for more than a century**, by linking people, spaces, places and memories on film culture[5].

Consequently, the "Cine Metro" app could easily be employed as an innovative digital tool for an extensive exploration of the "city-cinema" relationship established by Greek popular films, in which Thessaloniki maintained such a key role as highlighted by [7]. It could also pinpoint an extensive consideration of specific historical and cultural contexts, in which "Greek film production as a whole could be considered as a database, as an audiovisual archive, which is a significant source of the city's urban memory, regardless of any other cultural values it might possess"[6].

---

[3] The "CineMetro" app was designed and developed voluntarily for educational purposes by a student team of the Department of Informatics of the Aristotle University of Thessaloniki, under the guidance of professor Athena Vakali and in collaboration with film historian Dr. Angeliki Milonaki and film critic Yannis Grosdanis. More details on the team and the app are given at : http://oswinds.csd.auth.gr/CineMetro/.

[4] The profile of the Thessaloniki International Film Festival as outlined in its official website: http://www.filmfestival.gr/default.aspx?lang=en-US&page=586.

[5] The app is available for download in the following link in Google Play: https://play.google.com/store/apps/details?id=cinemetroproject.cinemetro.

[6] A. Poupou. "Cities shapes: Film prologues, introductory sequences and urban iconography" (in Greek). In: E. Sifaki, A. Poupou and A. Nikolaidou (eds). City and cinema. Athens: Nissos, 2011, p. 86.

Fig. 2. CineMetro app screenshots

4   Impact Potential and Adoption Feasibility

CineMetro is primarily addressing the interaction of citizens with a mobile app and ICT experiencing, but it further builds on the impact of cultural digitized content exposing in the ways people behave and interact in public spaces. City virtual spaces and leveraging trails by using simple and user friendly software modules design, enables future development of tools for data collection, various data threads integration, new smart cities installations cross-referencing etc.

In the collaborative economy, especially innovative ICT Services in smart cities have a strong business potential, as a result of the increasing urbanization which comes with a host of challenges for cities, local governments, businesses and citizens. The Cine-Metro extended apps can serve as a valuable tool for smart city application service providers (including SMEs) that are active in offering added-value to existing social, sensor networks and services to urban regions.

Such novel knowledge of a city's tales and trails will facilitate software tools implementation, intending to develop processes of data acquisition which will enable:

- **Social media monitoring and semantics analysis** through the collection of multimedia content shared through the CineMetro social networks and mapping of a number of different social networks to a single representation (to support hyperconnected virtual interactions cases).
- **Data collection from activities** to support, even in real time, events and experiences of life in public spaces, on the basis of the cultural content which is attached at the particular public spaces. As an example, an **online gaming experience** can be built on the basis of CineMetro and in relevance to cinematic city stories and their todays influence. The CineMetro mobile app will allow information gathering with respect to specifically chosen public spaces. The content gathered will then be classified and entered into a data repository, structured to provide specific keywords and ensure the necessary information to inform sociological analysis of complex social phenomena.
- **Physical sensing technologies** may additionally support highly useful capabilities, including: people participation dynamics, social proxemics (distances between people, people's trajectories, interactions' identification), as well as estimation of demographics (age, gender) and emotional state of people for some closed groups cases which will agree on particular purpose city experimenting. The targeted such advances may enable public spaces absolutely anonymous trajectory to maintain highest ethical standards and fully preserve privacy.

Table 1 indicates the level of CineMetro and its extension expected impact to each market stakeholder according to the segment targeted (top row covers the market segments) mostly relevant to the proposed work, and left column addresses the target stakeholders' categories. The level of impact is marked with a respective number of asterisks (*) correlating stakeholders and segments targeted (i.e. the more asterisks the higher impacted markets).

Table 1. CineMetro stakeholders impact

	Cities Services	City apps	Urban planning	ICT apps
SMEs	**	*	*	**
Startups	**	***	*	***
Entrepreneurs	**	***	*	***
Authorities; Policy Makers	***	**	**	*

Aristotle University has a strong academic audiences appeal since Aristotle University is the largest University in Greece and it covers all fields of study[7]. In CineMetro both computer scientists and film studies experts are involved and project's progress and results are already disseminated accordingly. All these academic audiences are encouraged to participate in any relevant events, activities and tasks. Aristotle University's groups OSWINDS which has implemented CineMetro, maintains close connections with other local non-academic organizations who promote innovative city experimenting and practices. As a proof of concept, in terms of its impact and future adoption, CineMetro has already been communicated to the next two Thessaloniki's popular SMEs/start-ups which are involved in the CineMetro dissemination plans:

- **Thessaloniki Walking Tours** team to discover the city of Thessaloniki, its history, its gastronomy, its people and their habits, its secrets and legends. This start-up invites you to experience the authentic aspects of the city through well-designed theme walks specifically aimed at providing the information and the means to spend a fascinating day in the life of this 2-thousand-year old city. Connecting past and present, we walk together through the streets of Thessaloniki, its sights, its markets, its neighbourhoods and its secret corners. With us, you will discover the human stories behind its important monuments, its art and its culture, the micro-history associated with the major events that shaped the complex character of Thessaloniki and its unique adventure through time (http://thessalonikiwalkingtours.com/)
- **Parallaxi** during the 24 years of its creative presence in the editing landscape of the country organized and continues to organize big events that alter the everyday life of the city. Like the exhibition of Greatest Kitsch at the Centre of Contemporary Art, the Cinema on the Street in five cities of the district, a rural summer cinema at ten characteristic monuments and neighbourhoods around the city and a giant urban experiment for a city in a Different View in June 2010 entitled "Thessaloniki Allios" (Differently). From the experiments of urban activists team "Thessaloniki Allios" was established and today still continues its engagement and action, with already 28 actions for design, architecture, environment, social inclusion etc. which have brought 150.000 citizens to its projects[8].

---

[7] Aristotle University of Thessaloniki https://www.auth.gr/en/uni.
[8] Thessaloniki Allios http://www.parallaximag.gr/thessaloniki/thessaloniki-allios.

5   Conclusions

In the collaborative economy, innovative ICT applications, tools, and services for smart cities have a strong societal and business potential, as a result of the increasing urbanization which comes with a lot of challenges for cities, local governments, businesses and citizens.

As it is evident from the proposed work, citizen ground up groups have already addressed the need to proceed on more humanizing the city efforts, integrating cultural content, todays actions, and people's perceptions. The proposed CineMetro mobile application materializes an approach which reveals city's film content relevance with city's spaces. Under an extended application further improvements involving analytics can be applied, offering a valuable tool for smart city application service providers (including SMEs) which are active in offering added-value to existing social networks and services.

Acknowledgments. The CineMetro android and iOS mobile applications were developed by the Informatics Department students : George Haristos, Kaltirimidou Effrosyni, Paniskaki Kyriaki, Papazoglou Christos, Syrtari Charikleia, Vaena Paraskevi whom the authors thank for their high quality developing code skills and their valuable contribution.

References

1. Cranshaw, J., Schwartz, R., Hong, J.I., Sadeh, N.M.: The livehoods project: utilizing social media to understand the dynamics of a city. In: International AAAI Conference on Weblogs and Social Media (ICWSM 2012) (2012)
2. Chesñevar, C.I., Maguitman, A.G., Estevez, E., Brena, R.F.: Integrating argumentation technologies and context-based search for intelligent processing of citizens' opinion in social media. In: Proceedings of International Conference on Theory and Practice of Electronic Governance (ICEGOV 2012), pp. 166–170 (2012)
3. Esuli, A., Sebastiani, F.: SENTIWORDNET: a publicly available lexical resource for opinion mining. In: Proceedings of LREC 2006, the 5th Conference on Language Resources and Evaluation, pp. 417–422. Genova (2006)
4. Guardian, by Poole, S.: The truth about smart cities (2014). http://www.theguardian.com/cities/2014/dec/17/truth-smart-city-destroy-democracy-urban-thinkers-buzzphrase
5. Johansson, F., Brynielsson, J., Quijano, M.N.: Estimating citizen alertness in crises using social media monitoring and analysis. In: Memon, N., Zeng, D. (eds.) 'EISIC', IEEE Computer Society, pp. 189–196 (2012)
6. Li, N., Chen, G.: Analysis of a location-based social network. In: Proceedings of The 2009 International Conference on Computational Science and Engineering (CSE 2009), vol. 4, pp. 263–270. IEEE Computer Society, Washington (2009)
7. Milonaki, Grosdanis, Y. (eds.): Cine thessaloniki. In: Stories for the City and the Cinema (in Greek), pp. 9–10. University Studio Press, Thessaloniki (2012)
8. Nikolaidou, A.: Cinematic uses of the athenian monuments. In: Cities in Film: Architecture, Urban Space and the Moving Image. International Conference, School of Architecture, University of Liverpool, 26–28th March 2008
9. Re.Work (2014). Cities Summit. London (2014). https://www.re-work.co/events/cities-2014/

10. Rösler, R., Liebig, T.: Using Data from Location Based Social Networks for Urban Activity Clustering. Geographic Information Science at the Heart of Europe, pp. 55–72. Springer, Heidelberg (2013)
11. Scellato, S., Mascolo, C., Musolesi, M., Latora, V.: Distance matters: geo-social metrics for online social networks. In: Proceedings of the 3rd Conference on Online Social Networks (WOSN 2010). USENIX Association, Berkeley (2010)
12. Scellato, S., Mascolo, C.: Measuring user activity on an online location-based social network. In: Proceedings of Third International Workshop on Network Science for Communication Networks (NetSciCom). Colocated with Infocom (2011)
13. Vakali, A., Anthopoulos, L.G., Krco, S.: Smart cities data streams integration: experimenting with internet of things and social data flows. In: Akerkar, R., Bassiliades, N., Davies, J., Ermolayev, V. (eds.) 4th International Conference on Web Intelligence, Mining and Semantic (WIMS 2014). ACM (2014)
14. Vakali, A., Angelis, L., Giatsoglou, M.: Sensors talk and humans sense Towards a reciprocal collective awareness smart city framework. In: International Conference on Communications (ICC) Workshops, pp. 189–193. IEEE (2013)
15. Zheng, Y., Xie, X., Ma, W.-Y.: Mining interesting locations and travel sequences from GPS trajectories. In: Proceedings of the 18th International Conference on World Wide Web (WWW 2009), pp. 791–800. ACM, New York (2009)
16. Zheng, Y., Zhang, L., Xie, X.: Recommending friends and locations based on individual location history. ACM Trans. Web 5(1), 44 (2011). Article 5

Privacy Through Anonymisation in Large-Scale Socio-Technical Systems: Multi-lingual Contact Centres Across the EU

Claudia Cevenini, Enrico Denti, Andrea Omicini$^{(\boxtimes)}$, and Italo Cerno

ALMA MATER STUDIORUM–Università di Bologna, Bologna, Italy
{claudia.cevenini,enrico.denti,andrea.omicini,italo.cerno}@unibo.it

Abstract. Large-scale *socio-technical systems* (STS) inextricably interconnect individual–e.g., the right to privacy–, social–e.g., the effectiveness of organisational processes–, and technology issues—e.g., the software engineering process. As a result, the design of the complex software infrastructure involves also non-technological aspects such as the legal ones—so that, e.g., law-abidingness can be ensured since the early stages of the software engineering process.

By focussing on *contact centres* (CC) as relevant examples of knowledge-intensive STS, we elaborate on the articulate aspects of *anonymisation*: there, individual and organisational needs clash, so that only an accurate balancing between legal and technical aspects could possibly ensure the system efficiency while preserving the individual right to privacy. We discuss first the overall legal framework, then the general theme of anonymisation in CC. Finally we overview the technical process developed in the context of the BISON project.

Keywords: Socio-technical systems · Contact centres · Anonymisation · Privacy

1 Introduction

Socio-technical systems (STS) are those systems where "the infrastructure is technology, but the overall system is personal and social, with all that implies" [12]. Large-scale STS [9] are nowadays typically characterised by a large number of participants and components, as well as by a huge amount of available data—recorded, produced, and used by the system activities.

Among the most relevant cases of large-scale STS are *contact centres* (CC)—in particular in Europe, where they involve nearly 1 % of its active population. CC are clearly *knowledge-intensive* systems, since they typically produce a wealth of spoken data, which are mined either manually or by rudimentary technical means. Spoken data in large-scale European CC are often *multilingual* and involve *multiple countries*, meaning that both national and EU laws and regulations on personal data and privacy have to be taken into account. In particular, processing of personal data is only performed when necessary, and by prior obtaining the *data subject*'s consent for the specific processing purpose.

© Springer International Publishing AG 2016
F. Bagnoli et al. (Eds.): INSCI 2016, LNCS 9934, pp. 291–305, 2016.
DOI: 10.1007/978-3-319-45982-0_25

Typical technology issues of CC as STS involve (1) basic speech data mining technologies with multi-language capabilities, (2) business outcome mining from speech, and (3) CC support systems integrating both speech and business outcome mining in user-friendly way. Scaling up to big (i.e., massive) data processing clearly scales up also the privacy and data protection issues. Moreover, when industrial research is performed, a distinction has to be made between the research phase – when software and technologies are being developed and tested – and the subsequent market stage—when real customer data are processed. These are the very motivations behind this paper: that is, how complex legal issues at both national and international level can be dealt with while building a complex software infrastructure for CC—both in the development and in the subsequent business phases. So, first of all, this paper aims at investigating how complex software infrastructures for CC may be developed and marketed in the full respect of the data protection legal framework.

The legal analysis should thus necessarily complement and support the technical work since the very early stages, acting as an enabler rather than an obstacle, by providing the legal framework within which a CC system may be developed and used. The analysis should: identify and analyse the legal requirements of speech data processing systems; investigate the relevant legal framework; determine the impact of legal and ethical issues on the deployment of the CC infrastructure; examine how such a system should be designed and used so as to comply with the applicable legal framework, while identifying the barriers that could potentially affect its design and deployment; and, finally, keep an eye over the procedures for data collection, storage, protection, retention, and destruction, so that they comply with national and EU legislation.

In this paper we focus on *anonymisation* [7] as a fundamental tool to deal with the potential conflict between opposite rights and needs, especially in the research and development phase of a large-scale, knowledge intensive STS. Conceptually speaking, anonymising amounts at defining *which* and *how much* information should be removed for some data to be acceptedly considered as anonymous—i.e, not de-identifiable [1] with "normal" means; technically speaking, an effective anonymisation process needs to suitably balance the effort for anonymising data with *(a)* the value of the resulting data, and *(b)* the purpose for which they are collected and used. In fact, while in principle the total anonymisation of personal data would obviously address the users' desire for privacy, the full availability of (spoken) data is often essential for the organisation to fulfil its goals—and at least useful for the efficiency of its processes.

As a result, the need for a suitable compromise between law-abidingness (and privacy needs), on the one side, and system and process efficiency, on the other, is a relevant goal not just for the legal analysis, but for the whole engineering process that leads to the construction of the CC infrastructure, so that a potential conflict of interests becomes *composition* of interests, and the law-abidingness requirement can be exploited as a *success factor* instead of being perceived as a possible source of delays and overheads—an issue that goes well beyond the (noteworthy) case study discussed here.

In the remainder of this paper we first recall the main legal issues (Sect. 2), then perform a socio-legal-technical analysis aimed at identifying the most relevant principles (about data protection and processing, about security measures, and others) and the consequent technological requirements (Sect. 3) as a pre-requisite to frame and discuss in depth the anonymisation process—first in general terms (Sect. 4), then in the specific context of the BISON project (Sect. 5). There, the specific goal is to understand how to structure the anonymisation process during the industrial research phase, yet without compromising the quality of development and testing, which is based on the data used, allowing the resulting STS to eventually deal with the proper amount of data when it reaches the business operation phase.

2 Legal Framework

Data protection is a fundamental human right, recognised by Council of Europe Convention – Treaty 108 [11], the first legally binding international instrument for data protection, by the Treaty on the Functioning of the European Union [6], and by the Charter of Fundamental Rights of the European Union [2].

The legal framework at EU level is laid down by Directive 95/46/EC (Data Protection Directive, or *DPD* [4]). Brought into force by all EU Member States national law, the DPD contains key principles for the fair and lawful processing of personal data, together with the technical and organisational security measures designed to guarantee that all personal data are safe from destruction, loss, alteration, unauthorised disclosure, or access, during the entire data processing period. Data processing requires even more care when it involves *large amounts* of personal and/or sensitive data: in particular, people should be given the possibility to manage the flow of data relating to them across massive, third-party analytical systems, so as to have a transparent view of how information data will be used, or sold.

The *data transfer from and outside the EU and cloud services* is therefore a particularly hot topic, since non-EU countries might provide an insufficient level of protection to personal data. This is why the flow of personal data is free between EU Member States, whereas the DPD sets restrictions on the export of personal data to third countries not ensuring an *adequate* level of data protection. Adequacy of data protection in a third country means that the main principles of data protection are effectively implemented in the national law of that country. Therefore, when there is no specific consent to the data transfer outside the EU given by the data subject, and when the level of data protection of the recipient's country is not deemed adequate, the data controller may be required – before exporting personal data – to contractually bind the recipient to set up enough security and organisational measures to grant adequate protection of the personal data—e.g., through standard contractual clauses, binding corporate rules.

2.1 Personal Data and (de-)identification

Personal data consist of any information relating to a natural person, who can be identified, either directly or indirectly, by reference to one or more factors specific to his/her physical, physiological, mental, economic, cultural, or social identity. It should be noticed that if the link between an individual and his/her data never occurred, or, it is somehow broken and cannot be rebuilt in any way (as in the case of anonymised data), the DPD rules no longer apply: this is why anonymisation turns out to be a fundamental tool to simplify both the industrial research process and the processing system design and development—clearly, in as much as the data value is not compromised. With respect to this issue, it is worth recalling the Explanatory Report to Convention 108 [11], which states that

- "identifiable person" means a person who can be *easily* identified: it does not cover identification of persons by means of 'very sophisticated methods' (Article 2, Sect. 28);
- 'the requirement appearing under litterae concerning the time-limits for the storage of data in their name-linked form does not mean that data should after some time be *irrevocably separated* from the name of the person to whom they relate, but only that it should not be possible to *link readily* the data and the identifiers' (Article 5, Sect. 42).

As concerns the collection of personal data (the very first processing operation), the DPD sets out some basic definitions and principles for lawful processing.

First, the DPD identifies two distinct roles: the *data controller* and the *data processor*. The former is in charge of personal data processing and takes any related decision—e.g. selection of data to be processed, purposes and means of processing, technical and organisational security, etc. The latter, instead, is a legally separate entity that processes personal data on behalf of a controller, in force of a written agreement and following specific instructions. In other words, the controller processes data on its own behalf, while the processor always acts on behalf of a controller, from whom it derives its power and range of activity. For instance, a company acts as a controller in processing its own customers' data, whereas the CC entrusted with the same processing acts as a processor on behalf of the company.

Personal data must be obtained and processed lawfully, be collected for *explicit* and *legitimate* purposes, and *used accordingly*. The processing purposes must always be clearly declared, primarily to the data subject, who has to be specifically informed of all elements related to the processing, before the processing itself is started: the data subject has then to provide his/her free, specific, and unambiguous *consent*. Any processing for undefined purposes is not law-abiding, and the consent given in such cases is not deemed valid. The same applies when the data subject is asked for just one consent in view of a plurality of purposes.

The data controller must not use the data collected for a given purpose to pursue a different one, also at a different time (i.e. after the declared purpose has been achieved): any further use of personal data for other purposes requires an additional legal basis if the new purpose of processing is incompatible with the original one. Furthermore, the data collected must be *strictly consistent* with the declared purposes: it is unlawful to collect more data than necessary (a.k.a. *principle of necessity*). Any data transfer to third parties is also a new purpose, potentially requiring additional legal support.

Personal data must also be *relevant* and *not excessive* in relation to the purposes for which they are collected and processed: only the specific data which are actually necessary to achieve a given purpose may be collected and processed, any wider collection resulting in law infringements. Personal data must also be *accurate* and *up to date*: whenever pieces of information on a given data subject turn out to be wrong, or need to be changed, the personal data must be consequentially corrected.

From the timing viewpoint, data may be retained *only for the period needed* to achieve the specific purposes for which they are being processed: then, they should be erased. However, it is possible to continue the data processing beyond the originally declared purposes if personal data are anonymised—that is, they cannot be linked back to an individual in any way. So, the anonymisation process may be regarded to as the last authorised operation of the processing of personal data, before they cease being personal data to become simply "data".

2.2 Accountability and Security Measures

According to the *accountability* principle, data controllers have to implement adequate measures to promote and safeguard data protection in their processing activities. Controllers are responsible for the compliance of their processing operations with data protection law, and should be able to demonstrate compliance with data protection provisions at any time. They should also ensure that the practical measures implemented to comply with data protection principles are effective. In case of larger, more complex, or high-risk data processing, the effectiveness of the measures adopted should be verified regularly, through monitoring, internal and external audits, etc.

Technical and organisational *security measures* should be adopted to protect personal data, during all the processing period, against the risks related to the integrity and confidentiality of data, in particular where the processing involves the transmission of data over a network, and against all other unlawful forms of processing. The level of data security requested by the law is determined by different elements, such as the nature (sensitive/non-sensitive) of the collected data, the concrete availability in the market of adequate security measures at the current state of the art, and their cost—which should not be "disproportionate" with respect to the necessity.

It is worth pointing out, however, that security measures are not limited to technical remedies, but also include organisational rules and procedures that should be strictly observed by all the subjects involved in data processing. The overall quality of the security measures is the result of a case-by-case evaluation, which must be performed before starting new personal data processing operations, and then implemented and adapted, when needed, during the whole data processing period, with regard to the technical solutions, and the human factor, too.

2.3 Big (Speech) Data

A CC infrastructure involves *speech recordings*, that is, processing biometric data (as in the case of the analysis of the tone, pitch, cadence, and frequency of a persons voice) for determining whether a person is who he/she declares to be.

From a data protection perspective, biometric technologies are closely linked to physical, physiological, behavioural, or even psychological characteristics of an individual—and some of them may be used to reveal sensitive data. Biometric data may also enable automated tracking, tracing, or profiling of persons: as such, their potential impact on privacy is quite relevant. Moreover, biometric data are by nature *irrevocable*: a breach concerning biometric data threatens the further safe use of biometrics as identifier, as well as the right to data protection of the concerned persons for whom there are no chances to mitigate the effects of the breach.

Therefore, the processing of biometric data is not only subject to the express consent of the data subject, but may also depend on the authorisation by Data Protection Authorities, and is submitted to strict rules on security measures that must be adopted to protect data: for instance, biometric information should always be stored in encrypted form; decryption keys should only be accessible on a need to know basis; the system should be designed in a way that allows the identity link to be revoked, either in order to renew it or to permanently delete it if the consent of the data subject is revoked; etc.

In this context, large-scale STS such as CC deal with big data because of the huge amount of data they collect, even though from a limited number of sources. Two main issues are worth highlighting:

– Big data analytics can involve the *repurposing* of personal data.
  If an organisation has collected personal data for one purpose and then decides to start analysing it for another one (or to make it available for others to do so), data subjects need to be informed of this novelty, and a new, specific consent is usually needed.
  This is particularly important if the organisation is planning to use the data for a purpose that is not apparent to the individuals because it is not obviously connected with their use of a service.
– Big data may intrinsically contrast with the principle of data minimisation and relevancy: the challenge for organisations is to focus on what they expect to learn, or, to be able to do by processing big data before the beginning of processing operations, thus verifying that these serve the purpose(s) they are to be collected for, and, at the same time, that they are relevant and not excessive in relation to such aim(s).

3   Socio-Legal-Technical Analysis

The current legal framework foresees a set of essential principles that should inspire the design and development of any law-abiding information system processing personal data. While some of such principles directly derive from the DPD – namely, from the "Principles relating to data quality" –, others concern the security measures that should be adopted, particularly with reference to the "Security of processing". These principles are further strengthened and detailed in the "General Data Protection Regulation" (GDPR) [8].

3.1   Relevant Principles

Relevant principles can be conceptually organised in three major categories, which are shortly detailed in the following:

(a) principles about data processing
(b) principles about security measures
(c) other relevant principles

Principles About Data Processing

1. *Principle of lawfulness and fairness*: any processing of personal data must be lawful and fair to the individuals concerned.
2. *Principle of relevance and non-excessive use*: personal data must be adequate, relevant and not excessive in relation to the purposes for which they are collected and/or further processed.
3. *Principle of purpose*: personal data must be collected for specified, explicit and legitimate purposes and not further processed in a way incompatible with those purposes.
4. *Principle of accuracy*: data must be accurate and, where necessary, kept up to date; every reasonable step must be taken to ensure that data which are inaccurate or incomplete, having regard to the purposes for which they were collected or for which they are further processed, are erased or rectified.
5. *Principle of data retention*: data must also be kept in a form which permits identification of data subjects for no longer than is necessary for the purposes for which the data were collected or for which they are further processed; Member States shall lay down appropriate safeguards for personal data stored for longer periods for historical, statistical or scientific use.

Principles About Security Measures

1. *Principle of privacy by design*: the protection of the rights and freedoms of data subjects with regard to the processing of personal data requires appropriate technical and organisational measures, at the time of the design of the processing system as well as at the time of the processing itself, particularly in order to maintain security and thereby to prevent any unauthorised

processing; these measures must ensure an appropriate level of security, taking into account the state of the art and the costs of their implementation in relation to the risks inherent in the processing and the nature of the data to be protected.

2. *Principle of appropriateness of the security measures*: the controller must implement appropriate technical and organisational measures to protect personal data against accidental or unlawful destruction or accidental loss, alteration, unauthorised disclosure or access, in particular where the processing involves the transmission of data over a network, and against all other unlawful forms of processing. Having regard to the state of the art and the cost of their implementation, such measures shall ensure a level of security appropriate to the risks represented by the processing and the nature of the data to be protected.

3. *Principle of privacy by default*: the controller shall implement mechanisms for ensuring that, by default, only those personal data are processed which are necessary for each specific purpose of the processing and are especially not collected or retained beyond the minimum necessary to achieve those purposes, both in terms of the amount of the data and the time of their storage. In particular, those mechanisms shall ensure that by default personal data are not made accessible to an indefinite number of individuals.

Other Relevant Principles

1. *Principle of least privilege*: in a given abstraction layer of a computing environment, every module (process, user, program) must be able to access only the information and resources that are necessary for its legitimate purpose.

2. *Principle of intentionality in performing any critical action*: examples include granting access to a wider set of users, selecting lower security settings, exporting data, reducing the number of anonymised items, etc.

3.2   Consequent Technological Requirements

The above principles translate into actual system requirements ranging from the system configuration to the user management, the way data are processed, and the security measures in general. Abstracting from any specific technical solution, the following issues can be identified and should be accounted for:

1. Different user groups and different users, with different privileges and access rights, so that each user is granted only the minimum set of rights that is necessary for his/her task, coupling maximum flexibility with maximum security; this asks for role-based authentication model, fine-grained set of user rights, adequate authentication mechanisms.

2. Default user profiles with the minimum set of rights, so that any addition of user rights giving access to data is always intentional.

3. Default anonymisation configuration corresponding to the maximum level of anonymisation, so that any custom configuration implying a decrease of

anonymisation level is always explicitly authorised, and therefore intentional and security-checked before proceeding.

4. Adequate support of detailed customised anonymisation levels, so that the system settings can be fine-tuned to the specific customer necessity—and no further.

5. Multiple levels of security, with proper warnings, whenever the default (i.e., maximum security) settings are lowered for any reason—e.g. during the custom configuration by authorised and suitably authenticated personnel.

6. Clear identification of the use-case scenarios when (authorised) personnel is allowed to access personal data—that is, non-anonymised copies of recordings, lawfully stored for authorised processing.

7. Severe restrictions on, or even denial of, transfer of non-anonymised data.

8. Immediate removal/wiping of non-anonymised copies of data recordings that may have been made during the processing, if required by the processing itself, as soon as their presence is no longer required.

4 Anonymisation Process

As detailed above, the legal framework allows personal data to be processed only to the extent they are needed to achieve specific purposes: whenever identifying data are not necessary, only anonymous data should be used.

As far as legal requirements for anonymisation are concerned, the DPD does not apply to data made anonymous in such a way that the data subject is *no longer identifiable*: it does not set any prescriptive standard, nor does it describe the de-identification process—just its outcome, which is a *reasonably-impossible* re-identification. The concrete application of such a general principle, however, is not easy: the main problem is to create a truly anonymous dataset, while retaining at the same time all the data required for a specific (organisational) task. On the other side, irreversibly-preventing identification requires data controllers to consider all the means which may likely reasonably be used for identification, either by the controller or by a third party.

The Directive 2002/58/EC (e-Privacy Directive) [5] also imposes anonymisation in certain cases. For instance, as far as subscribers' traffic data are concerned, processed within electronic communications networks to establish connections and to transmit data, it foresees that, when used for marketing communications services or value added services, personal data should be erased or made anonymous after the provision of the service. Besides, it imposes the providers of a public communication network or a publicly-available electronic communication service to erase or anonymise data no longer needed to transmit a communication.

The *Article 29 Working Party – Opinion on Anonymisation Techniques* (Article 29 WP henceforth) [3] is an important reference for compliance in anonymisation issues: it describes the main techniques used to anonymise personal data, and explains their principles, strengths and weaknesses, possible mistakes, and failures. The criteria on which Article 29 WP grounds its opinion on robustness focus on the possibility of:

- singling out an individual;
- linking records relating to an individual;
- inferring information concerning an individual.

Assuming that personal data have been collected and processed in compliance with applicable legislation, Article 29 WP on the one hand considers anonymisation as further processing, which generally speaking needs to comply with the compatibility assessment—e.g. data shall not be further processed for purposes incompatible with those specified at the moment of their first collection; on the other, however, it promotes the idea that anonymisation should be seen as further processing *compatible with the original purposes*, upon condition that the anonymisation process *reliably produces anonymised information*.

Again, a balance between different needs has to be found: although on the one hand removing directly the identifying elements in itself may not be enough to ensure the impossibility of re-identification, the principle set by Convention 108, on the other, states that identification *with very sophisticated methods* may not be relevant, as it should not be possible to identify a person *readily*.

Additional measures may often be needed to prevent identification, depending on the context and purposes of the processing for which the anonymised data will be used. In its Opinion 03/2013 on purpose limitation, Article 29 WP notes that

> Anonymisation is increasingly difficult to achieve with the advance of modern computer technology and the ubiquitous availability of information. Full anonymisation would also require, for instance, that any reasonable possibility of establishing a link with data from other sources with a view to re-identification be excluded. However, re-identification of individuals is an increasingly common and present threat. In practice, there is a very significant grey area, where a data controller may believe a dataset is anonymised, but a motivated third party will still be able to identify at least some of the individuals from the information released. Addressing and regularly revisiting the risk of re-identification, including identifying residual risks, therefore remains an important element of any solid approach in this area.

Adequate safeguards, whose strength should be proportionate to the adverse impact to the data subject of a possible re-identification, should also be considered if necessary—such as encryption, restrictions of access, etc.

More generally speaking, it is hard to assess in advance whether re-identification may or may not be possible, since it depends on how readily and how directly the link between the data and the identifiers is structured. Data controllers, for instance, should consider the *concrete means* that would be necessary to reverse the anonymisation process, their cost and know-how needed, as well as the likelihood and severity of implementing such means. They should conduct a data protection impact assessment to decide what data may be made available for reuse and at what level of anonymisation and aggregation: ideally, an impact assessment should be completed before disclosing information and making it available for reuse. Whenever controllers release anonymised datasets for use, the risk assessment should include re-identifiability tests (e.g. penetration tests). Finally, data controllers should keep into account that the risk of re-identification changes over time, with the evolution of technology: once-rare and sophisticated analytics techniques can quickly become commonly available, possibly even at low or no cost, or, new evidence could reveal accessible techniques for re-identification—like, e.g., in [13]. Thus, the data controller policies should be periodically reviewed in consideration of current and possible future threats.

Data processors, on the other hand, may process anonymised datasets communicated to them by data controllers: if they are not able to either directly or indirectly identify the data subjects of the original dataset, they will be acting lawfully with no need to consider data protection requirements. Still, before deciding if and how to use the anonymous data received by the data controller for their own purposes, they should evaluate the anonymisation techniques adopted by the data controller, because data processors may be held liable for consequences derived from their own data processing. Thus, if there is a risk of identification of the data subjects, the processing should be performed in compliance with the data protection law.

From the technical viewpoint, two aspects need be stressed: *(i)* different anonymisation techniques may be used, which may imply different levels of risk; *(ii)* anonymisation is necessarily defined with respect to some *threshold* about the easiness or probability of singling out, linking, or inferring an individual in a dataset—that is, when some data are considered sufficiently *de-identified*.

Article 29 WP – Opinion 05/2014 outlines the risks of singling out, linking, or inferring, with respect to some of the most used techniques (e.g. pseudonymisation, noise addition, substitution, etc.); in [1], a different approach, rooted in the USA, is discussed, based on the number k of *quasi-quantifiers* (i.e., identity-revealing traits) that must be cancelled in a dataset so as to ensure that the probability of re-identification is below a given threshold: this leads to the concept of *k-anonymisation* (there, a value of 3 is considered the minimum, with 5 being reasonably safe for the purpose). Also, this fits perfectly the other observation pointed out by Article 29 WP, i.e., that anonymisation can protect privacy and personal data only if anonymisation techniques are engineered and applied properly: context and objectives of the process must be clearly set to achieve the desired/required anonymisation level.

5   The Anonymisation Process in BISON

The issues discussed in this paper have been investigated in the context of the BISON project [10], aimed at developing an innovative tool for CC processing big speech data.

As it is common in industrial research, a fundamental distinction has to be pointed out between the *research phase* – when software and technologies are being developed and tested, but are not yet in actual production – and

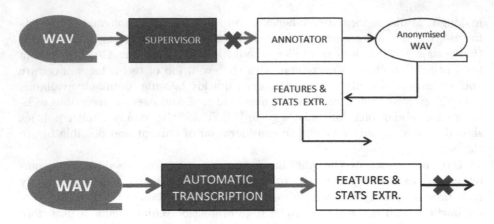

Fig. 1. Anonymisation during the Start-up stage and Research stage in BISON (Color figure online)

the subsequent, foreseeable *business phase* – when they actually deal with real customers' data.

Here, anonymisation is seen as the fundamental tool to set the industrial research phase free from the complex requirements imposed by the Data Protection rules, given that the DPD does not apply to anonymised data. At the same time, in the business phase that will follow the research project, the tool will have to deal with real user data, in compliance with applicable laws.

5.1   General Overview

In the first stage of the BISON research, data flow as in Fig. 1 (top): the anonymisation process is performed mostly with manual procedures, both because of the limited data size and because of the initial lack of automatic tools. The starting point is the audio file (WAV) of the call—which contains personal data. The call is examined by a *supervisor*, who is the only person authorised to access personal data, with the specific purpose of intercepting any personal data in the audio call and manually removing them by silencing (that is, by physically overwriting the relevant words with silence or some suitable "beep"). The result is a new audio file that contains no personal data: therefore, from this point in the data flow (red cross in the diagram), data can be considered as fully anonymised. The anonymised call can then be passed to the *annotator*, who is in charge of tagging the file with keywords, according to pre-defined technical specifications: the resulting annotated file is then further processed to extract statistics (features), which embed enough information to train the audio recognisers while making it impossible to reconstruct the original waveform and to trace back to the original personal data.

In the second stage of the BISON research, huge amounts of speech data need to be processed, which makes the manual annotation of personal data by the supervisor unfeasible: therefore, automatic transcription—for all the supported languages—has to be put in place. This change affects the stage where anonymisation takes place, since anonymisation is now performed on the original audio file (containing personal data) instead of on a manually pre-silenced audio file; it may also somehow reduce the reliability of the process, since no automatic transcriber can be considered 100 % effective in identifying terms and items related to any possible personal data: therefore, the "anonymised" file in this case may still contain some (hopefully and typically, a small amount of) personal data.

Of course, any effort should be made to reduce these errors to the minimum: thus, the automatic anonymiser should be designed and trained with the greatest possible care, and tested according to the best available practices. The subsequent feature extraction helps to deal with this issue, because the extracted statistics make it (mostly) impossible to reconstruct the original audio file. This is why the red cross in Fig. 1 (bottom) is conceptually placed only after the feature extraction step, instead of following the automatic transcription—although most of personal data are actually suppressed earlier in the chain. Features are finally exploited to feed the language recognisers with big data in many different languages (possibly under the requirement of signing a Non-Disclosure Agreement): in any case, data anonymisation takes place prior to the annotation of the recordings, which guarantees full anonymisation afterwards.

In a farther perspective, when the final system will eventually operate in the business context, the basic difference will concern *where* the extraction of statistics will be performed—namely, inside each CC, by the CC itself: so, no personal data will ever be delivered outside, and any processing will occur only provided that the appropriate consent, for the specific purpose, has been given, and in compliance with all applicable laws and regulations.

5.2 Technological Requirements

Despite the basic assumption of relying mostly on automatic anonymisation, some manual adjustments might still be necessary in the development and configuration phases—and possibly even at runtime, so as to capture any residual data that might have survived the previous checks. For this reason, automatic technologies should be coupled with an interactive tool, enabling the fine-tuning and (possibly live) control of the anonymisation process.

Such a tool should obviously adhere to strict security requirements: users' roles, rights, and restrictions should be tuneable on a fine-grain basis, and be further detailed case-by-case based both on the actual needs and the applicable national legal framework. Moreover, on-the-fly anonymisation should be available to deal with the case that some unexpected personal data are heard by the CC agent in charge of the call, requiring real-time anonymisation to be triggered.

5.3   Customisation and Future-Proofing

In the final state of the system (ready-to-market), users will need to be enabled to anonymise personal data whenever not needed for the specific purposes of the processing—and they should be able to do so in a *highly customisable* way. Customisation should be based on the specific CC requirements: for instance, it should be possible to enable anonymisation at different times (during/after the call), or based on the occurrence of specific situations, or as a feedback from speech analytics or data mining on text, etc.

A key challenge from this viewpoint is also to make anonymisation *future-proof* both with respect to a continuously-evolving legal scenario, as well as to the technology improvement, evolving even faster.

6   Conclusion

It is nowadays taken as understood that the practices of contemporary software engineering have to be extended to include non-computational issues such as normative, organisational, and societal aspects. This holds in particular for large-scale socio-technical systems: for instance, the law-abidingness of complex software systems including both human and software agents is quite an intricate issue, to be faced in the requirement stage of any reliable software engineering process.

In this paper we specifically address the problem of anonymisation of speech data in the case of contact centres, discussing the need for an accurate balancing between legal and technical aspects in order to ensure the system efficiency while preserving the individual right to privacy, and showing how the legal framework can actually translate into requirements for the software engineering process. By discussing the BISON approach, we show how the anonymisation process can be structured during the industrial research phase in order to make it possible for the resulting system to eventually deal with the amount of data actually required once it reaches the business operation phase.

Acknowledgements. This work has been supported by the EU-H2020-ICT-2014 Innovation action BISON – BIg Speech data analytics for cONtact centres (Grant Agreement no. 645323). Authors would like to thank all the partners of the BISON project for their invaluable cooperation in the development of the approach illustrated in the paper. Authors would also like to thank Dr. Silvia Bisi for her contribution to the project research.

References

1. Angiuli, O., Blitzstein, J., Waldo, J.: How to de-identify your data. Comm. ACM **58**(12), 48–55 (2015)
2. Charter of Fundamental Rights of the European Union. Official J. Eur. Communities **43**(C 364), 1–22 (2000)

3. Article 29 Data Protection Working Party - Opinion 05/2014 on anonymisation techniques, 0829/14/EN Wp. 216, 18 April 2014. http://ec.europa.eu/justice/data-protection/article-29/
4. Directive 95/46/EC of the European Parliament and of the Council of 24 October 1995 on the protection of individuals with regard to the processing of personal data and on the free movement of such data. Official J. Eur. Communities **38**(L 281), 31–50 (1995)
5. Directive 2002/58/EC of the European Parliament and of the Council of 12 July 2002 concerning the processing of personal data and the protection of privacy in the electronic communications sector (Directive on privacy and electronic communications). Official J. Eur. Communities **45**(L 201), 37–47 (2002)
6. Consolidated versions of the Treaty on European Union and the Treaty on the Functioning of the European Union. Official J. Eur. Communities 55(C 326), 1–390 (26 Oct 2012)
7. Fung, B.C.M., Wang, K., Chen, R., Yu, P.S.: Privacy-preserving data publishing: a survey of recent developments. ACM Comput. Surv. **42**(4), 14:1–14:53 (2010)
8. Regulation (EU) 2016/679 of the European Parliament and of the Council of 27 April 2016 on the protection of natural persons with regard to the processing of personal data and on the free movement of such data, and repealing Directive 95/46/EC (General Data Protection Regulation) (text with EEA relevance). Official J. Eur. Communities **59**(L 119), 1–88 (2016)
9. Omicini, A., Zambonelli, F.: Coordination of large-scale socio-technical systems: Challenges and research directions. In: Di Napoli, C., Rossi, S., Staffa, M. (eds.) WOA 2015 - From Objects to Agents. CEUR Workshop Proceedings, vol. 1382, pp. 76–79. Sun SITE Central Europe, RWTH Aachen University, Napoli, Italy, 17–19 June 2015. Proceedings of the 16th Workshop "From Objects to Agents"
10. The BISON Project. http://bison-project.eu/
11. Convention for the protection of individuals with regard to automatic processing of personal data, 28 January 1981
12. Whitworth, B.: Social-technical systems. In: Ghaou, C. (ed.) Encyclopedia of Human Computer Interaction, pp. 533–541. IGI Global (2006)
13. Zang, H., Bolot, J.: Anonymization of location data does not work: a large-scale measurement study. In: 17th Annual International Conference on Mobile Computing and Networking (MobiCom 2011). pp. 145–156. ACM, New York (2011)

The Butlers Framework
for Socio-Technical Smart Spaces

Roberta Calegari$^{(\boxtimes)}$ and Enrico Denti

Dipartimento di Informatica-Scienza e Ingegneria (DISI),
ALMA MATER STUDIORUM–Università di Bologna,
Viale Risorgimento 2, 40136 Bologna, Italy
{roberta.calegari,enrico.denti}@unibo.it

Abstract. Smart Spaces outline an intriguing application scenario where people are immersed in time and space in an augmented virtual environment, which suitably exploits ubiquitous computing technologies, space and time awareness, and pervasive intelligence. A number of technical, social, pragmatic challenges, arising from several perspectives and domains, need to be dealt with.

Moving from a Socio-Technical Systems approach, this paper first introduces the Butlers for Smart Spaces (BSS) architecture, which specialises the Butlers architecture originally defined for Smart Homes to the Smart Spaces scenario; then, it shows how BSS can be mapped onto the Home Manager platform, and discussed how a Smart Space can be designed and developed on its top—in general, and with regard to a specific example.

Keywords: Socio-Technical Systems · Smart Spaces · Home Manager · Home intelligence · Butlers architecture

1 Introduction

Smart spaces shape environments such as apartments, offices, museums, hospitals, schools, malls, outdoor areas, etc., referring to a lifestyle where computer systems seamlessly integrate into people's everyday lives, providing services and information "anywhere, anytime" [31,33] in a *pervasive, distributed, situated* and *intelligent* way [29]. A wide variety of scenarios have been explored over the years, addressing different physical spaces, applications and goals, but today the development of the Internet of Things [22,29,30,34] is providing the enabling technology to make such scenarios concrete: appliances and devices of any sort are being developed that embed network capabilities, often with some form of on-board intelligence. Not surprisingly, the major players are promoting their architectures and technologies – from Google's 'Works With Nest' [13] to Samsung SmartThings [32], Apple's Home Kit [2], Windows 10 IoT Core [22] – up to Amazon's Smart Home shop [1]. Personal assistants with natural language capabilities, like Google Now [14], Siri [3], Cortana [23], are also developing further capability of giving suggestions based on the user's current context and habits.

© Springer International Publishing AG 2016
F. Bagnoli et al. (Eds.): INSCI 2016, LNCS 9934, pp. 306–317, 2016.
DOI: 10.1007/978-3-319-45982-0_26

Socio-Technical Systems (STS), on the other hand, arise "when cognitive and social interaction is mediated by information technology, rather than by the natural world (alone)" [41]: they are by nature heterogeneous, distributed, made both of software agents, (sensors, actuators) and humans with their capabilities and social organisations [19,43]. Spread over a (potentially huge) number of autonomous components, with no centralised control, mixing humans and ICT components, STS raise several issues as concerns their analysis, design and implementation—from the capability of effectively coordinating the activities of so many decentralised components, to the time- and space-situatedness in dynamic, unpredictable, socio-physical environments.

In this paper we focus on Socio-Technical Smart Spaces (STSS)—the kind of STS which arise from Smart Spaces, in particular in the Smart Home context. Moving from an STS approach, we first introduce the *Butlers for Smart Spaces* framework as a possible reference architecture for STSS (Sect. 2), then present the Home Manager platform (Sect. 3) for the implementation of smart spaces in Smart Home contexts—namely, in IoT-aware environments. To show how it can be actually used, we take the case study of a Smart Oven and discuss its development and implementation on top of Home Manager (Sect. 4). Related work and conclusions are discussed in Sect. 5.

2 Butlers for Smart Spaces

The Butlers architecture [7] defines a technology-neutral framework made of seven conceptual layers, relating technologies, features and the corresponding value-added for users (Fig. 1). Although originally defined for the smart home context, it can be fruitfully specialised to STSS, and specifically to IoT scenarios. The bottom layers concern the enabling technologies, such as mono or bidirectional communication-enabled sensors, meters, actuators, etc.; in the middle, infrastructural/middleware layers aim to provide coordination and geographical information services. The top layers focus on specific aspects, like intelligence, sociality, gamification: as such, they are not necessarily to be taken in the sequence. The resulting map can be used both to locate a system based on its feature or, conversely, to identify unexplored market niches.

Most of today's smart/domotic devices (e.g. [1]), in particular, basically provide just remote monitoring or control facilities via some suitable Android/iOS app: so, they are conceptually located at level 2. The Butlers vision suggests that this is just the first chapter of the story: there's much more to be added to achieve real *Smart* Spaces. This is where the upper layers should come into play—and the conceptual map provides its key to what, how and why.

Leaving the full discussion to [7], what is relevant here is that a smart space (e.g. a smart home) can interact with its users not only to monitor (level 1) or remote-control (level 2) the environment (e.g. the home appliances), but – via a suitable coordination infrastructure (level 3) – more generally to provide an *immersed, smart experience*, taking into account the users' habits, behaviour, location, preferences (level 4) to reason on the overall situation (level 5) and

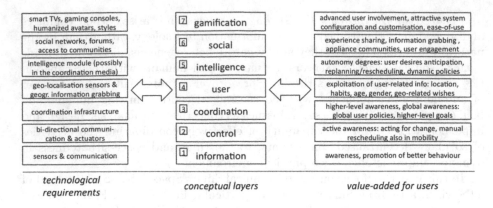

Fig. 1. Butlers multi-layer reference architecture, from [7]

possibly anticipate the user's needs. In this view, social networks (level 6) can be further sources of user-related information, while gamification (level 7) can be essential for technology acceptance—a crucial success factor in STS, where the human factor is at least as relevant as advanced technologies. *Butlers for Smart Spaces* (Fig. 2) is the contextualisation of the Butlers vision to Socio-Technical Smart Spaces. In this context, some lower-level functionalities are typically provided by the underlying infrastructure, while some envisioned upper functionalities are too far from the foreseeable future or from the current state of the art, so their layers can be collapsed/dropped. This is why information (1) and control (2) layers are grouped together in a single *Monitoring* layer—the ability to act on the environment being a fundamental property of the Smart Space notion itself. Smart Spaces also grab a lot of raw information, which needs to be pre-processed to become exploitable knowledge: since this activity is somehow in-between information retrieval (layer 1) and coordination (layer 3), a single *Services* layer is introduced on top of the Monitoring layer.

Moreover, since users and environment are the main protagonists of a Smart Space, coordination must necessarily take users into account, so coordination (3) and user-aware (4) layers can also be conveniently grouped. Such coordination is likely to be complex enough to justify a clear separation between (general and user-specific) goals and policies, so that different policies can be developed for the same goals. Accordingly two mid-layers, *Goals* and *Policies*, are introduced side-by-side at that level, making a step towards pro-activity and situatedness. Moving up, the very nature of a Smart Space suggests that the reasoning about the surrounding environment, which shapes the "Space", can be conveniently separated from the "more basic" reasoning layer—for both conceptual and practical reasons. The *Reasoning* and *Situated Reasoning* layers capture this separation, representing, respectively, the reasoning capabilities which exploit only the local/user knowledge, and which exploit also the surrounding environment. Gamification is left aside at this stage, as it is orthogonal to the tailoring of Butlers to the Smart Spaces context.

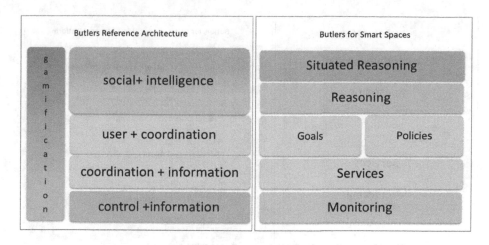

Fig. 2. Tailoring Butlers to the Smart Spaces context

3 The Home Manager Platform

Home Manager [8,9] is an open source platform [18] for STSS, inspired to the above architecture and explicitly conceived to be open, deployable on a wide variety of devices (PCs, smartphones, tablets, up to Raspberry PI 2), and – thanks to the underlying TuCSoN [26] agent infrastructure – suitable to accommodate "as much intelligence as the system needs, where the system needs".

Its purpose is to provide advanced services to users immersed in/interacting with the surrounding environment—in particular, the ability to reason on potentially any kind of relevant data, both extracted from the users's preferences and grabbed from other sources, so as to anticipate the users' needs.

3.1 Butlers on Home Manager

Figure 3 shows how the five layers of Butlers for Smart Spaces re-shape on the Home Manager platform. The TuCSoN infrastructure surrounds and encompasses all layers, as it enables the seamless integration of heterogeneous entities, bridges among technologies and agents' perceptions, and supports situated intelligence.

Each device is equipped with an agent, which acts as a sort of "proxy" to bring the physical device into the agent society that powers the Smart Space. At a basic stage, this agent enables the device monitoring and (possibly) remote control, grabbing the necessary information through TuCSoN sensors and probes, and acting on the environment via its actuators and transducers.

The Services layer takes the concrete form of *Service-Level Agents* and *Basic Policies*: the idea is that agents in this layer perform some information elaboration and possibly retrieval via mechanisms that, however, do not require sophisticated reasonings—for instance, grabbing information from weather web

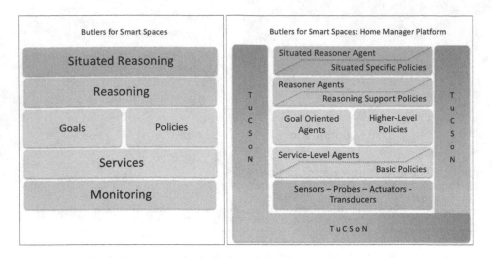

Fig. 3. Butlers for Smart Spaces on Home Manager

sites, or from selected Twitter pages, based on the selected basic policies, such as the user's preferred weather sites or followed Twitter people.

Analogously, Goals and Policies take the concrete form of *Goal-Oriented Agents* and *Higher-Level policies*, respectively. Policies at this stage typically concern everyday life habits and aspects – such as children not being allowed to set any Twitter policy, etc.; so, they are generally rather stable. Accordingly, the Goal-Oriented Agents are charged of autonomous decisions based on such policies: for instance, in the Twitter service case, the agent goal could be to retrieve suitable tweets from selected people and highlight the ones that, say, receive more than 100 likes, or refer to given topics, etc.

More complex, intelligent and fine-tuned behaviours call for further reasoning: *Reasoner Agents* are charged of potentially any kind of reasoning over user-related knowledge – typically the user's profile, habits, and preferences –, while *Reasoning Support Policies* encapsulate the corresponding rules. The top layer extends such capabilities towards situatedness, in time and space: *Situated Reasoner Agents* take into account the user location, movement, etc. to make situated deductions and perform real-time suggestions and pro-active actions: e.g., in the Twitter case, a reasoner could decide to include further Twitter pages if the user is moving to another city, assuming she might desire to receive travel/destination information (weather forecast, traffic, entertainments, etc.). *Situated Specific Policies* encapsulate the corresponding rules.

3.2   The Home Manager Technology

As shown in Fig. 4, devices participate to the agent society via their proxy agent. TuCSoN [26, 39] enforces the coordination laws to mediate among agents, governing both the agent-environment and the agent-agent interaction: in particular,

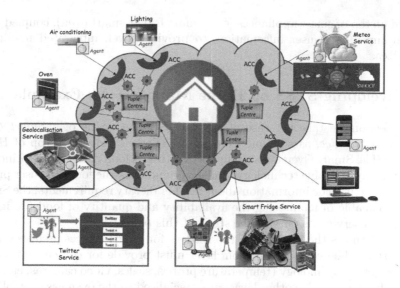

Fig. 4. The Home Manager Platform

its boundary artifacts, *Agent Coordination Contexts* (ACCs), define the agents' admissible operations and roles, while *tuple centres* – programmable, logic tuple spaces [25] – encapsulate the coordination laws and support situatedness [5]. This approach supports incremental evolution from a simulated environment to an "increasingly-real" system with some hardware devices, up to running "out of the box" on stand-alone platforms like the Raspberry PI 2. Legacy agents can also be integrated, exploiting the infrastructure to bridge the gap.

Currently, the Home Manager scenario models a smart house immersed in a smart living context, with several categories of independent devices (air conditioners, lights, etc.) and user categories [9]. At the basic operation level, the goal is to satisfy the users' desires (e.g. room light, temperature) while respecting some global constraints (e.g. energy saving, temperature range, etc.); at a higher level, however, the goal is more ambitious—to anticipate the user's needs by reasoning on the user's habits and on any user-related information, including the environment where he lives, travels, purchases goods, etc. The idea is to go beyond the mere monitoring and remote control of house appliances via app, as it is often found today in domotics system. So, its features include:

- exploit the user's location, tracked by the smartphone GPS, to enable an intelligent reasoner agent to take autonomous "situated" decisions;
- explore the environment around the user's location, extracting information about shops, services, etc., to be taken as a further reasoning knowledge base;
- retrieve information about the surrounding environment (e.g. weather) so as to tailor decisions to the user's habits and needs;
- interact with selected social networks (e.g. Twitter) to grab information that could later be exploited fur further reasonings;

– build on top of smart appliances (e.g., smart fridge, smart oven), coupled with environment and user information, to provide novel, integrated intelligent services.

4   Developing STSS on Home Manager: An Example

In this section we take the case of a Smart Kitchen, made of a *Smart Oven* and a *Smart Fridge*, as s simple example of STSS developed on top of Home Manager. The Smart Oven aims to support the user's food cooking—in principle, exploiting any available technology to identify and cook the food: the user profile is supposed to include information about his/her dietary requirements; the Smart Fridge is capable of monitoring the availability and quantity of food (designing its advanced services is outside the scope of this paper).

Figure 5 shows the tailoring of the Butlers for Smart Spaces architecture to the Smart Kitchen case. The bottom layer must provide for content awareness: any appropriate technology (temperature probes, scales, video cameras, etc.) can potentially be used. The other layers are specialised to the oven case, as follows:

– *Basic Oven services* include analysing the oven content, so as to warn the user (with a on-device screen, voice, notification, etc.) if the food exceeds the calories count of the day; to this end, the user profile, habits, preferences, diet plan, etc. should also be taken into account.
– *Basic Oven Agent & Policies* deals with aspects that do not require advanced reasoning skills: for instance, policies could require that children cannot set oven policies or modify the diet plan, that fish is cooked twice a week, etc.; the agent could guarantee that the proper amount of daily calories is introduced, while considering the user preferences; and so on.
– *Advanced Oven Agent* deals with reasoning on potentially any kind of user-related knowledge, to support more complex behaviour–e.g. making an exception to the general house policy that prevents cooking sweets in the weekdays if it is someone's birthday. Advanced techniques could be potentially exploited to learn how to cook new food, improve/adapt recipes based on users' feedback (want more chocolate, like it less sweet), etc.
– *Context-aware Oven* is charged of any opportunistic behaviour, situated both in time and space—such as suggesting a specific recipe based on the current fridge content, on food which is closer to the expiration date, etc.
– The two upper layers do not expose specific policies in this case.

The experimental prototype includes a Smart Oven (software only) and a Smart Fridge (both software-only and software+Raspberry PI hardware): the Raspberry PI 2 is coupled with GrovePI displays and an RFID tag reader to track the fridge content (Fig. 6, top). Fridge policies (under development) can range from guaranteeing e.g. that *(i)* at least 2 bottles of milk and 3 cans of beer are always present, *(ii)* the total of the shopping list reaches a minimum threshold to exploit free home delivery, *(iii)* the list is compared against multiple markets to find the most convenient, taking into account fidelity cards and special offers.

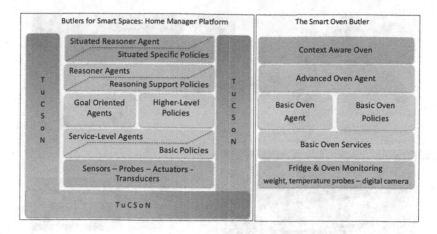

Fig. 5. Tailoring Butlers for Smart Spaces to the Smart Oven case

The Smart Oven is currently available only as a software simulation: the GUI supports recipe insertion/update/removal. Behind the scenes, it interacts with the Smart Fridge to check that the ingredients for the selected recipe are actually available, warning the user if this is not the case. The Oven reasoner monitors the fridge content and, based on predefined policies, compiles the shopping list and sends it via email to the selected supermarket for home delivery (Fig. 6, bottom). In a more complex scenario [8], the user can be geolocalised, switching on the oven automatically if he/she's buying a take-away pizza.

Overall, the key point is the chance of injecting intelligence and build applications that go well beyond the mere monitoring and remote control of appliances and devices via some Android/iOS app, as it is commonly found today: although this early prototype is clearly experimental, it highlights the many, challenging possibilities of integrating Smart Things in a pervasive, situated architecture.

Next steps aim to take Home Manager more and more "out of the box", exploiting the Raspberry both for the system core and to deploy smart appliances – smart fridge, smart oven, etc. –, thanks to the plethora of low-cost sensors, actuators, displays, cameras, etc. Looking at heterogeneity as an enabling value rather than an obstacle, we are building the software layer to support the interoperability between Windows 10 IoT Core, which can run on the Raspberry hardware, and Java, via a suitable bridge, so that TuCSoN agents can be designed in Visual Studio and integrated in the Home Manager ecosystem.

More in the perspective, we keep an eye on the evolution of the major players' technologies, looking for potential compatibilities—e.g. to allow Home Manager to exploit the many devices being developed for such frameworks, or, conversely, to possibly pack Home Manager as some sort of "plug-in" in such frameworks.

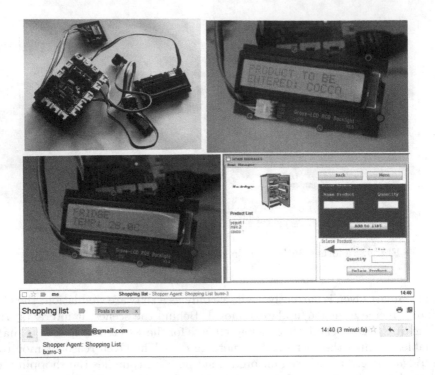

Fig. 6. Screenshots from our current prototype

5   Related Work and Conclusions

Both Socio-Technical Systems [19,21,27,41,43] and Smart Spaces [6,11,12,16,
17,20,34,36,38,40] have been widely explored in the literature for smart home,
offices, neighbourhood, streets, mobility, cultural heritage, health, etc. In the
Smart Home field, the Connected Home Platform and Development Framework
aims to interconnect home devices from different home control technologies [28],
while [15] discusses a control system, based on ZigBee sensors, for automatic
energy saving and user satisfaction. Ye and Huang [42] present a theoretical
framework for a cloud-based smart home, integrating home automation and
household mobility, while Song et al. [37] focus on WiFi and LTE coexistence in
IoT-based smart home systems.

Consumer electronics devices are also supporting remote access and control
via iOS/Android apps [10,24]. Google's "Works with Nest" aims to integrate heterogeneous apps and services in an unique framework [13], while Apple's "Home
Kit" [2] provides a set of APIs for integrating third party components and tasks:
[4] presents a smart home onto Apple's HomeKit Accessory Protocol, integrating iOS devices and accessories. Samsung SmartThings [32] is a combination of
hardware and software designed to make it easy to connect (Smart) "Things" to
the Internet and to each other. Control, monitor and security are ensured with
the SmartThings hub and sensor plugs from anywhere in environment.

Our approach has a different starting point: instead of moving from a specific need or application area, the Butlers for Smart Spaces (BSS) architecture introduces a technology-neutral reference for STSS in pervasive IoT contexts.

From the methodological viewpoint, the Butlers layers, tailored to the Smart Spaces scenario, can work both as design guidelines – to focus on the services required by an STSS application – and, conversely, as suggester for new application scenarios/niches, since the layered structure helps focussing on the conceptual location of today's devices and applications with respect to a whole stack of smarter possibilities, paving the way for advanced technologies and services.

Yet, the BSS contribution is not just theoretical: despite the technological neutrality, its layers can be mapped directly onto Home Manager, where the envisioned design finds its direct counterpart in the application structure. Developers remain free to code agents in a plurality of languages, delegating the underlying (TuCSoN) infrastructure for bridging among the diverse agents' ontologies, APIs, and knowledge representations—hereby also easing the integration of legacy components.

Despite its simplicity and its early development stage, the Smart Oven and Fridge example above means precisely to highlight these aspects, emphasizing how many intriguing possibilities can be opened once suitable intelligence chunks are injected where appropriate, providing for the construction of pervasive Smart Spaces where users interoperate immersed in time and space.

Future work will obviously proceed along this line, but also explore the multidisciplinary legal-ethical-technical issues that such deeply-pervasive IoT scenarios actually raise as concerns data privacy, device ownership, and so on [35].

References

1. Amazon: Smart home (2016). https://www.amazon.com/smarthome-home-automation/b?ie=UTF8&node=6563140011
2. Apple: Home Kit (2014). https://developer.apple.com/homekit/
3. Apple: Siri home page (2015). http://www.apple.com/ios/siri/
4. Bakir, A., Chesler, G., Torriente, M.: Program the IoT with Swift for iOS. In: Home Automation Using HomeKit, pp. 343–395. Apress, Berkeley (2016)
5. Casadei, M., Omicini, A.: Programming agent-environment interaction for MAS situatedness in ReSpecT. The Knowledge Engineering Review, January 2010
6. Chakrabarty, A.: Technology and Governance: Enabling Participatory Democracy. MPRA Paper 65231, University Library of Munich, Germany, June 2015. https://ideas.repec.org/p/pra/mprapa/65231.html
7. Denti, E.: Novel pervasive scenarios for home management: the Butlers architecture. SpringerPlus **3**(52), 1–30 (2014)
8. Denti, E., Calegari, R.: Butler-ising Home Manager: a pervasive multi-agent system for home intelligence. In: Loiseau, S., Filipe, J., Duval, B., Van Den Herik, J. (eds.) Proceeding of ICAART 2015, pp. 249–256. SCITEPRESS, Lisbon (2015)
9. Denti, E., Calegari, R., Prandini, M.: Extending a smart home multi-agent system with role-based access control. In: 5th International Conference on Internet Technologies & Society (ITS 2014), pp. 23–30. IADIS Press, Taipei, 10–12 December 2014

10. Efergy (2015). http://efergy.com/uk/products/rmpro/
11. Escobedo, L., Tentori, M., et al.: Using augmented reality to help children with autism stay focused. IEEE Pervasive Comput. **13**(1), 38–46 (2014)
12. Gal, C.L., Martin, J., Lux, A., Crowley, J.L.: Smart office: design of an intelligent environment. IEEE Intell. Syst. **16**(4), 60–66 (2001)
13. Google: Works with Nest (2014). https://nest.com/
14. Google: Google now (2015). https://www.google.com/landing/now/
15. Han, D.M., Lim, J.H.: Smart home energy management system using IEEE 802.15.4 and ZigBee. IEEE Trans. Consum. Electron. **56**(3), 1403–1410 (2010)
16. Helal, A., Moore, S., Ramachandran, B.: Drishti: an integrated navigation system for visually impaired and disabled. In: Proceeding of Wearable Computers, pp. 149–156 (2001)
17. Helal, S., Mann, W., El-Zabadani, H., et al.: The Gator Tech Smart House: a programmable pervasive space. Computer **38**(3), 50–60 (2005)
18. Home Manager. http://apice.unibo.it/xwiki/bin/view/Products/HomeManager
19. Jennings, N.R., Moreau, L., Nicholson, D., Ramchurn, S., Roberts, S., Rodden, T., Rogers, A.: Human-agent collectives. Com. ACM **57**(12), 80–88 (2014). http://doi.acm.org/10.1145/2629559
20. Kidd, C.D., et al.: The aware home: A living laboratory for ubiquitous computing research. In: Yuan, F., Hartkopf, V. (eds.) CoBuild 1999. LNCS, vol. 1670. Springer, Heidelberg (1999)
21. Li, T., Horkoff, J.: Dealing with security requirements for socio-technical systems: a holistic approach. In: Jarke, M., Mylopoulos, J., Quix, C., Rolland, C., Manolopoulos, Y., Mouratidis, H., Horkoff, J. (eds.) CAiSE 2014. LNCS, vol. 8484, pp. 285–300. Springer, Heidelberg (2014)
22. Microsoft: The internet of your things (2015). https://dev.windows.com/en-us/iot/
23. Microsoft: Cortana (2016). https://www.microsoft.com/en-us/mobile/experiences/cortana/
24. MyDlink (2016). http://www.dlink.com/uk/en/home-solutions/mydlink/
25. Omicini, A., Denti, E.: Formal ReSpecT. Electron. Notes Theoret. Comput. Sci. **48**, 179–196 (2001)
26. Omicini, A., Zambonelli, F.: Coordination for Internet application development. Auton. Agent. Multi-Agent Syst. **2**(3), 251–269 (1999)
27. Omicini, A., Zambonelli, F.: Coordination of large-scale socio-technical systems: challenges and research directions. In: CEUR Workshop Proceedings, WOA 2015 From Objects to Agents, vol. 1382, pp. 76–79. Sun SITE Central Europe (2015)
28. Papadopoulos, N., Meliones, A., Economou, D., Karras, I., Liverezas, I.: A connected home platform and development framework for smart home control applications. In: Proceeding of 7th IEEE International Conference INDIN 2009, pp. 402–409, June 2009
29. Ricci, A., Piunti, M., Tummolini, L., Castelfranchi, C.: The mirror world: preparing for mixed-reality living. IEEE Pervasive Comput. **14**(2), 60–63 (2015)
30. Rothensee, M.: User acceptance of the intelligent fridge: empirical results from a simulation. In: Floerkemeier, C., Langheinrich, M., Fleisch, E., Mattern, F., Sarma, S.E. (eds.) IOT 2008. LNCS, vol. 4952, pp. 123–139. Springer, Heidelberg (2008)
31. Saha, D., Mukherjee, A.: Pervasive computing: a paradigm for the 21st century. Computer **36**(3), 25–31 (2003)
32. Samsung Smart Things (2015). https://www.smartthings.com
33. Satyanarayanan, M.: Pervasive computing: vision and challenges. IEEE Personal Commun. **8**(4), 10–17 (2001)

34. Schaffers, H., Komninos, N., Pallot, M., Trousse, B., Nilsson, M., Oliveira, A.: Smart cities and the future Internet: towards cooperation frameworks for open innovation. In: Domingue, J., et al. (eds.) The Future Internet. LNCS, vol. 6656, pp. 431–446. Springer, Heidelberg (2011)
35. Schultz, J.: The Internet of Things we don't own? Commun. ACM **59**(5), 36–38 (2016)
36. Sittoni, A., Brunelli, D., Macii, D., Tosato, P., Petri, D.: Street lighting in smart cities: a simulation tool for the design of systems based on narrowband PLC. In: IEEE First International Smart Cities Conference (ISC2), pp. 1–6, October 2015
37. Song, Y., Han, B., Zhang, X., Yang, D.: Modeling and simulation of smart home scenarios based on Internet of Things. In: 3rd IEEE International Conference on Network Infrastructure and Digital Content (IC-NIDC), pp. 596–600, September 2012
38. Stojanovic, N.: On using query neighbourhood for better navigation through a product catalog: smart approach. In: IEEE International Conference on e-Technology, e-Commerce and e-Service, IEEE 2004, pp. 405–412, March 2004
39. TuCSoN (2008). http://tucson.apice.unibo.it/
40. Weilkiens, T., Lamm, J.G., Roth, S., Walker, S.: An Example: The Virtual Museum Tour System. Wiley, New York (2015)
41. Whitworth, B.: Socio-technical systems. In: Encyclopedia of Human Computer Interaction, pp. 533–541. Idea Group, Hershey (2006)
42. Ye, X., Huang, J.: A framework for cloud-based smart home, vol. 2 (2011)
43. Zambonelli, F.: Toward sociotechnical urban superorganisms. Computer **45**(8), 76–78 (2012). http://www.computer.org/csdl/mags/co/2012/08/mco2012080076-abs.html

Public Transportation, IoT, Trust
and Urban Habits

Andrea Melis[1]([✉]), Marco Prandini[2], Laura Sartori[3], and Franco Callegati[1]([✉])

[1] Department of Electrical, Electronic and Information Engineering,
University of Bologna, Bologna, Italy
{a.melis,franco.callegati}@unibo.it
[2] Department of Computer Science and Engineering,
University of Bologna, Bologna, Italy
marco.prandini@unibo.it
[3] Department of Political and Social Sciences, University of Bologna, Bologna, Italy
l.sartori@unibo.it

Abstract. The technological compound known as Internet of Things is enabling massive transformations in many fields. In this paper, we deal with one emerging scenario, Mobility as a Service, where the interplay between technical, regulatory and social aspects is intense. We advocate the need for interdisciplinary research, taking into account the different facets of a system which, in summary, aims at improving the quality of urban life by collecting personal data, tracking citizens' movements, correlating them with many other sources of information, and making the results widely available.

Keywords: Internet of Things · Mobility as a Service · Trust · Urban habits · Data collection · Governance

1 Introduction

Historically, infrastructures, administrations and citizens have been intertwined (to a certain degree) in a mutual, but non-linear, process of societal development. Transportation is a hot topic for big cities and regional districts in both developed and developing countries. Public transportation emerges as an even more crucial theme for future sustainability, especially in high density territories. If it is true that since the Fifties the time budget (in hours) allocated to urban transportation is stable while the number of kilometers steadily increased (the so-called Zahavi's conjecture), it means that at least part of the potential positive effects of transport innovations (alternative means of transport, integrated systems of mobility) has not been fully deployed and, thus, was lost. Now, a paradigm shift is happening from seeing mobility as a problem of infrastructure to designing mobility as a service (MaaS) to the community. The new paradigm offers a smart and sustainable optimization of urban mobility in complex and multidimensional metropolitan areas, but it requires pervasive, real-time data

© Springer International Publishing AG 2016
F. Bagnoli et al. (Eds.): INSCI 2016, LNCS 9934, pp. 318–325, 2016.
DOI: 10.1007/978-3-319-45982-0_27

collection and analysis, making it a prime application of the Internet of Things (IoT) model. Thus, it is important to start addressing both technical and social implications of IoT in order for potential benefits to be effectively realized.

In this paper we address technical and social implications of IoT in a specific setting: public transportation in urban contexts. By acknowledging the crucial role of the transport infrastructure for 'smart' territories and cities of tomorrow, we propose to start over and think about public transportation by changing the core premise: mobility cannot be anymore thought of just as an area of basic and standard public regulation, but as a Service. As such, mobility becomes an integrated framework for urban policy-making. From a technological PoV IoT allows for this innovative approach about Mobility as a Service (MaaS) (Sects. 2 and 3) connecting to stringent urban and social issues (Sect. 4). From a sociological perspective, we address three main points: the city as a whole, a pro-active regulatory approach and trust towards technology. Investigating mobility by a multidisciplinary eye has already served to pinpoint relevant areas for future research.

2   Public Transport in the IoT Age

In recent decades, public transport services steadily profited from the introduction of new technologies. The means of transport became faster, less polluting, more comfortable and accessible. The interaction of passengers with transport services became easier, as ticketing and payment systems went from paper to electronic and on-line planning and real-time information systems became available. Through more efficient exchange of information, transport operators began to see the interoperability between competitors as an added value (for example: coordination at transfer points can lead to better service, attracting more customers than it would happen by aggressively competing for the same route). The advent of IoT, however, potentially represents a real revolution for public transportation. Independent processes that required specific investments to deal with business needs (e.g., fleet management, fulfilment of quality of service obligations, route optimization, etc.) can be all seen as by products of a single platform, where thousand of autonomous objects can constantly acquire data captured from their surroundings, analyse them for local decision making and forward them to third parties [13]. These data are used to improve services' efficiency, and to make them smarter and more customized. In this picture, the citizens are not simply end users any more, but become components of the service itself and contributors to its development. Out of the many examples that confirm this claim, a few are provided hereinafter.

2.1   Case Studies

In various contexts, pilot projects as well as large-scale deployments have proven the value of the technologies that now compose the foundations for IoT.

Transport for London (TFL) has already implemented a huge network collecting data through devices, such as ticketing systems, sensors attached to vehicles, and traffic signals, but also by means of surveys, focus groups, and social media [6]. These data have manifold uses: for example, a single "journey mapping" application can aggregate anonymized data to allow the study of overall flows, or to produce real-time maps showing passengers the network status, as well as more individual analyses through personalized travel habits. Data analysis also helps TFL respond in an agile way when unexpected events occurs and travel data is also used to identify customers who regularly use specific routes and send tailored travel updates to them.

Similarly, the Kontron Intel architecture was used to create a map representation of traffic, depicting realistic vehicular mobility traces of downtown Portland, Oregon [5]. With this map, the authors were able to organize traffic by directing vehicles through the most suitable paths according to traffic congestion. This work focused on cars, but the principle could be adapted to all public transport, taking into account route constraints.

Other examples of the same kind can be found in adapted Markov chains, fed by real-time traffic information, to predict congestion trends on freeways [14], while in another case study [9] smart-phones embedded sensors are used to create a network of vehicles to track the availability of parking spaces, suggesting where to park.

Gubbi et al. [2] defined otherwise a cloud-centric version of an IoT network system contextualized for the design of traffic management in medium and big cities. They proposed a framework enabled by a mix of public and private clouds in order to provide the capacity to utilize the IoT. Following these results, Leng and Zhao [4] provided a practical implementation of this kind of IoT network for traffic management, realized with a cloud-centric monitoring system of vehicles. Viviani's work [16] describes a technical solution adopted in the Padova Smart City project, a proof of concept deployment created in collaboration with the city municipality, that pushes the same concepts further towards the idea of MaaS.

3   MaaS and the Future for Mobility

Mobility as a Service [10] is an innovative approach to the integration of public and private transport, made viable by the integration in a coordinated infrastructure of the technologies illustrated in the previous section. Born in the city of Helsinki, this paradigm is starting to spread throughout Europe and beyond, aiming to establish standards for the interoperability between different (even in terms of country) operators, and to encourage the creation of alternative solutions to the standard "mass transit/private car" duality, as both new technologies and social trends emerge [7].

Very briefly, the principle of MaaS is that as long as every detail of the demand and supply for transportation services is known in real-time, there is no need for passengers to commit on specific means. Instead, they will enjoy a broad spectrum of alternatives from which to choose, taking into account the needs of the moment. For example, one could specify a very strict set of constraints in terms of comfort and timing, likely to result in a choice of premium means, while another could simply express the need for reaching a destination at the best price, getting a virtual ticket, and receiving real-time instructions about which means to use to complete the trip. Many business models are possible. In the simplest form, a MaaS operator could simply be a smart broker for planning and paying trips on existing networks. A more innovative way would be selling mobility packages allowing travelers to use pre-configured amounts of usage of different means. From the transport operators viewpoint, a MaaS platform could be a great opportunity to leverage integration and to exploit unused capacity. For example, a taxi company exposing vehicle availability and position in real-time could offer lower prices during off-peak times, thus appearing as a good alternative to mass transit; data-mining could allow operators to foresee correlations between various conditions (events, weather, accidents) and transportation needs, to allocate materials in the best possible way. Ideally, the ICT infrastructure enables these models by tracking timing, position, and availability of trains, buses, subways, shared bikes, shared cars, taxis, Uber cars, Lyft cars, etc. in an overall effort of opening data and standardizing the interfaces to access them [15]. In short, in a mobility context, both the users and the operators can benefit from the smart definition of trips, provided enough availability of data and efficiency of processing is available. This is exactly the kind of challenge IoT architectures are up to [11].

Thus, the role of public administration can undergo a significant change. Some administrations could choose to play the role of a central MaaS operator, exerting a stronger control on the local mobility agenda. Others could leave the field to private companies, hoping to benefit from market-driven optimization of citizens' patterns of mobility. They could also accurately monitor citizens, using collected data to plan investments and direct incentives towards specific goals. The first scenario allows for a more respectful and regulated approach to citizens' privacy while the second leaves room for malevolent or misleading collection and use of data. One example of the latter is Google legal advisor David Drummond's defensive reply to the question about future uses of data collected by their driverless car [8]. According to authors, it is too early to regulate over the driverless-car about data collection and uses, because it is not yet foreseeable what is worth (implicitly 'for the company'). In any case, governments will need to face the challenges MaaS provide, and to think about needed regulatory changes to make it viable for innovative cities. The adoption of MaaS will sustain a transition from a public transport system traditionally coordinated by the government to a multi-faceted system where exert coordination through the help of other actors. For example, determining who is in charge of setting the standards will affect business and consumers in parallel with data protection policies.

It is worth noticing that, in many places, this change could introduce significant trust issues. The organization of transport infrastructures by public bodies guarantees (at least in principle) that travelers' interests are safeguarded. In a fuzzier scenario, it could be very difficult to verify how sensitive data are processed and by whom, as detailed in Sect. 4.3.

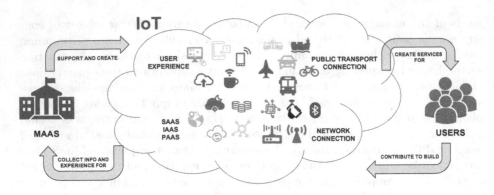

Fig. 1. Maas, IoT and user work-flow

Figure 1 illustrates a development scenario of MaaS. The central cloud is the concept of IoT architecture understood as a heterogeneous set of networks, technologies and experiences. The user enjoys the huge power of the IoT infrastructure and its vast datasets, yet at the same time he enriches the IoT by providing valuable information from mobile devices' sensors and from his own user experience, as well as contributing additional computational power ("fog computing").

The MaaS' role is twofold too, in that it uses infrastructural functions and user data to create many layers of value-added applications, which enlarge the set of available IoT services.

4   Social Implications of IoT and MaaS

Technology always takes a user to 'tango', even though seldom it is a learnt lesson. Social implications of IoT and its siblings, such as MaaS-enabling infrastructures, are still to be empirically investigated, but three issues are already worth mentioning.

4.1   Not only Mobility, It Is About the Whole City!

Rethinking mobility means thinking about the whole city (or territory) with all the social and economic complexity it brings along. Citizens live and experience, innovate or destroy the city, moving within it. Daily routines and social trends in public or private domains offer grounds to better conceptualize mobility. Moreover, in the last decades, users have been progressively accustomed to have more autonomy in the uses they make out of technology (e.g., from the more traditional mass media to the malleable family of ICTs). IoT-powered platforms kindly welcome and incorporate users' feedback in a potential virtuous circle. Since urban mobility is not the only aspect of urban life that users engage in, it is likely that feedback about 'mobility' already incorporate and reflect other daily routines (about work, leisure and family habits). Thus, MaaS could offer optimized solutions to mobility that are 'embedded' in social contexts in contrast with more top-down and thought-in-silos solutions. Thinking about the whole city and its dynamics (and not only about transportation) also offers a potential for nudging. A deep knowledge about urban life is essential to policies that sustain and encourage more sustainable behaviours (e.g., nudging and giving suggestions to include walking in daily routines, to combine "transportation", healthy behaviour, and street liveliness). Mobility is a fundamental piece of the city life that needs to be thought as a part of a more general urban agenda.

4.2   A New Pro-active Approach to Policy-Making

Successful policies show a deep understanding of the context they intend to regulate. Lately, bottom-up contributions to policy-making have been possible and emphasized through the power of ICTs, social media and so forth. Yet, a lot can still be done in a more traditional, but effective, way. When it comes to urban mobility combined with technological innovations, policymakers could make a first move engaging with relevant actors to collect all possible insights about existing needs (as explained in previous paragraphs). This pro-active approach reaches the goal of a deeper understanding of the context while listing also all possible existing constraints. For sure, there are regulatory constraints that, if not adapted and molded to the new context, could limit possible innovations. There is a need for a change in the regulatory framework that is thought to govern hundreds of transactions or services, not millions of connected things (and their data). What level of de-regulation and what kind of new rules policy-makers will come up with is a truly interesting area of research.

4.3   Trust

Why and to what extent should one trust IoT for mobility? Security directly relates to trust as users have to be confident that IoT applications are secure from cyberattacks or external misuses, like any other networked environment. From a technical point of view, IoT security should be thought of a device vulnerability in an evolutionary process supported by threat-and-solutions to emerging security issues [3]. Debating over security issues is not new in IT, but the scale to 'unique' challenges to IoT make it even more central [12]. From a sociological point of view, privacy and awareness arise as crucial aspects of security when the classic model of 'terms and conditions' acceptance does not apply to IoT. Users do not have direct knowledge and control over IoT devices in order to express privacy preferences in their daily uses of IoT applications or in the management of the data collection that either 'tap into' or 'feed into' the increasing big data analytics. Thus, it is quite difficult to seize and adapt IoT apps to the social norms underlying what is deemed to be private, public or shareable. If the Internet changed the boundaries of the right to privacy, IoT threatens to go even further. In this respect, future technical developments should look closely to privacy-by-design principles and privacy preferences in order to set the basis for higher fairness in data collection and use.

When it comes to mobility, security is linked to trust also because individuals need to be safe when using transportation means which are totally new (e.g., self-driving cars [17]) or which could be perceived as threatening (e.g., shared cars with unknown drivers). As an example, gender or age perspectives could help in designing mobility as a service considering different needs in terms of easiness and safeness in public spaces. By the same token, IoT apps could be effectively used for law enforcement and public safety (monitoring urban areas, recovering stolen cars, etc.). Yet, they could be also maliciously put at work to monitor single individuals that are no criminals as in the recent case of Banksy (where researchers used crime prevention analytics and mobility maps to match and uncover who could possibly be identified as the famous street artist [1]).

5 Conclusions

As soon as public transportation is acknowledged as a service it could become a smart tool for urban governance. Not only can urban mobility be updated through a mix of new and existing solutions, but it can be shaped along with emerging societal trends in public and private domains that affect urban life. It is important to accumulate and store urban knowledge not confined to urban mobility in order to proactively design smart policies. Moreover, IoT and MaaS could offer new chances for designing ways by which to build and consolidate users' trust and confidence towards technology.

Therefore, technological solutions could really empower and sustain innovative answers to mobility issues. Yet, it could be so, only if we share a perspective where social, political and urban needs meet technological opportunities.

References

1. Chandler, A.: Banksy unmasked? http://www.theatlantic.com/technology/archive/2016/03/banksy-unmasked/472152/
2. Gubbi, J., Buyya, R., Marusic, S., Palaniswami, M.: Internet of things (IoT): a vision, architectural elements, and future directions. Future Gener. Comput. Syst. **29**(7), 1645–1660 (2013). http://www.sciencedirect.com/science/article/pii/S0167739X13000241, includingSpecial sections: Cyber-enabled Distributed Computing for Ubiquitous Cloud and Network Services; Cloud Computing and Scientific Applications Big Data, Scalable Analytics, and Beyond
3. Rose, K., Eldridge, S., Chapin, L.: The Internet of Things (IoT): an overview. Techical report, The Internet Society (ISOC), October 2015
4. Leng, Y., Zhao, L.: Novel design of intelligent internet-of-vehicles management system based on cloud-computing and internet-of-things. In: 2011 International Conference on Electronic and Mechanical Engineering and Information Technology (EMEIT), vol. 6, pp. 3190–3193, August 2011
5. Marfia, G., Pau, G., De Sena, E., Giordano, E., Gerla, M.: Evaluating vehicle network strategies for downtown portland: opportunistic infrastructure and the importance of realistic mobility models. In: Proceedings of the 1st International MobiSys Workshop on Mobile Opportunistic Networking, MobiOpp 2007, pp. 47–51. ACM, New York (2007). http://doi.acm.org/10.1145/1247694.1247704

6. Marr, B.: How big data and the internet of things improve public transport in london (2015). https://www.linkedin.com/pulse/how-big-data-internet-things-improve-public-transport-bernard-marr

7. Finger, M., Bert, N., Kupfer, D.: Mobility-as-a-Service: from the Helsinki experiment to a European model? Technical report, Observer European Transport Regulation (2015)

8. Phippen, J.W.: Who drives a driverless car? http://www.theatlantic.com/national/archive/2016/02/google-driverless-car/462153/

9. Salpietro, R., Bedogni, L., Felice, M.D., Bononi, L.: Park here! a smart parking system based on smartphones' embedded sensors and short range communication technologies. In: 2015 IEEE 2nd World Forum on Internet of Things (WF-IoT), pp. 18–23, December 2015

10. Pippuri, S., Hietanen, S., Pyyhtia, K.: Maas finland. http://maas.fi/

11. Hietanen, S.: CEO, ITS Finland. Mobility as a service the new transport model? Technical report, MaaS Finland (2014)

12. Sicari, S., Rizzardi, A., Grieco, L., Coen-Porisini, A.: Security, privacy and trust in internet of things: the road ahead. Comput. Netw. **76**, 146–164 (2015). http://www.sciencedirect.com/science/article/pii/S1389128614003971

13. Yate Team: Connecting public transport to the internet of things (2015). https://blog.yate.ro/2015/08/27/connecting-public-transport-to-the-internet-of-things/

14. Tian, D., Liu, C., Wang, Y., Zhang, G., Xia, H.: A freeway travel time predicting method based on iov. In: 2015 IEEE 2nd World Forum on Internet of Things (WF-IoT), pp. 1–5, December 2015

15. Viviani, M.: Mobility as a service (2015). http://www.webnews.it/2015/08/07/mobility-as-a-service/

16. Zanella, A., Bui, N., Castellani, A., Vangelista, L., Zorzi, M.: Internet of things for smart cities. IEEE Internet of Things J. **1**(1), 22–32 (2014)

17. Zetter, K.: Researchers hacked a models, but teslas already released a patch. http://www.wired.com/2015/08/researchers-hacked-model-s-teslas-already

Author Index

Absalom, Richard   105
Alani, Harith   97
Alexiev, Vladimir   205

Badii, Atta   55
Bagnoli, Franco   19, 42, 55, 115, 148
Balazs, Balint   55
Berlinguer, Marco   27
Bodrunova, Svetlana   176

Calegari, Roberta   306
Callegati, Franco   318
Cardelli, Chiara   148
Carle, Georg   218
Castellani, Tommaso   55
Cecchini, Cristina   123
Cerno, Italo   291
Cevenini, Claudia   291
Coppolino Perfumi, Serena   148
Cordeiro, Edwin   218

D'Orazio, Davide   55
De Paoli, Stefano   74
Denti, Enrico   291, 306
Diani, Mario   257
Diplaris, Sotiris   74
Dupont, Anthony   239
Duradoni, Mirko   115

Ermoshina, Ksenia   244

Fernandez, Miriam   97
Ferri, Fernando   55
Filippov, Vladimir   176
Föls, Michael   97
Fumero, Antonio   89
Fuster Morell, Mayo   27

Gkatziaki, Vasiliki   74
Gkrosdanis, Ioannis   278
Grifoni, Patrizia   55
Guazzini, Andrea   42, 115, 123, 148
Guidi, Elisa   123

Hall, Wendy   161
Halpin, Harry   244
Hartmann, Dap   105
Herring, David   97
Heyman, Rob   133
Hirche, Philipp   74

Imbimbo, Enrico   42

Jamin, Emmanuel   205

Kapadia, Vishal   74
Katmada, Aikaterini   3
Koltcov, Sergei   176
Koltsova, Olessia   176
Kompatsiaris, Ioannis   3, 205
Kompatsiaris, Yiannis   133

Liparas, Dimitris   205

McCutchen, Ethan   74
Melis, Andrea   318
Menichinelli, Massimo   189
Mills, Richard   74
Milonaki, Angeliki   278
Montero, Calkin Suero   189
Morales, Stephanie   239
Musiani, Francesca   244

Nenadic, Iva   231
Niedermayer, Heiko   218
Nikolenko, Sergey I.   176

Olszowski, Rafał   66
Omicini, Andrea   291
Ostling, Alina   231

Pacini, Giovanna   19, 55
Papadopoulos, Symeon   74, 133
Pavan, Elena   257
Petkos, Georgios   133
Piccolo, Lara   97
Prandini, Marco   318

Prasad, Srivigneshwar R.   74
Puigbo, Marti   205

Raumer, Daniel   218
Romero, Inés   89
Rueda, Yolanda   89

Salcedo, Jorge L.   27
Sartori, Laura   318
Satsiou, Anna   3
Scharl, Arno   97
Schwellnus, Nikolai   218
Serban, Ovidiu   55
Sessa, Carlo   239

Simeonov, Boyan   205
Siow, Eugene   161
Skaržauskiené, Aelita   105
Spyromitros-Xioufis, Eleftherios   133
Stefanelli, Federica   42

Tiropanis, Thanassis   161
Turk, Žiga   239

Vakali, Athena   278
Valente, Adriana   55
Voigt, Christian   189
Vrochidis, Stefanos   205

Printed in the United States
By Bookmasters